BLACK HOLES, QUASARS, AND THE UNIVERSE
Second Edition

Harry L. Shipman   University of Delaware

HOUGHTON MIFFLIN COMPANY   BOSTON

Dallas   Geneva, Illinois   Hopewell, New Jersey   Palo Alto   London

Printed in the U.S.A.

Library of Congress Catalog Card Number: 79-49834

ISBN: 0-395-28499-6

CONTENTS

PREFACE

In the last decade, astronomers have discovered that violent, explosive phenomena play a vital role in cosmic evolution. For example, the end of a star's life cycle often results in violent collapse of the core. Some stellar cores become so compact that nothing, not even light, can escape from them, and they become black holes. The nuclei of galaxies contain explosive phenomena as well. The exploding nuclei of some galaxies are the most luminous objects in the universe: the quasars. In the last few years, as observations have provided confidence in our picture of cosmic evolution, there has been renewed interest in cosmology, the study of the evolution of the entire universe.

This book covers a dynamic field; the fast pace of the research frontier has called for a second edition. I am happy to report that there is little in the first edition that can be regarded as downright wrong, but there are many places in which it is now quite incomplete. I have updated all chapters and have essentially rewritten some of them in order to keep this book current.

I believe this book can fill three needs. First, many interested non-astronomers read magazine articles on black holes, cosmology, or other topics treated here and want to know more. Next, more colleges offer a course on modern astronomy, addressed to the non-science major, that goes beyond or supplements the traditional one-semester survey courses. Third, the survey courses can use this book as an enriching addition to the standard introductory texts which generally devote relatively little space to the topics treated here.

In revising the first edition, I left no chapter untouched. The treatment of pulsars in Chapter 3 is considerably expanded. Chapter 5 introduces the new black-hole candidates V 861 Scorpii and Circinus X-1, and presents recently uncovered facts regarding the prima donna of all black-hole candidates, Cygnus X-1. Chapter 6 has been completely rewritten. Most of what were classified as the frontiers of black-hole research in the first edition are now facts, and have been moved to Chapter 4. A host of *new* frontier areas have appeared: globular cluster x-ray sources, x- and gamma-ray burst sources, evaporating black holes, and the detection of possible gravitational radiation from a pulsar in a double star system.

Part 2 talks about many new phenomena related to quasars: the discovery that BL Lacerta objects are definitely active galaxies, the association of some quasars with galaxies of similar redshifts, clarification of the source of infrared emission in many objects, the discovery of radio jets, and the increasing prominence of black-hole models for the energy source. I rewrote Chapter 12, which deals with the nature of quasar redshifts, in view of recent work that shows that the redshifts of some quasars are produced by the expansion of the universe.

In Part 3, I broadened the scope of Chapter 13 to include brief treatments of areas of cosmic evolution that are of great importance to our own evolutionary history: the production of the atoms we are made of, the star we orbit around, the planet we live on, and life itself. The remainder of Part 3 presents new observations on cosmology. Particularly noteworthy is the discovery of our motion relative to the motion of the rest of the universe, shown by the large-scale variation in the intensity of the background radiation emitted by the million-year-old universe. Throughout the book I included mention of the continuing stream of research results that fill out our picture of cosmic evolution.

People have written me to ask for suggestions for additional material that could be used along with this book in a one-semester course. In my own course, I add a week and a half on relativity (between Chapters 3 and 4). To supplement Chapter 13, I add some material on the history of cosmology and the origin of life. Other possible additions would be a discussion of various aspects of space astronomy (technology, the physics of rockets, the teamwork mode of operating space observatories) to supplement, say, Chapter 6; supergravity and other attempts to unify the fundamental forces of nature (Chapter 6); cosmic geometry (Chapter 13), or a quick review of general relativity (Chapter 4). I added some of the new material in Chapter 13 with the idea of opening doors that lead to many other astronomical topics. The Bibliography suggests possible readings on these topics.

I thank all my astronomy mentors, particularly Owen Gingerich, Jesse Greenstein, Bev Oke, and Steve Strom, for inspiring me to work ever harder in my career. Stimulating discussions by colleagues too numerous to mention have contributed greatly to this book and to the quality of astronomical life. Mike Zeilik, Dick McCray, Ken Brecher, and Remo Ruffini helped materially by reading early drafts of the first edition; Greg Shields and Stu Shapiro gave similar help with the second edition. Tom Williams, Dave Shaffer, and Mike Simon provided useful comments on individual chapters. Students at Pierson College, Yale University, first suggested that I teach a course on these topics, and thereby generated the idea for this work. Pierre Mali, Donald Richards, and the English Department at Kingswood School contributed by teaching me how to write an English sentence. My wife Wendy has put up with my spending many long hours on this project. All astronomers are grateful to federal and private agencies for providing financial support to astronomy. I personally thank the National Science Foundation, the Research Corporation, the University of Delaware Research Foundation, and the National Aeronautics and Space Administration for supporting my own work, which corroborates many of the ideas expressed here.

Harry L. Shipman
Newark, Delaware

PRELIMINARY: BASIC ASTRONOMICAL TERMINOLOGY

This book really starts in Chapter 1, not here. The Preliminary section contains background material that you may or may not want to read, depending on your background in astronomy and your self-confidence. I suspect that you really picked up this book because you wanted to learn about black holes and quasars, and not about parallaxes and magnitudes. However, in reading about these exciting astronomical objects, you will occasionally need basic astronomical knowledge. Since readers' backgrounds vary, I included all the basic material in this section, apart from the main body of the book. This section is intended as a reference, which you may want to peruse from time to time as you read.

The basic physics needed for astronomy, which underpins some of the forthcoming material, is here. It includes the building blocks of matter—electrons, protons, and neutrons; the relation between light and energy; the spectra of stars and their uses; and the ways astronomers and physicists measure radiation from the stars.

If you want to go on to Chapter 1, please do so. Or if you really want to read this section first, I shall not stop you.

Matter

If you divide matter—any form of matter—into smaller and smaller pieces, you discover that all matter is organized on three main levels. Sugar, for example, is composed of sugar molecules (see Figure P-1). Each sugar molecule has the properties of sugar; It is sweet, easily digested, soluble in water, and has food value. If you try to divide a sugar molecule, you do not have sugar any more. You have atoms.

Atoms are a lower level of organization of matter. All molecules are composed of atoms, and the number and arrangement of different types of atoms within a molecule determine the properties of that molecule. Sugar contains 12 carbon, 11 oxygen, and 22 hydrogen atoms. All atoms of a particular chemical element are similar. Any of the oxygen atoms of sugar

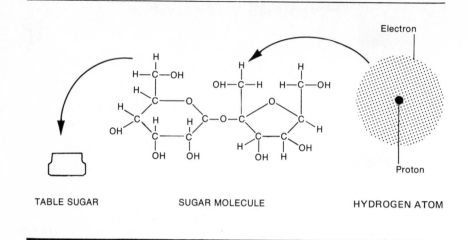

TABLE SUGAR   SUGAR MOLECULE   HYDROGEN ATOM

FIGURE P-1   Sugar is composed of sugar molecules, each of which possesses the properties of sugar. The molecules are composed of atoms, which in turn are composed of protons, electrons, and neutrons.

could be replaced by any other oxygen atom in the universe and you would still have sugar.

If you could examine atoms closely, you could see what they are made of. Just as molecules are made of smaller units —atoms—atoms are also made of smaller units. Any atom contains a nucleus surrounded by a number of electrons. While the electrons orbit the nucleus, they are moving so fast that they are correctly visualized as a blur around the nucleus. Figure P-2 shows a photograph of a molecule: The electron clouds are clearly visible as blurs surrounding the invisible nuclei. The nucleus, at the center of the atom, contains protons and neutrons, which are much heavier than the electrons. Protons and electrons have equal charges, which are of opposite sign, and neutrons (as their name implies) have no charge. An atom in its normal state has equal numbers of protons and electrons, so it too has no charge as a whole. The number of electrons in an atom governs how the atom interacts with other atoms in forming molecules; thus the number of electrons in an atom determines what element that atom is.

Subatomic particles, like electrons, protons, and neutrons, can interact with each other in differing ways. The simplest interaction occurs when a neutron outside of an atomic nucleus decays into a proton and an electron. Other interactions are more complex. Collision between two protons starts a chain of reactions, the first step of which is the formation of heavy hydrogen, or deuterium, when one of the protons changes to a neutron. Such interactions make the sun shine, and are central to the evolution of stars and other astronomical objects.

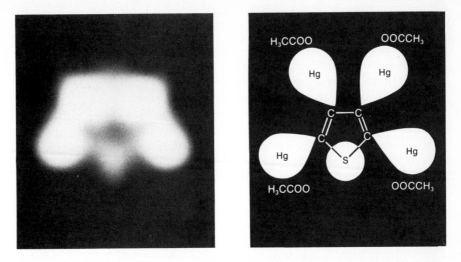

FIGURE P-2   *Left:* A photograph of a molecule of 2,3,4,5-tetraacetooxymercurithio-phene. Only the mercury and sulfur atoms can be seen in this photograph; the other atoms are too small to be seen. *Right:* A drawing of the structure of this molecule, showing how the mercury and sulfur atoms are bound in the molecule. (Photographs taken from "Image of a Sulfur Atom," by F. P. Ottensmeyer, E. E. Schmidt, and A. J. Olbrecht, *Science,* vol. 179, pp. 175–176, January 12, 1973; photograph copyright 1973 by the American Association for the Advancement of Science. Courtesy, *Science* and F. P. Ottensmeyer.)

Is there any level of organization below protons, neutrons and electrons? This question lies at the frontier of particle physics. Most physicists studying high-energy phenomena believe that there is a way that a proton, for example, can be taken apart to reveal its building blocks; these building blocks have been named quarks. However, since no one has yet found a quark, the theory remains unproven.

Light and energy

Almost all of our information about astronomy comes to us through light and other forms of electromagnetic radiation such as radio waves, x rays, and infrared radiation. No one has yet brought us a piece of a star, although pieces of the moon have made their way to earth. Since light is the source of virtually all our knowledge of astronomy, its nature is fundamental to any study of the subject.

Light is a form of energy. It is difficult to precisely define the term *energy.* It is easier to develop a mental picture of the energy concept by asking what energy does. The usual definition of energy is "the ability to do work," an accurate—if dry—description. The type of work that energy does varies with its form. Light energy can illuminate the printed page; heat energy can keep us warm; kinetic energy (or energy of motion) can be used to move something from one place to another. All activity can be thought of as involving manipulation of energy—changing the character of energy from one form to another.

Whatever happens in the universe can neither create nor destroy energy. Thus a fundamental question about any astronomical phenomenon is, What is the energy source? The sun bathes the earth in energy, as its rays deposit 1,388,000 ergs of energy per second on each square centimeter of surface area on top of the atmosphere. (Ergs and all other units of measurement are defined at the end of this preliminary chapter.) This energy comes from nuclear reactions taking place in the center of the sun, and is transformed from nuclear energy into light energy by the solar furnace. We intercept and use some of this energy, changing it to different forms. But whatever happens, energy is neither created nor destroyed. This principle is so important that it can be called a physical law, the First Law of Thermodynamics: Energy is neither created nor destroyed.

The energy contained in light comes in little packets. Each packet, a photon, carries a certain amount of energy. The amount of energy depends on the character of the photon; an energetic x-ray photon has trillions of times as much energy as a low-energy radio photon.

Light is a form of electromagnetic radiation. Electromagnetic radiation can be understood most easily, perhaps, by thinking of radio waves. An electron in a radio or a TV antenna feels forces pushing it along the antenna, first one way and then the other. These forces come from the electric and magnetic fields carried by the radio wave. All radiation contains these electric and magnetic fields, reversing direction at periodic intervals.

The forms of electromagnetic radiation differ in how quickly the reversal of direction takes place. Reversals come much less often in a radio wave than in a light wave. The rate at which these reversals take place determines the frequency of the wave. The frequency equals the number of times the electric and magnetic fields go through complete cycles in one second. Frequency is measured in hertz, or cycles per second. For example, if your radio is tuned to 1100 kilohertz, in the middle of the AM dial, the radio is picking up the particular electromagnetic waves in which the fields reverse direction 1,100,000 times per second.

Closely related to frequency is wavelength. If you measure the distance that a light wave travels in one cycle of the change in field direction, that distance is one wavelength. If the frequency of the wave is high, the electric field reverses direction rapidly and the wave does not have to travel far for the field to go through a complete cycle. Thus high-frequency waves have short wavelength and low-frequency waves have long wavelength.

Electromagnetic radiation comes in all varieties, from very energetic gamma rays of very short wavelength and high frequency on one end of the spectrum to radio waves of long wavelength, low frequency, and low energy on the other. Historically, several different names have emerged to characterize different parts of the spectrum, and these names are listed in Figure P-3, along with wavelengths, frequencies, and energies of typical photons. (Powers-of-ten notation, used in this figure, is explained later in the section.) The scale at the bottom of the spectrum illustrates what a small part of the spectrum visible light is. Below the scale are examples of the kinds of instruments used to pick up the different types of radiation. The atmosphere blocks a great part of the electromagnetic spectrum from our view, since only visible light and radio waves can travel unimpeded to the ground. For some parts of the spectrum, like the infrared, radiation is blocked by water vapor in the lower part of the atmosphere. To see this radiation you must fly above the lower atmosphere in an airplane or balloon. (There are a few infrared frequencies that are not blocked by water vapor.) For other parts of the spectrum, such as the ultraviolet, satellites are necessary to see what objects in space are doing. (See Figure P-4.)

The astronomer is interested in the great extent of the electromagnetic spectrum because different types of objects emit different types of radiation. There are several ways in which light can be emitted from an object. Here we consider the most common—radiation from hot objects.

Thermal radiation is the usual term for characterizing light emission from hot objects. Anything that is hot radiates, generally producing photons with roughly the same amounts of energy that is present in the individually moving particles. Heat comes from the motion of molecules in an object, and the temperature of the object indicates how much energy each molecule in the object possesses. Consider a particular part of the electromagnetic spectrum, red light. This type of radiation is emitted by objects whose moving molecules have the same energy as a photon of red light. Such objects are at temperatures of 5900 Kelvins, which happens to be the temperature of the sun. Infrared photons, lower-energy photons, are emitted by cooler objects. The right-hand column in Figure P-3 lists the temperatures of objects that emit the types of radiation listed. Hot objects emit radiation over a wide range of wavelengths, producing a *continuous spectrum* or *continuum*. (Figure P-3 lists the temperature at which the peak number of photons is emitted at the indicated wavelength. Further, the temperatures given are for objects without intrinsic color, technically called *blackbodies*.)

The last column of the figure provides some insight into the types of objects that are likely to be found by astronomers studying various parts of the spectrum. Stars, for example, have temperatures roughly between 1000 and 100,000 K. Thus they are found by astronomers studying the infrared, optical, and ultraviolet regions. X-ray astronomers see hotter

RADIATION TYPE	WAVELENGTH (METERS)	FREQUENCY (HERTZ)	PHOTON ENERGY (ELECTRON-VOLTS)	CORRESPONDING TEMPERATURE (KELVINS)
GAMMA RAYS	1.2×10^{-13}	2.5×10^{21}	10^7	3×10^{11}
X-RAYS	1.2×10^{-9}	2.5×10^{17}	10^4	3×10^7
ULTRAVIOLET	1.5×10^{-7} (1500 Å)	2×10^{15}	8.3	24,500
BLUE	4×10^{-7} (4000 Å)	7.5×10^{14}	3.1	9150
RED	6×10^{-7} (6000 Å)	5×10^{14}	2.0	5900
INFRARED	2×10^{-6}	1.5×10^{14}	0.62	1850
MICROWAVE	3×10^{-4}	10^{12}	0.004	9.3
RADIO	1	3×10^8	1.2×10^{-6}	0.0035

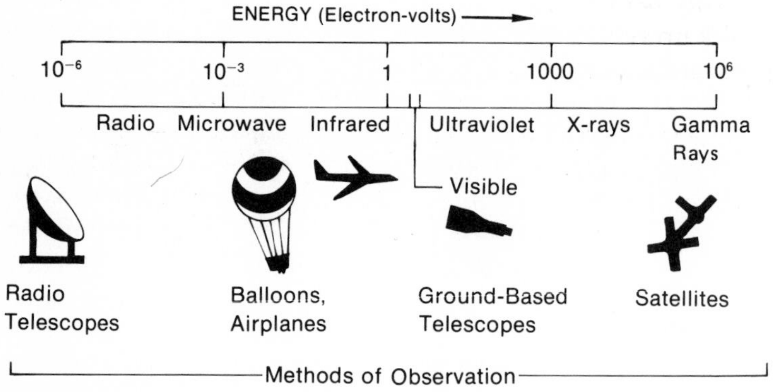

FIGURE P-3   Different types of radiation (see text).

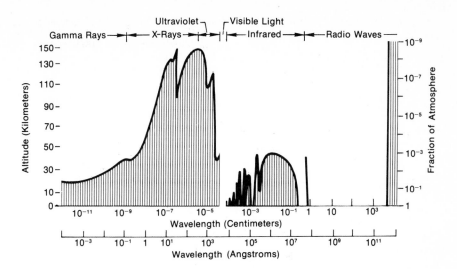

FIGURE P-4   Absorption of radiation in the atmosphere. The upper boundary of the dark area specifies the altitude where the intensity of celestial radiation is cut down to half its original value by the atmosphere. Ultraviolet and x-ray observations must be made from high-altitude rockets or satellites. Infrared observations require high-altitude mountain observatories in dry climates or airplanes. (From Leo Goldberg, "Ultraviolet Astronomy," *Scientific American,* June 1969. Copyright © 1969 by Scientific American, Inc. All rights reserved.)

objects, such as streams of gas falling into a black hole, heated up by the hole's gravity before the gas falls in. Astronomers studying the microwave region pick up very cool objects, such as the radiation left over from the Big Bang at a temperature of a few degrees. There are no objects in the natural universe with temperatures of 0.0035 K, so the radio astronomer does not pick up thermal radiation. Instead, he (or she) picks up nonthermal radiation, which in quasars comes from fast electrons spiraling in a magnetic field (see Chapter 8).

Stars and their properties

Before 1950, astronomers obtained virtually all their information about the universe from visible light, a tiny fragment of the electromagnetic spectrum. Most visible light in the universe is emitted by stars, and astronomy had so far concerned itself primarily with stars. The extension of our view to other parts of the electromagnetic spectrum has introduced astronomers to other types of objects, such as quasars and perhaps black holes. Stars do come into the picture in a supporting role; thus a brief review of the properties of stars is in order.

A star is a ball of gas that produces light. Exactly how much light? An answer involves measuring the star's brightness as seen from the earth and the star's distance from the earth. These are separate tasks.

Stellar magnitudes

Deferring to historical custom, astronomers measure the apparent brightness of stars through the magnitude scale, a very confusing way. This scale originated when Hipparchus made the first star catalogue in the second century B.C. He classified the stars according to their brightness, labeling the brightest stars first magnitude, the next-brightest second magnitude, and so on down to fifth magnitude. Since his catalogue was so useful, astronomers kept the magnitude scale rather than adopt a more logical one. Much confusion about the magnitude scale can be alleviated if you remember that it goes backward: big (positive) magnitudes mean faint stars.

The magnitude scale has been made more quantitative since the days of Hipparchus, but the main features are still with us. Magnitudes can be measured to two decimal places by putting a photometer (a device that acts essentially like the automatic light meter in a camera) at the back end of a telescope and seeing how much light is hitting the photometer. Stars that differ in brightness a hundredfold differ in magnitude by 5. [Thus the magnitude m equals -2.5 log (amount of light) plus a constant.]

Stellar distances

One must know distances of stars if one is to determine whether a given bright star is a big one far away or just a nearby one. For example, Rigel, in Orion's foot, and Sirius are both very bright stars in the winter sky. (Figure 2-2 shows where to find these stars.) Sirius is quite close, while Rigel is much farther away. If these two stars were at the same distance from the earth, Rigel would be much brighter.

The distances to nearby stars are measured by *triangulation*, or *parallax* (see Figure P-5). You examine the position of a given star from opposite ends of the earth's orbit. If the star is close, it subtends a large angle as viewed from the two ends of the orbit. If it is far away, the angle is smaller. The size of the angle thus determines the distance of the star. Now you can tell whether it is a bright star that is close or a still brighter star far away.

The intrinsic brightness of a star, or how much energy it puts out into space, is the star's *luminosity*. To measure this, you measure the star's magnitude as seen in the sky, or its *apparent* magnitude, and its distance from you. If you know that distance, you can then calculate how bright it would be if it were some standard distance, such as ten parsecs from the

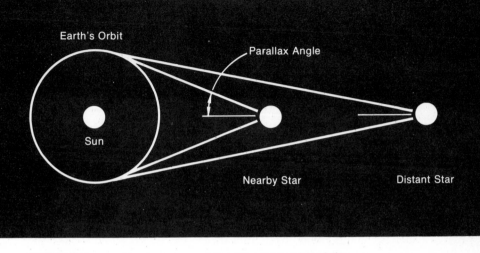

Earth's Orbit

Parallax Angle

Sun

Nearby Star

Distant Star

FIGURE P-5   How to measure stellar distances by triangulation

earth. A star's magnitude at a distance of ten parsecs from the earth is its *absolute magnitude.*

The sun has an absolute magnitude of 4.72. Were it only ten parsecs away—nearby by stellar standards—it would be so faint that it would be barely visible in a light-polluted suburban sky, hard to pick out amidst the illumination from city lights reflected off dust particles in the air.

Absolute magnitude is related to luminosity. For example, if a star has an absolute magnitude of 9.72, it would be five magnitudes (or a hundred times) fainter than the sun would be at that distance. Since the sun has a luminosity of 3.9×10^{33} ergs/sec, this star would have a luminosity only one one-hundredth as large, or 3.9×10^{31} ergs/sec.

Stellar temperatures and spectra

Once you know the luminosity of a star, you want to know whether it is luminous because it is large and cool or small and hot. You want its surface temperature. One way to determine a star's temperature is to look at its color; hot stars are blue and cool stars are red. The electromagnetic spectrum figure, Figure P-3, shows this result. Another way to determine the temperature of a star is to look at its spectrum.

If you spread the light of a star out into a rainbow and photograph it, you have a spectrum of the star. Figure P-6 shows a spectrum of Xi Aquarii, with the colors going from blue to red. The spectrum is just an image of the star in various wavelengths. The spectrum is spread out or widened from top to bottom so that it can be analyzed more easily.

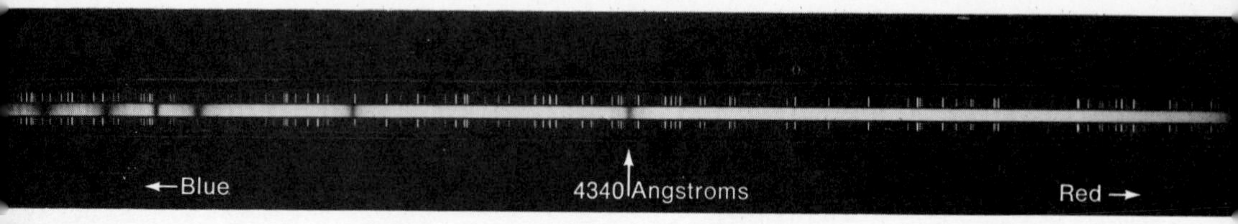

←Blue   4340 Angstroms   Red →

FIGURE P-6   The spectrum of Xi Aquarii. (Obtained by the author at Kitt Peak National Observatory.)

The most interesting feature of the spectrum is the series of dark vertical lines crossing it. Wherever such a dark line appears, the star is not emitting much light at that particular wavelength. *Spectrum lines* exist whenever there is a sudden change in the amount of radiation at a particular wavelength. They can be dark lines or *absorption lines* when radiation is lacking, and bright lines or *emission lines* when, somehow, radiation is concentrated at a particular wavelength. Emission lines are a hallmark of quasar spectra. We'll talk about them later, in Chapter 9, when you need to know what they are.

The dark absorption lines in the central spectrum of Figure P-6 carry a message that, correctly interpreted, provides an enormous amount of information about a star. For the present, focus on the temperature of the star, and for definiteness, focus on one particular line, at a wavelength of 4340 angstroms. (One *angstrom unit,* abbreviated Å, corresponds to 10^{-8} centimeter.) What goes on in the star to produce this shortage of photons at 4340 angstroms?

As radiation travels from the surface of a star toward space, it must travel through the obstacle course of atoms in the star's outer envelope. This obstacle course is filled with thief atoms, which steal energy from the radiation. These thieves are selective in what they steal. Let's follow the paths of three photons through the outer envelope, with wavelengths 4340, 4330, and 4350 angstroms—one at a line center, and two on either side of the line. These photons pass by hydrogen atoms (Figure P-7). These atoms are the thieves. They can exist in a number of different energy states, corresponding roughly to different distances of the electron from the nucleus. It happens that the amount of energy the 4340-angstrom photon carries is exactly the right amount necessary to kick a hydrogen atom from its first excited state into a higher energy state, the fourth excited state. A hydrogen atom in the first excited state acts like a thief, stealing the energy of the photon so it can move up to a higher energy state. The hydrogen atom has little use for the 4330- or 4350-angstrom photons, since they carry either too much or too little energy to move the atom to a higher energy state. Thus the obstacle course of the star's outer envelope has highly selective thieves in it, who steal the energy from only one type of photon. Some photons can travel through easily, whereas others have trouble. Photons that are absorbed by the outer envelope, like the 4340-angstrom one, create gaps, or dark lines, in the star's spectrum.

Whether a photon can make it through the obstacle course or not depends on the types of atoms in the star's outer envelope. The 4340-angstrom photons can be absorbed only by a hydrogen atom in the first excited state. If the star is cool enough that there are relatively few hydrogen atoms sitting around in this excited state, these are not enough available thieves waiting to steal these photons and there is no gap or line in the spectrum. If the star is somewhat hotter than the sun, there are many hydrogen atoms in the first excited state and the gap is very large, as it is in the spectrum of Xi Aquarii in Figure P-6.

Thus you can determine the temperature of a star by examining the strength or weakness of certain spectral lines, such as the 4340 line (called Hγ, or H-gamma). Spectral classification is now a well-developed art. Someone good at this work can just look at the spectrum of a star and obtain a fairly good idea of its temperature, its luminosity, and any peculiarities it might possess. Spectra are grouped into different classes, denoted by letters of the alphabet. Historical reasons dictate that the order from hottest to coldest is not straightforward: O, B, A, F, G, K, M.

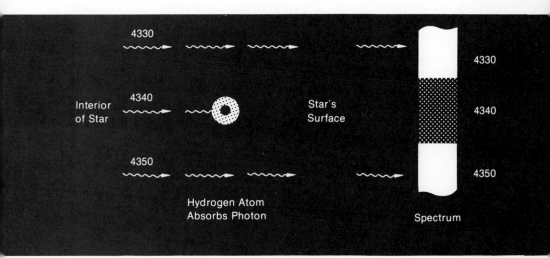

FIGURE P-7   Formation of a dark line in a stellar spectrum. Radiation flows from left to right, from the stellar interior through the surface to the space outside of the star. The 4340-angstrom photons are absorbed by hydrogen atoms, leaving a gap in the spectrum, called an absorption line.

Units of measurement

Many aspects of astronomy involve measuring physical quantities and creating theoretical models to explain these measurements. The measurements are quantitative, expressed in numbers that must refer to some kind of unit.

Length

Most units of length are familiar, but astronomical distances are so large that special units are often used. Distances between stars are measured in *parsecs*. A star one parsec away has a parallax angle (Figure P-5) of one second of arc, and is 206,265 times farther from the earth than the sun is. One parsec is 3.085×10^{13} km. Parsecs are convenient units for measuring stellar distances because stars are roughly one parsec apart in our part of our galaxy. One parsec is a very long distance; it is difficult to visualize exactly how long. People try analogies: If the earth were 6 inches from the sun, the nearest star would be 28 miles (1.33 parsecs) away. But somehow these analogies do not convey a true impression of how vast and empty the universe is.

Energy

Probably the most familiar unit of energy is the kilowatt-hour, a feature of your electric bill. Ten 100-watt light bulbs shining for an hour use up one kilowatt-hour's worth of energy. Most of the energies quoted in this book are in *ergs*. One erg is not much energy; a kilowatt-hour is 3.6×10^{13} (36,000,000,000,000) ergs. The energy of motion (or kinetic energy) of a two-gram insect crawling along at a speed of one centimeter per second is one erg. (In English units, the bug weighs 1/14th of an ounce and covers an inch in 2.5 seconds.) *Electron volts* are also used to measure energy; one electron volt is 1.60207×10^{-12} erg. Photons of visible light, with energies of a few electron volts, carry very little energy in terrestrial terms.

Temperature

All temperatures in this book are expressed in Kelvins or degrees Celsius (centigrade) above absolute zero.

Angles

All positions on the sky are measured in angles, because you can measure only the angle between light rays coming from different objects.

Angles are measured in degrees, minutes, and seconds of arc, where 360 degrees equal one circle, 60 minutes equal one degree, and 60 seconds equal one minute. The moon and sun are half a degree across in the sky. Telescopes enable astronomers doing high-precision work to measure positions of stars with an estimated error of 0.01 second of arc, which equals the diameter of a dime 200 miles away. Such precision can be achieved only by repeating the measurement many times.

Powers-of-ten notation

Astronomers and also other scientists often have to deal with very large or very small numbers. To save effort, a shorthand system has been developed. A number like 3,000,000 is expressed as 3×10^6. Where does this come from? The form of a number is $X.XXX \times 10^n$, where n is the number of zeros to be added to the number, or the number of places the decimal should be shifted to the right. If the exponent is negative, the decimal point should be shifted to the left. Thus 3×10^3 is 3000; 6×10^{-2} is 0.06; and 3×10^{12} is 3,000,000,000,000. (Be sure you can figure these out.) If you prefer to think in words, the following equivalents are useful:

$$10^3 = \text{thousand}$$
$$10^6 = \text{million}$$
$$10^9 = \text{billion}$$

Further, you can extend units of measurement by using prefixes. The metric system uses these extensions a great deal, in common with measurements in astronomy. In this book I shall often talk about kiloparsecs (1 kiloparsec = 1000 parsecs) or megaparsecs (1 megaparsec = 1 million parsecs). If you like large numbers, you can use *giga-* (10^9) and *tera-* (10^{12}), but at this point we usually use powers-of-ten notation. Other prefixes are *centi-* (10^{-2}), *milli-* (10^{-3}), *micro-* (10^{-6}), and *nano-* (10^{-9}). Thus the gross national product of the United States can be thought of as exceeding one trillion dollars, one teradollar, or 10^{12} dollars. If you really want to impress people, tell them about googols and googolplexes. A googol is 10^{100}, and a googolplex is 10^{googol}, that is, $10^{10^{100}}$!

You can find more information about the terminology of astronomy in the Glossary or in an elementary astronomy textbook. "Suggestions for Further Reading" at the end of the book lists a few appropriate selections.

1 INTRODUCTION: THE VIOLENT UNIVERSE

Quasars and black holes are two inhabitants of the new, violent universe that the astronomy of the 1960s revealed. Beginning with the discovery of quasars in 1963, observations revealed that explosions and catastrophic collapse are important stages in the evolution of some astronomical objects. These explosions accelerate electrons to speeds close to the speed of light; some galaxies release as much energy in one second as our sun releases in ten thousand years. The discovery of this violent universe has changed astronomy, producing what several people in the field have called a golden age. The pace has quickened: New discoveries and new interpretations appear every month—sometimes even weekly. Each new advance poses still more questions, sometimes making the universe more puzzling, not less. We do have some ideas of what quasars and black holes are, though important details are still unclear. What we know about them makes them the most exotic and exciting objects in the universe. What we still have to learn about them could lead to a revolution in physics.

A black hole represents the ultimate triumph of gravity in its role as the regulator of a star's life cycle. When a massive star dies, gravity becomes so strong that the star cannot possibly hold itself up. What happens next? We do not know for certain. One possible end to a massive star's life is total collapse. Its tombstone is a black hole. Outside, we can still feel the gravity from the hole, but we can't see it because gravity prevents light from escaping from the star's surface.

If we can't see a black hole, how do we tell what it looks like? The calculations of the theoretical physicist help us here. We know (or think we know) how gravity works, and we can use this knowledge to give us a good idea of what a black hole does to its surroundings. Holes provide explorations of the outer limits of Einstein's theory of gravity. But black holes are more than the products of the fertile imaginations of theoretical physicists. X-ray astronomers overcome the limitations of the earth's atmosphere by launching rockets and satellites to observe radiation that probably comes from heated gas, emitting its last gasp of x rays before it is sucked into a black hole. In the last decade, knowledge of the life cycles of stars has progressed, and black holes have appeared a likely end to the lives of massive stars.

Although black holes contain the strongest gravitational forces in the known universe, quasars are probably the most energetic objects in the universe. The astronomer who examines a photograph of a quasar sees a small, starlike dot, indistinguishable at first from an ordinary star. Closer analysis of the light from quasars reveals that in all cases, this light is much redder at the earth than it was when it left the quasar. This *redshift*, most people believe, means that the quasars are very far away, more distant than the most distant galaxies we can see. Thus the quasars, whose central energy sources are not much bigger than the solar system in some cases, must be emitting more light than the average galaxy. This cosmological interpretation of quasars is supported by recent discoveries of galaxies that are exploding as violently as the quasars. The cosmological interpretation raises the question, Where does all the energy come from?

Investigation of quasars is not the only probe of the distant universe. In the past several years, a wide variety of observations has provided strong evidence that cosmic evolution started with the *Big Bang*, a tremendous explosion that occurred 20 billion years ago. The most important of these observations was the discovery of radio waves left over from the Big Bang. In 1978, Arno Penzias and Robert Wilson, the Bell Laboratory scientists who made this discovery, won the Nobel Prize for their efforts. The big cosmological question now is, Will the expansion of the universe go on forever? If it eventually stops, the universe will someday collapse, draw itself together under the influence of its own gravity, and evolve as a giant cosmic black hole. *Cosmology*, the study of the evolution of the entire universe, explores the outer limits of Einstein's theory of gravitation in the same way that the study of black holes does.

The study of black holes, quasars, and cosmology is part of the golden age of astronomy. A definitive picture of these objects is not immediately available, for the violence in the astronomical universe is somewhat mirrored by turbulence in the astronomical community. The astronomical controversies are interesting in themselves nevertheless, for the story of black holes and quasars illuminates the methods that astronomers use when trying to interpret the universe. I hope incidentally to communicate some of the flavor, excitement, and frustration of this scientific undertaking as I describe the violent universe.

Controversies and changing interpretations of nature are common in the scientific world, despite the popular image of science as some sort of Delphic oracle. Unless you read the research literature, which is published in highly technical, condensed journal articles, most of your scientific knowledge probably comes from textbooks. Textbooks usually seek to encapsulate an area of knowledge in a neat, shiny package. As a result, the textbook reader often gains the impression that the research that led to the knowledge was equally neat and shiny. It wasn't. A researcher must find his or her way through a tangled maze of uncertain observations and contradictory interpretations to try to fill in the total picture.

The scientific process

Two basic conflicts underlie the so-called scientific method, or the way that scientific research is done. In our immediate story, these conflicts provide a background for the study of black holes and quasars. The first conflict is between theory and observation. It involves the attempt to match a mental model of the natural world with the evidence our senses gather. The second conflict is between the desire to uphold the currently accepted basic physical laws and the need to change them by a scientific revolution when conflict between theory and observation becomes overpowering.

Models and reality

J. L. Synge, in *Talking about Relativity,** describes the conflict between theory and observation as an interplay between two worlds—the M world, or model world (as I shall call it), and the R world, or real world. The real world is the world we live in: It contains apples, oranges, stars, black holes (maybe), and other such things. Scientists seek to understand the real world by matching it to a model world.

The model world exists only in people's minds. It contains castles in the air, created from the basic laws of science and mathematics. Also needed is some basic picture of what some object is like to build a model. For example, consider a model of the sun. You start with the relevant laws of physics, or prescriptions of the behavior or matter under certain conditions. You then add the basic picture: The sun is a ball of $1,989 \times 10^{33}$ g of gas, which is 70 percent hydrogen, 27 percent helium, and 3 percent heavier elements. Using mathematics and often a computer, you then discover certain properties of the object you are trying to model. The computer produces a value for the solar luminosity, for example, from the model. The real world now enters, for the model's luminosity should be the same as the sun's real luminosity, within the limits of error.

The connection between the model world and the real world comes from observations, but this connection is not always straightforward.

---

*Bibliographical details for works mentioned in the text can be found in "Suggestions for Further Reading" at the end of the book.

Remember that the world of the observer is a kind of model world by itself. A dot formed by the blackening of a few grains of silver bromide on a photographic plate is not a quasar; it is only an image of a quasar. You then need to interpret the image. As controversies about interpreting photographs that seem to depict luminous bridges connecting quasars and bright, nearby galaxies demonstrate, interpretation of an image is sometimes uncertain.

The matching and modeling process also takes place in other disciplines, some nonscientific. Science is unique in that the matching is quantitative. The model produces some numbers that can then be compared with observed ones. Sometimes these numbers are not completely certain, since experimental error does exist. Models themselves have some uncertainty because of the approximations that scientists make in computing them.

Matching theory to observation is sometimes called prediction. That is, a theorist tells an observer what he or she should see and the observer then goes out and sees if the theorist's idea is true. Some people argue that matching real and model worlds *always* involves prediction, and that prediction is the essence of science. To me, the word *prediction* is too narrow a term, especially in astronomy. We astronomers can only observe the universe, we can't do experiments with stars. As a result, theorists spend a great deal of time trying to model or explain existing patterns in the real world, predicting in reverse. I believe that a more appropriate word is *understanding*. When a model matches reality, we think we know what causes the real world to behave the way it does. (For a fuller discussion about prediction in science, see Stephen Toulmin's *Foresight and Understanding*.)

The progess of science involves both theory and observation. It is often possible for theorists to forget that they are only model-builders. The models become so fascinating that they become real, like the statue of Pygmalion. Synge, a theorist himself, calls fascination with models the Pygmalion syndrome. You get caught up in your own work and become oblivious to the fact that you're only dealing with pencil marks on paper (or ten-foot-high piles of computer output) and not real stars. We astronomers have a particular difficulty here because we can't touch a star. All we know is what our telescopes reveal. As a result, the astronomer can develop great faith in a theoretical picture of, say, a quasar, which has no basis in reality. You have to keep talking to your colleagues or go and make some observations yourself to remind yourself that you are ultimately trying to prove something about the real world. This means not simply indulging yourself in an elaborate mathematical exercise, thinking it represents the real world. The Pygmalion syndrome is moderately prevalent among black-hole theorists, so watch out for it. There is nothing wrong with mathematical speculation so long as you recognize it as such.

Scientific revolutions

But how does science advance? How are obsolete models replaced by new ones? There are always some anomalies, some areas in which the model and observations do not quite match. Sometimes all you have to do is modify the model world in a very small way. You build a new room on the castle in the air to make it look more like the real world. Less often, the change in the model must be more fundamental, since the basic laws of science, the foundations of the model, are at fault. When these laws are replaced, a scientific revolution occurs. Thomas Kuhn, in *The Structure of Scientific Revolutions,* has given a fairly complete and instructive model of what happens during a scientific revolution. I think that his model applies to the revolutions that may take place as a result of the discoveries in astronomy that this book recounts.

We seek a scheme to distinguish between big changes, or scientific revolutions, and minor ones. A big change involves a change in the foundations of the model world, or the paradigm. The *paradigm,* a crucial concept in Kuhn's scheme, is in its narrowest sense a set of rules. These rules are the physical laws, or the foundations of all theoretical models. "You cannot exceed the speed of light," "Energy is neither created nor destroyed."

But a paradigm is more than just a set of rules written down in textbooks. Behind the rules lies a philosophy or world view. During the Copernican revolution, the essence of the change in the paradigm was the realization that the earth, the home of human beings, was not the center of the universe. No one except the astronomers really objected to changing the rules of planetary motion. Only when Galileo made nonscientists, particularly theologians, realize that changing the rules also meant removing the earth from its central location did the hierarchy of the Catholic church realize the magnitude of the paradigm change.

Most of the time—when science is not undergoing a revolution— the processes of normal science are at work. Theoreticians are busy building and refining the castles in the air of the model world, using the existing paradigm, as they seek to match model and observation. Observers approach the matching process from the other end, discovering new patterns that need explaining and testing the models. It is like a picture puzzle, in a way. Observations tell you, for instance, that the image of Mars is always accompanied by two little dots. You expand your model of the solar system to include two moons of Mars; you name them Phobos and Deimos. Nothing basic has changed. You are still using the same foundation, the same paradigm. You have added a new room to Castle Solar System, and filled in a piece of the puzzle. There is always work of this sort to be done, as the world of science always contains a few anomalies, where theory and observation do not quite agree.

Occasionally the processes of normal science, of eliminating anomalies by building more and better models and obtaining more observations, do not work so easily. Sometimes it is impossible to build a model, based on the existing paradigm, that can explain a bothersome observation. Often the situation is not clearly drawn; you can stretch the model to explain the observations but you may be stretching it too far. When you take a reasonably simple model, a small, neat castle in the air, and patch on extra rooms and staircases and cupolas and porticos, the whole structure can begin to look awful. The paradigm becomes badly distorted by this excess architecture. Maybe the paradigm is wrong and needs to be changed. The principle that a scientific explanation must be neat, involving a minimum number of arbitrary assumptions, has proved useful in the past. It has been given the name "Occam's Razor," after a fourteenth-century philosopher.

Paradigm change completes the picture of a scientific revolution. To review: An anomaly develops and grows, as model and reality refuse to be brought together. When the anomaly has resisted repeated attack, a crisis develops. The paradigm must be changed, since the foundations are at fault. A chaotic period follows, during which a new paradigm is sought. When a new one is found and accepted, the revolution is complete and the new order takes over. The new paradigm must be able to explain the anomaly that caused all the trouble in the first place, and it must be consistent with all other known facts.

Black holes and quasars: a scientific revolution?

The objects discussed in this book—black holes, quasars, hyperactive galaxies, and the universe itself—are always interesting, sometimes strange, and occasionally totally baffling. An easy response to a baffling phenomenon is to postulate a scientific revolution, to seek some new physical laws in order to explain it. Yet such an approach ignores the success of current models in explaining what we know already and the elegant simplicity of a universe that is governed by a few universal laws that explain how matter interacts. Is the understanding of black holes, quasars, and the evolution of the universe likely to lead to a scientific revolution? A revolution is improbable, but the possibility is always there.

Start with black holes. A classic scientific revolution occurs when observations present puzzles that theory cannot explain. We are not yet positive that black holes even exist, though the evidence for their presence in the real world is pretty strong. However, the questions that even the theoretical investigations unveil are sufficiently profound that it is possible—though not probable—that a deeper and perhaps radically different understanding of the nature of gravitation will emerge from current research.

A scientific revolution is more likely in the field of quasars. Here observational astronomers have discovered some startling objects. Although theoretical understanding has, so far, been able to explain the principal features of the observations, many details remain to be worked out. Astronomers can explain— more or less—the production of prodigious amounts of energy in galactic nuclei; they explain by the use of theoretical models. Some models involve black holes of galactic dimensions, collapsed objects containing millions of solar masses. The events in these galactic cores seem so outlandish when compared to the quiescent cosmic evolution taking place in the galactic fringes, near the sun, that new laws of physics might seem necessary. But no, these are new phenomena—new ways that Nature has put matter together—that have appeared. We don't need new physics yet.

Our picture of the evolution of the universe is startlingly complete. The 1960s and 1970s saw the appearance of many new facts that have fit quite well into the classic Big Bang cosmology. The picture of the universe expanding from a primeval explosion for 20 billion years is almost too comfortable. No revolutions seem imminent in this field.

This book discusses some strange, unusual objects. But these objects can evolve in ways that require no new physical laws. The same stuff that you and I are made of, atoms and molecules, has probably just been put together in an unusual way to produce black holes, exploding galaxies, and Big Bang universes. The current paradigm is not being destroyed. Although revolutions *are* possible, they are still a little way off, and may never come at all. In science, the odds are against scientific revolutions. Far more incipient revolutions have petered out than have succeeded.

The remainder of this book will explore black holes, quasars, and cosmology, as I examine the possibilities of a scientific revolution and the relations between the model world of the theorist and the real world of the observer. Black holes, the subject of Part One, are primarily theoretical. Where do they come from? What are they? Have we observed any? Do they really exist? The scene then shifts to quasars and their potential first cousins, the exploding galaxies. Here the observers are ahead of the theorists; they provide observations and explanations sometimes follow. What is a quasar? What type of radiation does it emit? Where does the energy come from? Are the redshifts really cosmological? Quasars, which are probably the most distant objects yet observed, introduce cosmology, the subject of Part Three. Cosmology, the evolution of the entire universe, unifies the rest of astronomy. By studying parts of the universe, we seek to contribute to our understanding of the whole.

This book, like all books about an active research field, is in danger of being outdated by the fast pace of research. I shall identify well-accepted fact, informed opinion, and speculation as such in various parts of the book, and shall summarize these categories in tables at the end of each part.

Although I try to avoid speculations, I have found that nonastronomers are generally interested in the most speculative parts of the field. Black holes and quasars are strange objects, and you should keep in mind which of their attributes are well understood and which are fascinating products of theorists' imaginations.

1 BLACK HOLES

Nothing can escape from a black hole, not even light—that is why it is called a black hole. Black holes may form when stars are so overwhelmed by the force of their own gravity that they cannot keep from collapsing. A collapsing star shrinks. If its core is massive enough, it keeps shrinking; the surface keeps collapsing until the entire star has shrunk to a point. Surrounding this point is a volume of space where the gravity is so strong that any light trying to escape to the outside world is sucked back to the central point. No one inside a black hole could communicate with the outside world; he or she would be cut off from our universe by the event horizon.

Such is our model of the black hole. Is there any reason to believe that black holes are part of the real world, not just exotic inhabitants of the model world? Chapters 2 and 3 concern the known facts of stellar evolution. There are some indications that massive stars do end their lives as black holes, but not enough is known about the late stages of stellar evolution to allow us to say that black holes must exist. Chapter 4 presents a model of a standard, well-understood black hole. There are a few stars, we believe, that may have companions that are black holes. Chapter 5 discusses related bits of evidence that black holes exist in the real world as well as in the model world. The frontiers of black-hole research are treated in Chapter 6.

2 STELLAR EVOLUTION: TO THE WHITE-DWARF STAGE

Any star spends its entire life in a struggle against gravity. In the same way that we are pulled toward the center of the earth by gravity, every gas atom in a star is pulled toward the star's center. If the star cannot resist this pull by exerting some sort of pressure, the star will collapse as all the atoms respond to this relentless pull and fall inward. If the star kept on collapsing, all the atoms would eventually end up in the center of the star, and the star would become a black hole. A black hole is a dead star, for it will never become anything else. It just sits in space, dark and menacing, swallowing up any matter that comes too close. Fortunately, black holes are very small; there is virtually no chance that the earth will ever collide with one.

How can any stars exist at all, with gravity always trying to turn them into black holes? The interiors of stars must be, and are, able to exert a pressure to oppose the gravitational pull. If the star is to avoid the black-hole fate indefinitely, this pressure must sustain itself as the star cools off. Some stars are able to exert such a pressure and end their lives as white dwarfs or neutron stars. White dwarfs compact much of the mass of the star into a volume the size of the earth, while neutron stars are smaller still, only 20 kilometers across.

Do just any stars become black holes? If so, which? How does a star end its life? To further explore these questions, which bear heavily on the origin of black holes, we must consider the factors that govern a star's life cycle. These two factors are gravity and pressure. Different kinds of pressure are at work at different times of the star's life. A star that is fully mature but still robust, the sun, can be our model for the interior of a star.

The battle with gravity

Every atom in the sun is attracted to the sun's center. A balance between heat pressure and this gravitational force enables each atom to resist this attraction. To develop a mental image of this balance, imagine the sun as a ball of gas, 1.39×10^6 km (863,000 miles) in diameter, consisting of three layers: the visible surface, called the photosphere, a gaseous envelope containing most of the mass of the sun; and a small central core in which nuclear reactions occur. The entire weight of the envelope presses down on the central core; something must keep the core from collapsing under this tremendous weight.

The weight of the envelope is balanced by the pressure in the central core of the sun. This pressure is tremendous: 2×10^{17} dyn/cm^2, or 2×10^{11} times the air pressure at sea level on the earth. It is this pressure that holds the sun up. This pressure comes from the tremendous heat at the sun's center, where the temperature is about 15 million Kelvins.

But can this pressure be maintained? Heat is continuously leaking out of the sun as it shines. The envelope blankets the central core, slowing down the heat leakage, but the never-ending flow of energy from the hot core, through the envelope, to the photosphere, and eventually to outer space as the sun shines threatens to deplete the central core of the source of its pressure — heat. How does the sun cope with this heat loss?

The sun is continuously generating energy in its central core to replenish its heat. This generated energy balances the energy lost into space, so that the sun can continue to hold itself up (Figure 2-1). The generation of energy maintains the core temperature and the core pressure so that the sun can continue to hold itself up, to shine, and to keep us warm.

The structure of a star like the sun, or the way that it holds itself up, is governed by these two requirements: It must have a central pressure high enough so that the weight of the envelope can be supported, and it

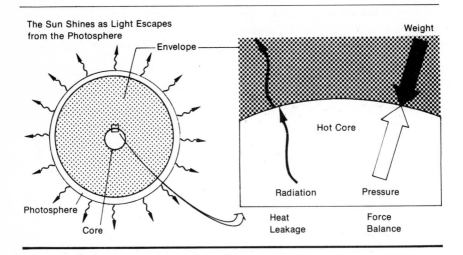

FIGURE 2-1   How the sun keeps from collapsing. Excess pressure in the hot core (white arrow) counterbalances the weight of the envelope (solid arrow). The interior constantly loses energy to the envelope and ultimately to outer space because of the flow of radiation from the core, through the envelope to the photosphere, and to space as the sun shines.

must have some source of energy in the central core so that this central pressure can be maintained in the face of a continuous loss of energy as the star shines. The energy source is crucial; without it, the weight of the envelope would cause the star to collapse. The life cycle of a star is governed by the evolution of this energy source; the energy source opposes the force of gravity.

One way in which the core can provide energy to keep itself hot is to contract gravitationally. As the core shrinks slowly, with gravity forcing it to occupy a smaller volume of space, gas in it is compressed. When any gas is compressed, it heats up. Heat provided from compression can maintain the central pressure necessary to hold the star up. Yet this energy source is not a long-lasting one.

When a star obtains its energy from gravitational contraction of the core, the star as a whole undergoes changes in its structure because its insides are shrinking. Gravitational contraction cannot supply the energy needed to keep the sun in balance for long, because there is not enough energy to be had from contraction alone. If contraction were the sun's only energy source, the sun would exhaust its energy supply in 15 million years. The earth has been warm enough to support life for a few billion years, and the only source of heat for the earth is the luminosity of the sun. Thus a simple, seemingly irrelevant fact—the existence of life on earth a few billion years ago—shows that something other than gravitational contraction must supply the sun's energy. Gravitational contraction just doesn't last long enough.

Nuclear fusion is the energy source that keeps the sun alive. If hydrogen gas is hot enough, hotter than 4×10^6 K, fusion reactions like those that occur in a hydrogen bomb can produce energy. In a sense, the sun is one great big slow-burning hydrogen bomb. Following a chain of nuclear reactions, four hydrogen atoms in the sun's core coalesce to form one helium atom and liberate energy in the process. This fusion is a very efficient process and produces a great deal of energy; the conversion of one kilogram (2.2 pounds) of hydrogen to helium yields enough energy to keep a 100-watt light bulb burning for one million years! The heat energy from this reaction in the sun's core keeps the interior hot. As long as hydrogen is being converted to helium in the sun's central core, the sun will remain stable and continue to keep the earth warm.

Evolving stars

How long can hydrogen fusion last? This process of energy generation, like any process of energy generation, requires fuel, and the sun's supply of hydrogen fuel is limited. Once the hydrogen runs out, other energy sources come into play; the sun, and other stars like it, go through a number of stages during their life cycles. Stars of different masses follow different evolutionary paths once they quit the hydrogen-fusion stage, called the *main-sequence* stage. Before we pass to the fascinating drama of the late stages of stellar evolution, let us pause to examine a variety of main-sequence stars. As examples, I pick a few representative main-sequence stars from the winter sky, shown in Figure 2-2.

The main sequence

Most of the stars in the sky are main-sequence stars, and inside are much like the sun. They keep themselves from collapsing by burning hydrogen in their cores. Different stars, of different masses, may have different detailed structures, but their general properties are basically the same. (They differ, for example, in how they transport their energy to the surface and in the details of the fusion reactions.)

The critical factor in distinguishing between various main-sequence stars is mass. Some stars, such as Sirius, the brightest star in the sky, are more massive than the sun. These massive stars are hotter and brighter than the sun. Sirius, with its mass of 2.2 solar masses, is 21 times as luminous as the sun. If it were where the sun is now, it would be fearsomely bright, and the earth would be too hot to live on. The biggest and brightest main-sequence stars are about 60 to 80 times as massive as the sun and 300,000 times as luminous.

Finding stars less massive than the sun is not so easy, since they are also less luminous than the sun, and therefore fainter and harder to see in the sky. One small main-sequence star, Epsilon Eridani, can be found in the winter sky by tracing out the sinuous outline of Eridanus, the River, southwest of Orion. You need a dark sky and a bit of patience to find it, for Eridanus contains no bright stars visible from midnorthern latitudes. Epsilon lies at a bend in the river, at the western edge of the constellation; its position is shown in Figure 2.2. It is not much further away than Sirius, but since it has only 0.7 solar mass compared with Sirius's 2.2, it is much fainter, with a luminosity that is 0.25 the sun's. It is so faint that it does not deserve a proper name; it is called *Epsilon* because it is the fifth-brightest star in the constellation of Eridanus. (*Epsilon* is the fifth letter in the Greek alphabet. The brightness ordering was done by eye estimates and is not always precise.)

In the winter sky we can see powerful, massive, hot main-sequence stars like Sirius and feeble, small, and cool main-sequence stars like Epsilon Eridani. Why? The luminosity of a main-sequence star depends almost entirely on its mass. In a massive star, such as Sirius, the interior must

FIGURE 2-2   The southern winter sky as it appears from midnorthern latitudes. The brighter stars are shown, along with guideposts to the fainter stars in Eridanus.

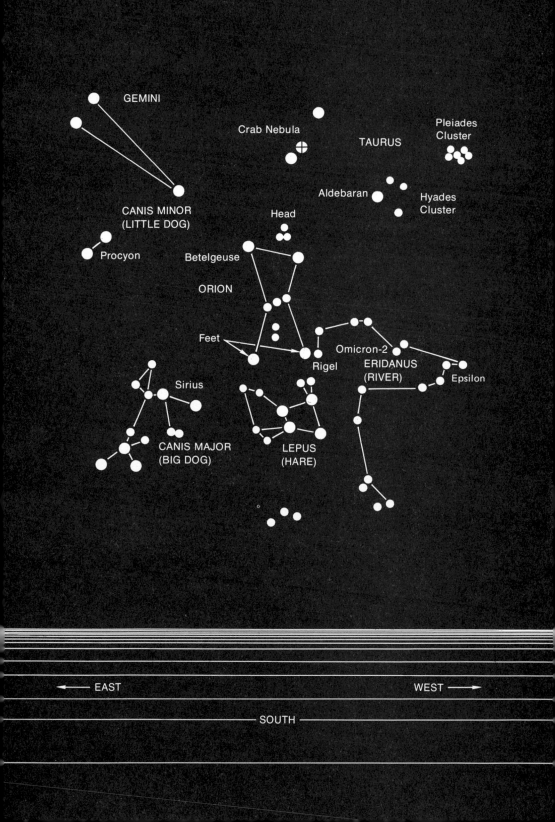

support the weight of the envelope by maintaining a very high pressure. The interior must be quite hot so that the pressure will be kept high enough. As a result the interior contains much heat energy that leaks out into space and makes the star shine brightly. The nuclear furnace must burn rapidly to replenish this leaking energy, and the hot core produces fast-burning, nuclear reactions. In a less massive star, like Epsilon Eridani, the interior is cooler than Sirius's, since there is not so much weight to support. The pressure is less, the temperature is lower, the nuclear reactions proceed more slowly, there is less energy to leak out into space, and the star is dimmer.

This picture of a galaxy of main-sequence stars—some massive, hot, and luminous, some small, cool, and less luminous—does not depict the whole story of stellar evolution. Stars stay on the main sequence only as long as they can burn hydrogen at their centers. Sooner or later, as a star burns all its hydrogen to helium ash in its central regions, its core will run out of hydrogen fuel. Massive stars, in which the central nuclear reactions proceed rapidly, will run out of fuel long before their less massive counterparts. The sun will remain on the main sequence for 5 billion years, and Sirius, a more massive star, will leave the main sequence 1.5 billion years after its birth. All massive stars are young; massive stars burn themselves out to the point of invisibility before they can become old. But what happens in these stars that leave the main sequence when they run out of hydrogen at their centers?

Red giants

Before stars die, they go through some very interesting evolutionary stages. A star that has run out of hydrogen fuel at its center leaves the main sequence and turns to the only other available source of central energy, gravitational contraction of the core. As the helium core contracts, it heats up, thus producing energy to replenish the energy lost as the star shines. Paradoxically, as the core and interior of the star shrink and heat up, the envelope expands and cools. (The details of this process are complex and fortunately peripheral to the story.) The star's surface cools to less than half of its main-sequence surface temperature, and the star swells to tens or hundreds of times its main-sequence size, while the interior is shrinking and increasing in temperature. The star becomes a red giant.

A red giant is continuously readjusting itself to maintain the necessary balances as the core shrinks. Eventually the core becomes hot enough that another nuclear reaction can begin at the center: The helium nuclei, the ashes left over from the earlier main-sequence hydrogen burning, now become fuel as they fuse to become carbon nuclei. (In a red giant, hydrogen is still being fused to helium in a shell, but the shell cannot provide energy to hold the inner parts of the star up.) The core must be hotter than

10^8 K to make the helium-to-carbon reaction go. But once the star starts fusing helium, it has found another way to keep its outer envelope from collapsing, and the contraction of the core stops, for a while at least.

One of the best known red-giant stars is Betelgeuse, the bright star in Orion's right shoulder. Orion is an easily recognized constellation, a bastion of the winter sky. (Figure 2-2 shows where it is. Yes, the name Betelgeuse is pronounced "beetle juice." It comes from the Arabic *Ibt Al-Jauzah,* meaning "Armpit of the Central One.") The surface of Betelgeuse is cool, only "red" hot (about 2300 K), whereas the sun is "yellow" hot (5760 K). Betelgeuse varies in size, expanding and contracting irregularly over periods of a year or so. It is truly a giant star. If you should put the center of Betelgeuse where the sun is, the earth would always be inside Betelgeuse's surface and would be vaporized (Figure 2-3). The sun, too, will become a red giant in 5 billion years, when it exhausts its hydrogen fuel. It will not be as large as Betelgeuse, but the earth will be uninhabitable nevertheless, since the sun will be too bright. We have 5 billion years to prepare for this fate, so the human race should be able to find a new place to go by then.

The center of Betelgeuse is much hotter than the center of the sun. The helium fuel in Betelgeuse has probably been exhausted, leaving a core of carbon. It is probably burning helium in a shell surrounding the hot, shrinking carbon core, with a central temperature of several hundreds of millions of degrees, in contrast to the sun's comparatively mild 15 million degrees. The sun, a middle-aged star, still has fuel left for 5 billion more years on the main sequence. Betelgeuse is old, doing its best to survive on

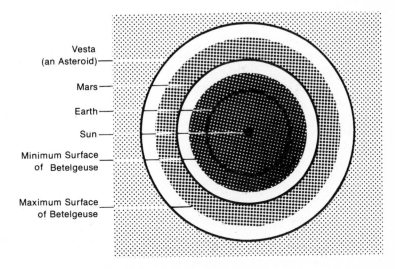

Vesta (an Asteroid)
Mars
Earth
Sun
Minimum Surface of Betelgeuse
Maximum Surface of Betelgeuse

FIGURE 2-3   The size of Betelgeuse, compared with the solar system

the remaining fuel. It is rapidly approaching the day of reckoning when it will totally exhaust its nuclear energy supply.

What happens to stars like Betelgeuse as they near the end of their stellar lives? During the red-giant stage, the star's core is contracting under the influence of gravity. Every once in a while, new nuclear reactions can begin as the center becomes hot enough. Helium burns to form carbon and oxygen; carbon fuses into neon and magnesium; oxygen burns to silicon and sulfur; and neon, magnesium, sulfur and the rest fuse in a series of partially understood reactions to form iron. Once you make iron, no more energy can come from fusion. If you wish to fuse iron with any other atom, you must supply energy; you cannot use such reactions to supply the energy needed to hold a star up. This contraction is shown in Figure 2-4; the star's evolution is indicated downward on the diagram, showing increasing compaction of the center, or higher central density, caused by gravity. The star is held up every once in a while on one of the lines as a reaction takes place that temporarily delays the collapse.

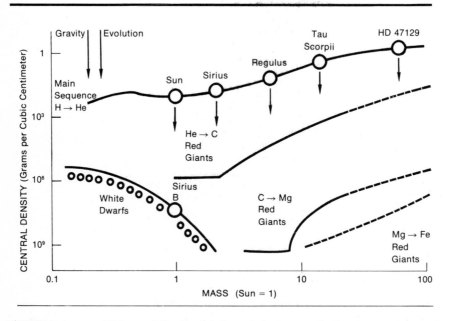

FIGURE 2-4   Stellar evolution through the red-giant stage. Gravity always compresses the core of any star, carrying it downward on the diagram. If the star does not lose mass, it follows a straight line downward. Most stars lose some mass and tail off toward the left. As a star reaches one of the solid or dashed lines, it pauses in its evolution, as energy liberated from fusion of nuclear fuels temporarily halts the collapse. Degeneracy pressure can permanently halt collapse for those stars that become white-dwarf stars.

The details of stellar development during the late red-giant stage are still poorly understood. But the overall picture is clear. Gravity sends orders to the core: The core must contract. Execution of these orders can be temporarily postponed when nuclear reactions can supply energy to the core and keep the interior hot enough for the weight of the envelope to be supported. However, any fuel that the star turns to as an energy source is eventually exhausted. The core again contracts, and the star's evolution proceeds inexorably to the final state.

Real stellar evolution

We thus know a model for the life of a star from main sequence to red giant. How much confidence do we have that real stars actually evolve this way? Does this model of stellar evolution, which sounds so neat, have anything to do with reality? We must bring the observations in. The model has been checked extensively and generally holds up well; representative examples are convincing.

The model must provide that the mass of a star determines the star's evolution, at least through the red-giant stage. In particular, a massive star spends less time on the main sequence than a less massive one. To check this provision, consider two clusters of stars in the constellation Taurus, seen in the western sky in winter. These are the Pleiades, or Seven Sisters, and the Hyades, which surround the bright star Aldebaran. (Aldebaran is not a member of the Hyades cluster; it just happens to lie in the same direction in space. See Figure 2-2 for the location of these objects.) The Pleiades are young stars, and the cluster contains many bright, blue, high-mass main-sequence stars. Such stars do not exist in the older Hyades; they have burned out. The observations fit the model.

Another expectation of the model is that massive stars should be more luminous than less massive stars. Earlier, we compared the sun and Sirius. Figure 2-5 shows how theory and observation compare for a larger number of stars. A number of diagrams like Figure 2-5 have been prepared in recent years, based on different methods of determining stellar masses. The agreement between the theoretical models and observations of real stars is good; this gives us confidence in the models.

There is a potential time bomb ticking away on the shelves of the stellar evolutionists. Theory indicates that there should be a detectable number of neutrinos coming from the sun. These neutrinos are subatomic particles produced in nuclear reactions at the sun's center, and they interact with matter so rarely that they pass right through the sun from center to surface. (A neutrino beam would have to pass through several hundred light-years of lead to be completely stopped.) Seeing these neutrinos would give us clear evidence about the sun's center.

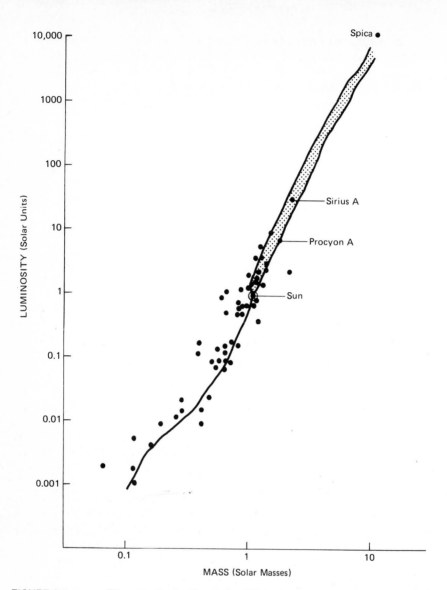

FIGURE 2-5   The mass–luminosity relation. The solid curve and shaded area represent theoretical calculations; the dots represent observations. The agreement between the trend of the dots and the trend of the theoretical curves supports the theory. (Data and theory from E. Novotny, *The Physics of Stellar Atmospheres and Interiors,* Oxford, England, Oxford University Press, 1972, and G. Veeder, *Astronomical Journal* 79 (1974), page 1056)

Dr. Raymond Davis has attempted to find these neutrinos by intercepting them with a swimming-pool-sized vat of cleaning fluid located deep in a South Dakota gold mine. So far, he has had no success. His results indicate that the number of neutrinos from the sun is less than one-fifth of what the theory provides for. Whether there is some minor deficiency in the theory or whether there is a significant weakness in the model is not known at present.

The observational tests of the theory described above all relate to main-sequence stars. It is not so easy to check the theory against red-giant stars, since there are fewer red giants in the sky. Further, most of the interesting changes in a red giant take place in the giant's invisible core, and their effect on the surface of the star is not straightforward. However, red giants in clusters, as far as they can be checked, have the luminosities and temperatures that theory says they should have.

The fate of stars beyond the red-giant stage is still somewhat mysterious. Eventually, a star will finish its nuclear burning and become a corpse, an object that is generating no energy. Scientists have observed two types of stellar corpses: white dwarfs and neutron stars. A third may exist, black holes. The tour of the stellar graveyard begins with the white-dwarf stars, stars that are about the same size as the earth.

Stellar corpses: white-dwarf stars

We left the red giant as its core was continually following the orders of gravity to contract, stopping occasionally to ignite and complete a nuclear reaction in its core. Must the orders of gravity be followed? If so, the star must continue to contract indefinitely, becoming a black hole. As the star reaches the end of its life, heat pressure eventually loses the battle with gravity—the energy needed to sustain the heat has been lost into space as the star shines. When the fuel inevitably runs out, the star's interior cools to a point at which heat pressure is no longer important. But if a small enough core is left after the red-giant stage, a star can find a final resting place as a white-dwarf star, a star no larger than the earth (Figure 2-6).

The first white dwarf ever discovered, and the brightest known today, is Sirius B, also called the Pup, Sirius being the Dog Star. Although Sirius looks like a single star to the naked eye, a good telescope will show that it is a system of two stars orbiting each other. The brighter of the two, called Sirius A or often merely Sirius, is the main-sequence star mentioned earlier. It is difficult to see the companion, known as B since its feeble light is overwhelmed by A; Sirius A is ten thousand times brighter than B. Sirius B is faint because it is small, packing into an earth-sized volume the same

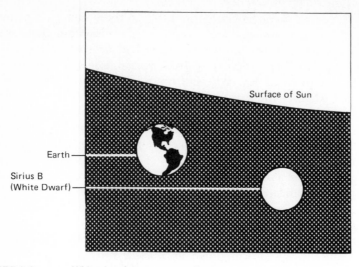

Earth ─────

Sirius B
(White Dwarf) ─────

Surface of Sun

FIGURE 2-6   White dwarfs are very small.

mass that the sun contains in a much bigger volume. Thus Sirius B is much denser than the sun. A matchbox full of solar matter would weigh about 15 grams (half an ounce), while the same volume of Sirius B matter would weigh about 10,000 kilograms (10 tons) if it were weighed here on earth (Figure 2-7).

Holding white dwarfs up

Sirius B, the Pup, has already passed through the red-giant stage. How has it managed to defeat the relentless command of gravity? In the interior of Sirius B, matter is compressed to a state in which it exerts pressure just because it is compressed, not because it is hot. This type of pressure is called *degeneracy pressure,* an unfortunate name, because the word degenerate can be applied in other contexts to the moral character of someone or something. Morality has nothing to do with the interior of a white-dwarf star. It is the high density of matter that produces this temperature. The matter in Sirius B can cool slowly as energy leaks out to space, and the degeneracy pressure will still support the weight of the star against the force of gravity.

Degeneracy pressure exists because the matter in the interior of white-dwarf stars like Sirius B is compressed to an extraordinary state. In an ordinary gas, such as the gas in the center of the sun (Figure 2-8 left), atoms are far apart. They fly about freely and bounce off other atoms,

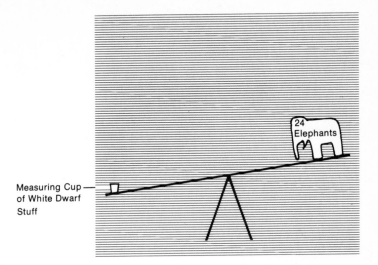

FIGURE 2-7   White dwarf stuff is very dense; a cupful would outweigh two dozen elephants.

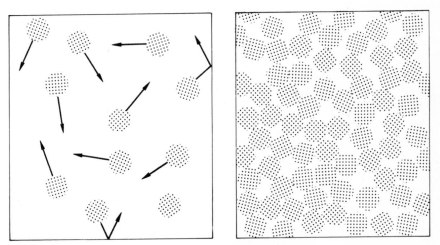

Ordinary Gas   **Degenerate Gas**

FIGURE 2-8   In an ordinary gas (left), atoms are free to move about, since they are much smaller than the distances separating them. When the gas is so dense that the electrons become degenerate, the packing of the electrons exerts pressure (right).

exerting pressure because they move. In the interior of Sirius B, the gas is compressed so much that the electrons touch each other. Their close packing exerts pressure—degeneracy pressure. Degeneracy pressure plays a role only when gases are compressed to very high densities.

The important aspect of degeneracy pressure, from a stellar evolutionist's viewpoint, is that it exists no matter how cool the gas is. The temperature of an ordinary gas is related to the speed with which the atoms are moving. The higher the temperature, the faster the atoms move. Cool an ordinary gas, and the atoms move more slowly and exert less pressure because they collide less often and less forcefully. But the pressure from a degenerate gas comes from the close packing of the material, not from its motion. Whether the degenerate gas is hot or cold does not affect the pressure. The gas can cool off to any degree, and the degeneracy pressure will remain.

Structure of a white dwarf

So how does degeneracy pressure work in the interior of a white-dwarf star, a sphere of degenerate matter? At any point in the star, some kind of pressure must support the weight of the layers on top of that point to prevent the star from collapsing. This pressure is supplied by degenerate electrons in a white-dwarf star. This degeneracy pressure acts no matter how cool the white-dwarf star is. The star then evolves by cooling. Ultimately it becomes so cool that it cannot be seen. It takes about five billion years for a typical white-dwarf star to cool from its highest temperature to the temperature of the sun.

Degeneracy pressure acts like a barrier, preventing a white dwarf from further collapse (see Figure 2-4). Once a star has become a white dwarf, it can collapse no further. Not all stars become white dwarfs, because degeneracy pressure has its limitations. If a star with more than 1.4 solar masses attempts to stabilize itself as a white dwarf, electron degeneracy will not be strong enough to hold it up and the star will collapse further. The existence of this limit, known as the *Chandrasekhar limit,* means that stars that are to become white dwarfs must end their lives with less than 1.4 solar masses of material. Although rotation could, in theory, stabilize a heavier white dwarf, known white dwarfs do not rotate fast enough to allow rotation to be significant.

The white-dwarf model can be checked. If you know the mass of a white dwarf, the theory tells you its density (it can be read off a graph like Figure 2-4). Knowing its density tells you how big the white dwarf should be: High-density white dwarfs are small; low-density ones are large. You can then go and measure sizes and masses of white dwarfs and see how well they conform to the theory. These measurements have been made and vindicate the theory.

Origin of white-dwarf stars

We have lately examined a stellar corpse—a white dwarf, and a star just about to die—a red giant. How are these two stages of stellar evolution connected? Unfortunately, the late evolution of red giants is a complex phenomenon and all the factors are not understood. But there is a theory, which the Polish astronomer B. Paczynski and the American William Rose developed, for the origin of white dwarfs. I present this theory in the remainder of this chapter and then discuss the evidence for it. To this point, this chapter has been factual: well-tested theories and observations. From here to the end, it is informed opinion: ideas that a number of astronomers (including me) believe to be true but that have not been sufficiently well proved to be accepted unequivocally.

Let us consider the past life of Sirius B, the Pup, in the Paczynski-Rose scenario for the evolution of white dwarfs. At one time the Pup was a red giant, considerably larger and brighter than Sirius A. The Pup would then have been brighter than Venus at its brightest—bright enough to cast a shadow here on Earth. It then had more mass than Sirius A now has, probably 2.5 to 3 solar masses, contrasted with A's 2.2 solar masses. The Pup's core then contracted as it successively used up various nuclear fuels. Eventually the core became so dense that degeneracy pressure from electrons could prevent the further collapse of the star. At this point, the Pup still looked like a red giant, but was basically a very hot white-dwarf core, with a mass of one solar mass, surrounded by a gigantic, very tenuous envelope, which contained the remaining mass.

Eventually, in Paczynski's picture, the core lost its grip on the envelope, as pressure from radiation from the core overbalanced the pull of gravity. The envelope drifted off into space, becoming visible as a gas cloud surrounding a small, hot white star. We can see such gas clouds, known as planetary nebulae, around other stars; one of these is shown in Figure 2-9. Rose believes that planetary nebulae are formed somewhat more forcefully, as explosions in the hydrogen-burning shell of the star cause the envelope to blow away. In either case, the red giant loses its outer envelope, violently or nonviolently, leaving behind the core, which contains roughly half the mass that the star started with.

The envelope then dissipated in a few thousand years, a very short time as far as stellar lives are concerned, and the white-dwarf core, the star we now know as the Pup, was left. The Pup cooled as its atoms and electrons gradually lost their energy of motion. However, since the Pup's core was supported not by the motion of the electrons (heat pressure) but by electron degeneracy pressure, the Pup did not contract any more and remained constant in size. It is now quite hot, some 29,000 K (three times as hot as Sirius A), but very small—smaller than the earth. In the future, it will continue to cool. Several billion years from now, it will be too cool to be seen. It will be an invisible black dwarf, a hazard to interstellar navigation.

FIGURE 2-9   NGC 7293, a planetary nebula in Aquarius. (Hale Observatories)

The sky is full of white-dwarf stars. Tens of thousands of white-dwarf suspects have been listed. More than 500 have been examined in detail and are known to be these tiny stars, stars as small as the earth and a million times denser than rock. By coincidence, Figure 2-2 displays the location of three of the four white-dwarf stars nearest the earth, though the stars themselves are invisible to the naked eye. Each of these stars can be located because it is the companion of a brighter star, a star that we *can* see. We have already discussed Sirius B, the companion to Sirius, the brightest star in the sky. Procyon, in the constellation Canis Minor (Little Dog), also has a white-dwarf companion. Omicron-2 Eridani is a triple-star system, containing two main-sequence stars that are cooler than the sun and one white-dwarf star.

Do all stars become white dwarfs? If so, black holes would exist in the model world only, as gravity could not make any star smaller than a planet. Black holes would then pass into the world of esoteric mathematics and out of our immediate consideration. But is is only the small stars that become white dwarfs.

The reason that all stars may not become white dwarfs is that some stars are too massive. Remember, degeneracy pressure can be asked to hold up only a star smaller than 1.4 solar masses. Eventually in larger stars the electrons are squeezed to the point that degeneracy pressure can no longer cope with gravity. Gravity can force a star around the white-dwarf barrier to collapse (shown in Figure 2-4) if the star finishes its life with a mass of more than 1.4 suns. Sirius B could never have become a white dwarf if it had not unloaded 1.5 to 2 solar masses of gas when it shed its envelope as a planetary nebula. Thus the only stars that can become white dwarfs are those small enough to end the red-giant stage with cores smaller than the magic figure "1.4 solar masses."

But saying that a white dwarf forms from a core of less than 1.4 solar masses begs the question. We seek to connect the red-giant stage with the white-dwarf stage a little more securely and really to answer the question, What stars become white dwarfs? Paczynski's calculations indicate that small stars, stars with less mass than some critical mass, become white dwarfs. Latest results indicate that this mass limit is between 2 and 6 solar masses, with a best guess of 3.5.

Unforeseen complications might invalidate this conclusion that all low-mass stars become white dwarfs. Several factors have not yet been included in calculations like Paczynski's. Magnetic fields and rotation may be important. Stars in close binary systems probably evolve quite differently. But it does seem probable that at least some stars follow the hypothetical sequence outlined above for Sirius B: red giant → planetary nebula → white dwarf.

Observational support for this scheme comes from two facts. The number of white dwarfs in our galaxy is roughly equal to the number of stars smaller than 3.5 solar masses that have evolved past the red-giant stage. Stars at the center of planetary nebulae resemble hot white dwarfs, and theoretical calculations indicate that they should eventually become white dwarfs. The sequence low-mass star → red giant → planetary nebula → white dwarf is one possible end to the life of a star.

Stars are gigantic spheres of gas. A star's structure is determined by its need to hold itself up by some sort of internal pressure. Evolution of a star is governed by the changing nature of the processes that maintain this pressure. A main-sequence star obtains its energy from the conversion of hydrogen to helium in its center. When the star runs out of central hydrogen, it becomes a red giant as its core contracts and its surface expands.

The red giant turns to other nuclear reactions in the center as it searches for a way to hold itself up, but it eventually runs out of all possible fuels. If the core is small enough, the core can stabilize itself as a white-dwarf, since degeneracy pressure from electrons can hold the star up without involving heat. This white-dwarf stage is one possible end to stellar evolution: a star with roughly the mass of the sun squeezed into a volume no larger than the earth's. Do all stars end their lives as white dwarfs? No, for we have observed another endpoint to stellar evolution: the neutron star or pulsar, in which one solar mass or so of gas is compressed into a tiny sphere 20 kilometers across.

3 SUPERNOVAE, NEUTRON STARS, AND PULSARS

In the first year of the Shih-huo period, in the fifth moon, on the day of Ch'ih Ch'iu [July 4, 1054], a guest star appeared several inches southeast of T'ieng Kuang [a star in what is now called the constellation of Taurus].
— *Sung-Shih* (History of the Sung Dynasty)

I make my kowtow. I observed the phenomenon of a guest star. Its color was slightly iridescent. Following an order of the Emperor, I respectfully make the prediction that the guest star does not disturb Aldebaran [the brightest star in Taurus, Figure 2.2]; this indicates that . . . the country will gain great power. I beg to store this prediction in the Department of Historiography.
— Yang Wei-T'e, Imperial Astronomer, 1054

So did the Chinese court record the appearance of the most spectacular event of stellar evolution, a supernova. They saw a star appear where no star had been seen before. What was this—the birth of a new star? No, there *had* been a star there; it was merely too faint to be seen with the naked eye. Supernovae mark the death, not the birth, of stars. Suddenly this dying star became much brighter—as bright as the whole Milky Way galaxy. It would be seen in daylight for several months, and in the nighttime for a year or so. "Eventually it faded, and became invisible."[1] (Notes are found at the end of the book.)

The Chinese chronicles mark the first link in a story that culminated some 900 years later in the discovery of neutron stars—objects whose extreme density can be understood by imagining the entire mass of the sun packed into a volume the size of the earth's crust under a typical U.S. county, only 20 kilometers across. We have identified the Crab Nebula (Figure 3-1; Figure 2-2) as the debris of this explosion that the Chinese, Japanese, and Koreans recorded in 1054. This cloud of glowing gas is about three parsecs across and is filled with electrons gyrating around magnetic lines of force at speeds close to the speed of light. Near the center of the nebula is this neutron star, which was recognized as such only in 1968. Neutron stars make white dwarfs look almost normal, as they are 10^9—one billion—times as dense as this other form of dead star.

FIGURE 3-1   The Crab Nebula, Messier 1. This is the remnant of the 1054 supernova. (Hale Observatories photograph)

Neutron stars are the second exhibit in the gallery of stellar corpses. The existence of such objects was first proposed in the 1930s, shortly before it was realized that black holes could also exist. Neutron stars and black holes both remained in the speculative fringes of the model world until the 1960s, when the discovery of pulsars brought the neutron stars into the real world of discovered astronomical objects. They are part of the violent universe, since they are the debris of a supernova explosion. In order to ask whether there is still a third type of stellar corpse—the black hole—we must first find out what neutron stars are and where they come from.

The tortuous tale of how these objects were discovered illustrates the interplay between the model world and the real world, and the story may be repeated with the discovery of black holes. The neutron stars themselves may tell us something about the types of stars that created them. When we have established the neutron-star story, we can extend the evolutionary scenario into a grand scheme for the death of stars, the current working hypothesis for the end of stellar evolution. Although this scheme may not be correct, it is a useful way of organizing what we do and do not know about the way that stars die.

Exploding stars

The Chinese chronicles quoted at the beginning of this chapter referred to a "guest star," or one that appeared where no star had been seen before. The appearance of a guest star is quite a surprise to anyone who is accustomed to looking at the sky. The familiar stars in their familiar patterns—the constellations—have seemed to be immutable. Year after year, the same stars rise at their accustomed time. You always see the Pleiades, part of Taurus (Figure 2-2), rise around midnight in July, and they are followed by the bright star Aldebaran. Thus it always was and thus it will always be, or so it seems. Occasionally, though, the familiar patterns of the stars are disrupted by the appearance of a new star—a bright one, out of place. One such new star was the Chinese "guest star" of 1054, in Taurus. Tycho Brahe, one of the key figures of the Copernican revolution, made his reputation by carefully observing one of these celestial interlopers, the supernova of 1572. His book *De Nova Stella* ("On The New Star") gave Tycho his reputation and these new stars a name: novae, or new stars. Two classes of "new stars" are now recognized, the *novae* and the much brighter *supernovae*.

Novae and supernovae

The names *novae* and *supernovae* are misleading, since these stars are not "new" stars at all but stars that brighten spectacularly as they leap to the stellar graveyard. What is the difference between these two classes of new stars—novae and supernovae? The answers to these questions deepen our probe into the late stages of stellar evolution.

The ordinary novae are much more common but much fainter than the supernovae, which are more spectacular. Novae that are visible to the naked eye occur every decade or so, and they rarely become as bright as the brightest star in the sky. If you look for a nova, you just look for an extra star in some constellation. The luminosities of novae are comparable to the luminosities of the brightest stars in the galaxy—up to 10^6 times the luminosity of the sun.

Supernovae are much rarer; the last one seen by men on earth was Kepler's supernova of 1604. They are much more powerful than the novae; they become as bright as an entire galaxy—billions of times as luminous as the sun. Historic supernovae in our galaxy have been visible in the daytime for as long as two months.

The ordinary novae are related to the white-dwarf sequence of stellar evolution. Some of them are recurrent; they have been known to flare up more than once. The nova phenomenon generally occurs in a binary system that has as one member a very hot white dwarf. A detailed theoretical model of novae that agrees with all observations does not exist yet, but it is generally supposed that gas falling on the surface of the white dwarf triggers the nova outburst. This gas is pulled from the other star by gravity. Although there are many unanswered questions about novae, it seems probable that the nova phenomenon is one branch of the white-dwarf sequence. The supernovae are similar to the novae, in that the same general pattern of a star's becoming vastly brighter occurs. Yet they are different: Once a star has become a supernova, it has died.

Supernovae in history

Historic observations of supernovae are few, because the phenomenon occurs infrequently. Most of the supernovae in our galaxy were recorded by the Chinese, whose court astronomers, careful observers of the sky, recorded the appearance of unusual objects. Their thoroughness is attested to by the fact that they did not miss any appearances of Halley's comet in the last 2000 years. You go back through their chronicles and see that faithfully every 76 years the comet's arrival is recorded. They did distinguish between comets and "guest stars," or novae and supernovae. Somewhere around half a dozen supernovae have been recorded in the chronicles. We have gained much information about the brightness of supernovae from the chronicles, but sometimes it is a little difficult to figure out exactly what the chronicles mean. For instance, the A.D. 185 supernova is described: "Kheihai of the second year of Chung-P'ing [early Han dynasty], a guest star appeared in Hang Mang, about the size of half a mat. It was of five colors. . . ."[2] What is the visual magnitude of half a mat?

But despite the difficulties of interpreting these chronicles in modern terms, the data are invaluable. We have identified the gas clouds that are the remnants of the outer layers of these stars that exploded and died in antiquity. A knowledge of when these gas clouds were ejected helps us enormously in interpreting their nature.

We also ask, How often does a supernova explode in our galaxy? Careful inspection of the astronomical records of the Chinese imperial government, supplemented by analysis of the Japanese and Korean records, show that since the birth of Christ, supernovae appeared in A.D. 1006, 1054 (the Crab Nebula), 1572, and 1604. The guest stars of A.D. 185 and 1181 were probably supernovae, and there were two in A.D. 386 and 393 that *might* have been supernovae.[3]

With eight supernovae in two millennia, it would seem that they are relatively rare events. But most of the Milky Way galaxy—the giant star system that we live in—is hidden from us, obscured by enormous clouds of interstellar dust. Radio astronomers have discovered a radio source called Cassiopeia A. Analysis indicates that Cas A is the remnant of a supernova that would have been seen two or three centuries ago, were it not hidden by interstellar dust. Allowance for the obscuring of the galaxy by dust indicates that a star in our galaxy dies as a supernova every two decades.[4]

Sooner or later one of these supernovae will be close enough so that we can see it, record its light, and further probe this spectacular form of star death.

Modern work on supernovae did not begin until the 1930s, when Fritz Zwicky, the pioneer of supernova research, realized that he must look toward other galaxies to see supernovae with any substantial frequency. In 1933, he began examining nearby galaxies with a 10-inch refractor to see whether any stars had flared up. In 1936, his chances for finding supernovae improved when the first Mount Palomar telescope, an 18-inch Schmidt, was installed to begin the Palomar supernova search program. The telescope still patrols the sky, examining galaxies to see if any stars in the galaxies suddenly become brighter. In 1939, the search program found the brightest supernova of the twentieth century. This supernova, shown in Figure 3-2, was a ninth-magnitude star at maximum. The brightness of this supernova allowed Caltech astronomer Rudolf Minkowski to obtain a magnificent series of spectra of this supernova. Through these and similar series of spectra, we have some understanding of the supernova phenomenon.

Supernovae and stellar evolution

What causes a star to become a supernova? Why should a star suddenly become as bright as an entire galaxy? A supernova explosion is a stellar funeral, marking star death. The star's inner core collapses, for the star has burned all its nuclear fuel and is unable to hold itself up. The collapse of the core releases vast amounts of energy, which causes the envelope to expand. Most of the details of this process are still mysterious, but the general outlines are understood empirically even if we cannot model them.

Violent events in the interior of a supernova, which we do not understand, cause the star's outer layers to expand very rapidly, producing a sudden increase in the brightness of the star. Calculations of what happens to the star's outer layers indicate that these gas clouds should expand at speeds of 5000 km/sec and have temperatures of 10,000 K. The fact that real supernovae have been observed to show such speeds and temperatures makes astronomers more confident that our general picture of supernovae as exploding stars is indeed correct.

But not all details are explained. Supernovae are of two types, called—not surprisingly—types I and II. Theoretical models can explain type II supernovae. But the message of light from type I supernovae is written in a language that is very difficult to understand. Type I supernovae concentrate their light at particular wavelengths; but what atoms emit energy or light at these wavelengths? Disputes continue; no one really knows. Why don't the gas clouds cool off faster than they do? What chemical elements are in the clouds of gas that make up supernova remnants?

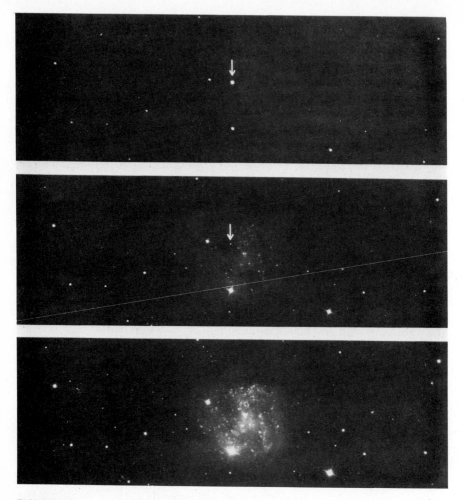

FIGURE 3-2   The supernova in the distant galaxy IC 4182, the brightest supernova of the twentieth century. *Top:* At its maximum brightness, the supernova was more luminous than the galaxy; a photograph showing the star does not show the galaxy. *Middle:* Supernova and galaxy are comparable in brightness. *Bottom:* The supernova has faded and is no longer visible. (Hale Observatories photograph)

These are the types of unanswered questions that make contemporary astrophysics exciting. We more or less understand supernovae, but it is tracking down the remaining uncertainties that deepens our understanding, and occasionally tells us that the working models are wrong.

Eventually the expanding gas clouds of supernovae become large enough to become visible as distinct gas clouds, clouds like the Crab Nebula (Figure 3-1). These clouds expand rapidly; the Crab has become perceptibly bigger in the last 40 years. After thousands of years, these clouds become several parsecs across. An increasingly popular model postulates that a large fraction of the gas between the stars in the Milky Way Galaxy is composed of the shells of ancient supernovae.

We know less than we might about supernovae because the ones we observe now are located at great distances, and therefore, are relatively faint. Powerful telescopes are needed to observe these objects, and powerful telescopes are not always available at the moment that a supernova in a distant galaxy happens to explode. Were one to explode in the Milky Way galaxy, or even in a relatively nearby galaxy, we could observe it more carefully and more precisely. We could even refine theories to a degree that is presently impossible. Once the supernova has exploded, the gas shell expands and drifts off into interstellar space. This shell mingles with the rest of the interstellar gas, possibly compressing it in ways that are currently being analyzed, and ceases to affect the evolution of the central core.

What is left over from a supernova? In some cases at least, not all the star is blasted off into the depths of interstellar space. Theoretical analyses are uncertain. In some models, the star explodes completely; in others, a very massive core is left. Observations show a visible core that remains from one of the historic supernova explosions—the explosion of 1054 that produced the Crab Nebula. That core is a *neutron star,* an object still stranger than the white-dwarf star. Nature has found a second way to defeat the orders of gravity and halt gravitational collapse. In neutron stars, the collapse of a stellar core is halted just short of the point of no return, the point at which gravity overcomes all forms of pressure and the star shrinks indefinitely.

Pulsars: the neutron star discovered

Supernova pioneer Fritz Zwicky and his colleague Walter Baade, in 1934, put forth a far-reaching but speculative suggestion: "With all reserve we suggest the view that supernovae represent the transitions from ordinary stars into *neutron stars,* which in their final stages consist of extremely closely packed neutrons."[5] Five years later, in 1939, J. Robert Oppenheimer (who later became well known in connection with the atomic bomb) and his student George M. Volkoff showed that yes, indeed, neutron stars *could* exist. These objects are about 20 kilometers across, the size of a large city (Figure 3-3). At their centers they are 10^{15} times as dense as water—as dense as the nucleus of an atom.

But all this work was purely theoretical. It is one thing to suggest the possible existence of something that packs the mass of the sun into a volume 20 kilometers across, and another thing to actually *find* such an object. The work of the 1930s showed only that neutron stars could be inhabitants

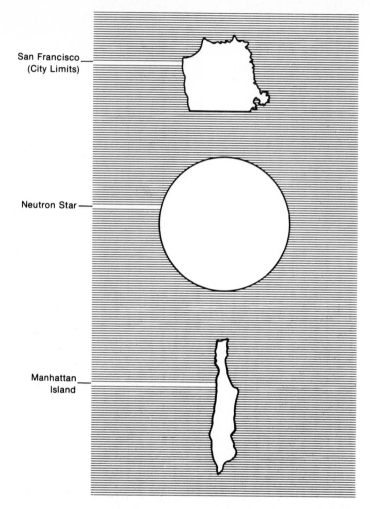

San Francisco __
(City Limits)

Neutron Star —

Manhattan __
Island

FIGURE 3-3   Neutron stars are very small.

of the model world; before they could enter the real world, they had to be found.

Not much happened in the neutron-star field until the 1960s. Supernova work was continuing, of course. During World War II, when Dutch astronomy was enduring the ordeals of Nazi occupation, Jan Oort and his friend the orientalist Duyvendak realized the importance of the "guest star" in the Chinese chronicles. Zwicky and his colleagues in California were discovering more and more supernovae and trying to unravel their story. But the neutron-star idea remained just that: an idea.

In the 1960s, interest in neutron stars revived somewhat. X-ray astronomers were just beginning to send sounding rockets into the upper atmosphere to discover what x rays were being emitted by various celestial objects. Theoreticians began trying to guess what kinds of objects might emit x rays, and the neutron star was suggested as a possible candidate. Neutron stars, being small, would be expected to be very hot as a result of their compression. Hot objects emit high-energy radiation like x rays, so a neutron star might be an x-ray source. This idea was stillborn, however; from what we now know about neutron stars, one of them would have to be closer than the nearest star to be detected by current x-ray telescopes if it were just shining. The seed had been planted, and neutron stars appeared once more in the literature.

Pulsars

In 1967, neutron stars were finally discovered in an unexpected way—as *pulsars,* or pulsating radio sources. Most radio sources in the sky emit a hiss of radio noise, like static on a radio. Pulsars emit their radio-frequency radiation in regular bursts, or pulses. Listen to them, and they tick each time a pulse comes through. It is now generally agreed that these pulsars are neutron stars. But how were they found?

It was at Cambridge University, England, that the appropriate equipment for discovering pulsars was set up. Jocelyn Bell and Anthony Hewish were not trying to discover pulsars. How could they—no one had thought that neutron stars would emit pulsed radio-frequency radiation. Bell and Hewish were on a different track—trying to determine the size of radio sources by watching to see whether the sources twinkled as their radio waves passed through the interplanetary medium. Just as stars twinkle and planets do not on a cold, sharp winter night, small pointlike radio sources would twinkle (or scintillate) as their radiation passed through the thin, wispy gas between the planets, while larger sources would remain steady. The scientists were able to discover the pulsars because they were looking for time variations in the strength of radio sources.

The discovery of pulsars came in the summer of 1967, as Bell noticed something rather odd on the weekly 400 feet of charts the Cambridge telescope produced. What seemed to be bursts of radio emission appeared on the records around midnight. At first, she wondered if these bursts might be caused by some source of interference on the earth. Radio astronomers have to untangle celestial signals from radio emissions from those of more prosaic objects. Radar installations, automobile ignitions, snowmobile engines, and even some electric motors, for example, emit a few radio waves. Bell did not want to announce the discovery of something new in the sky, only to discover that it was just the radio waves from a refrigerator in the lab. In the late summer of 1967, she and Hewish continued to track down the source of the signals, seeking to eliminate the possibility that terrestrial interference was all they were looking at.

By the end of September, though, it became clear that the source was extraterrestrial, for it passed overhead earlier and earlier each night, just as the stars do. On November 28, the source came in very strongly, and the Cambridge astronomers were finally able to pick up the pulses. Further analysis indicated a very remarkable source: an extremely short pulse of 0.016-second's duration arrived every 1.33730115 seconds, and the pulses came in very regularly. Bell then searched through several miles of chart records, and soon three more pulsars were found. Up to this point, the Cambridge group had kept these discoveries secret, but by February 9, 1968, they were ready to announce their discovery to the world. The search now began: What were these pulsars?

The primary focus of the search was some astronomical object that would so *something* — pulsate, rotate, or finish an orbit — very regularly with periods of seconds. We needed a good pulsar clock, since the pulses from all four pulsars were evenly spaced, with accuracies of one part in ten million. (A watch that lost one second per month would be that accurate.) At one point it was thought facetiously that these pulsars might be interstellar navigation beacons for some advanced civilization, and they were jokingly dubbed LGM's (for Little Green Men). Sadly, a few tabloid newspapers got hold of this under-the-table gossip, and the *National Enquirer* blazoned forth that we astronomers had really discovered another civilization. Most definitely we had not, but such sensationalizing can occur at the time of an exciting discovery. The serious question, however, was, What are the pulsars?

The answer was not long in coming. The year 1968, following the discovery of pulsars, saw a vast amount of observational and theoretical work on these objects. Papers appeared frequently in the journals, and sometimes people were so impatient to announce their results that they used the *New York Times* to announce the discovery of a new pulsar. Yet with all this activity, it was October 1968 when the solution was discovered, at the National Radio Astronomy Observatory (NRAO) in Green Bank, West Virginia. David Staelin and Edward Reifenstein, staff members of NRAO, found a pulsar in the middle of the Crab Nebula. The Crab was known as the remnant of the 1054 supernova. The link between pulsars and some other astronomical phenomenon was forged by this discovery, since it was certain that at least one pulsar was a supernova remnant.

Theorists had been busy, too, during these hectic months. One answer to the possible nature of pulsars was the spinning-neutron-star idea that Thomas Gold of Cornell proposed. Others had hypothesized that the pulsars were white dwarfs or peculiar double stars. The combination of the theory that a neutron star was a supernova remnant with the presence of a pulsar in the middle of the Crab strongly supported the neutron-star hypothesis. Furthermore, the only serious competitors to this idea could not explain the very rapid speed of the Crab pulsar: it pulses 30 times every second, the fastest pulsar speed. White dwarfs simply would not pulsate

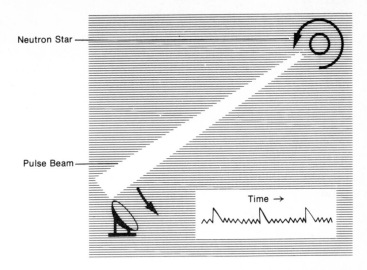

Neutron Star

Pulse Beam

Time →

FIGURE 3-4   The pulses of pulsars come because the radio emission of pulsars is directed in a very narrow beam like the beam of a lighthouse. As the beam sweeps by the radio telescope, a sudden increase of the radio noise, and then decrease, is observed.

that fast. Numerous other arguments have now completed the matching of the neutron-star models to real pulsars. Although we are still refining models and making additional observations to cement the junction between the neutron-star model and the real world, the early vision of Baade, Zwicky, Oppenheimer, and Volkoff was confirmed. Neutron stars really were the corpses left after a supernova explosion.

Understanding pulsars

Pulsars are neutron stars. What have observations of pulsars—observations that have continued since pulsars were discovered in 1968—taught us about the properties of neutron stars? Less than one might expect. A major problem is that no one has yet come up with a reasonably detailed mechanism to account for the emission of pulses by a rotating neutron star. Yet even with those limitations, we have learned something.

All models of the pulse mechanism start from the same general concept, illustrated in Figure 3-4. Somewhere in the vicinity of the pulsar, a beam of radiation is produced. As the pulsar rotates, this beam sweeps the sky in the same way that a lighthouse beam sweeps around the horizon. When the beam points toward a radio telescope, the telescope registers a pulse of radiation. When the beam rotates a little farther around, the telescope does not register radiation from this pulsar. Each time the neutron star rotates, its beam sweeps by the telescope and we see a pulse. A few pulsars have *two* emitting regions, which produce two beams that sweep by the telescope each time the star rotates. One consequence of this beaming mechanism is that some pulsars are unobservable from our place in the universe. If the beams, sweeping around the sky, happen to miss the earth and its battery of radio telescopes, our civilization does not discover this pulsar.

The basic pulsar model of Figure 3-4 was part of Thomas Gold's original suggestion that pulsars are neutron stars. But fleshing out this skeletal model with detailed calculations, trying to provide some concrete theoretical picture that can be tested by future observations, is hard. Near the surface of the gyrating neutron star, rapidly rotating magnetic fields produce electrical fields that can accelerate charged particles like electrons and protons to high speeds. These high-speed particles emit radiation when they whiz through the strong magnetic fields near the pulsar. But how near the pulsar is the radiation produced? Is it right near the surface, where the fields are 10^{12} gauss (10^{13} times—or ten trillion times—the magnetic field at the earth's surface), or is it farther from the neutron star?

Pulsar timing

Because we don't understand how the pulses are produced, we derive little information about the neutron stars from analysis of the pulses themselves. But the pulses do provide a very precise measurement of the rotation rate of the pulsar. Measure exactly how often the pulses arrive, and you know how fast the neutron star is spinning. Do this over a period of time, and you can see the changes in the rate of rotation that occur when the pulsar evolves.

Almost all pulsars slow down very gradually. Radio astronomers can time the pulses so precisely that they can detect period changes of 1 part in 10^{15}, a period increase of one quadrillionth of a second, each time a pulse is emitted. As an example, the first pulsar ever discovered, PSR 1919+21, has a period that lengthens by 0.116 nanosecond, or 0.116 billionths of a second, every day.

Pulsars slow down because they emit energy in the form of high-speed particles. The oldest pulsars are those with long periods, or periods that change very slowly. There are a few pulsars with periods that change so slowly that, apparently, they are more than a billion years old. But these are exceptional. Most pulsars cease to emit pulses after ten million years.

Once a neutron star stops emitting pulses, it just sits there in space, too small and faint to be detected. Once this tiny ball was an enormous, luminous red-giant star, easily seen. A little later it was a spectacular supernova, outshining an entire galaxy. Later still, it produced pulses of radio radiation. But now, and for billions of years in the future, it is just a tiny, slowly spinning, invisible little neutron star, surrounded by a few drifting wisps of gas.

Pulsar irregularities

The last few paragraphs have outlined the history of pulsars in the simplest form. A newly born pulsar gyrates madly, whipping up a storm right near its surface, and accelerating particles to high speeds. These fast particles somehow produce radio waves that sweep around the sky. Soon this frenzied rotation slows down, and after a few million years the pulses fade away. The neutron star continues to exist forever as a tiny, 10-km ball, 10^{15} times as dense as water. Is this the whole story?

The overall pattern—pulsars slowing down and then fading away— is modified by various types of irregular behavior in the youngest pulsars. A year after pulsars were discovered, Paul Reichley and George Downs were using the 210-foot Goldstone antenna of NASA's Deep Space Network to monitor some pulsars. In February 1969, the pulsar in the constellation Vela, PSR 0833−45, suddenly sped up (Figure 3-5). Two more similar events, called *glitches*, occurred later, in 1971 and 1976. What happened?

The pulsar in the middle of the Crab Nebula is a very young pulsar, and it, too, exhibits glitches. Since the Crab pulsar is fairly bright and emits optical pulses (in addition to radio pulses), it can be monitored with relatively small telescopes. One definite glitch occurred on September 28 or 29,

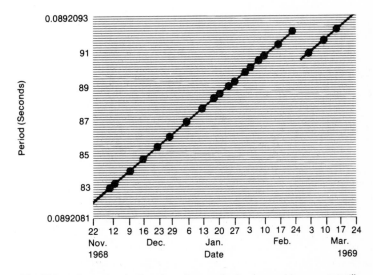

FIGURE 3-5   The Vela pulsar was slowing down in a perfectly reasonable manner until sometime between February 24 and March 3, 1969, when its period decreased by 134 nanoseconds. This speedup, observed once again in this pulsar and twice in the Crab pulsar, is called a glitch. (Adapted from *Nature* 222, p. 229, 1969.)

1969, but this period change was only one one-thousandth as great as the changes observed in the Vela pulsar. Other period changes in the Crab pulsar have been interpreted as smaller glitches, or even as continuous period irregularities.

Thus the well-behaved pulsars slow down relentlessly. Some other pulsars do not conform to this rule and speed up occasionally. A few pulsars change their periods irregularly in very small, almost imperceptible ways. Before relating these changes in period to the neutron star itself, we must take a look inside the neutron star and see what neutron star matter is like.

Structure of a neutron star

Observations of pulsars continued through the 1960s and 1970s. One early pulsar worker, Anthony Hewish, shared the Nobel Prize in 1974—the reward for an important scientific discovery. During this time period, theorists renewed their attacks on the problem of the structure of a neutron star. Material in a neutron star is incredibly dense, 10^{15} times as dense as water. A bit of neutron star matter one one-hundredth the size of a pinhead would be able to balance the two dozen elephants shown in Figure 2-7. We understand the behavior of matter under such conditions only incompletely, and base our models of neutron stars on this under-standing.

Theoretical models of neutron stars vary in detail, but all have the same general structure, shown in Figure 3-6. The behavior of the surface layer, which is a few meters thick, is regulated by the pulsar's magnetic field. The crust, below the surface, is a kilometer-thick layer of material that is 10^{17} times stiffer than steel, with densities that range from 3×10^5 to 10^{14} g/cm^3. Next is a superfluid core, in which circulation currents flow freely, without resistance. The center may or may not contain a solid core of strange subnuclear particles, particles that do not normally exist under less extreme conditions.

Neutron stars are supported by degeneracy pressure. Just as elec-trons in a white-dwarf star resist being compacted by gravity, so do the neutrons in neutron stars resist the gravitational squeeze. Neutrons are far smaller than electrons, so neutron stars are much smaller than white-dwarf stars (recall Figure 3-3).

A great hope of pulsar researchers is that pulse-timing observations can act as probes of the interiors of the neutron stars, and thus test the behavior of matter at densities that are characteristic of the nuclei of atoms. Yet the observations are few. The Vela pulsar has had three glitches, and the Crab pulsar has many tiny irregularities. The tiny irregularities are thought to come from cracking or crumbling of the crust under strain, although they could also result from readjustment of the circulation cur-rents in the superfluid interior. The large glitches in the Vela pulsar are far

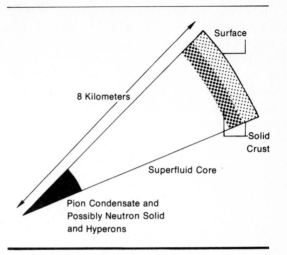

FIGURE 3-6   A slice of a neutron star. The significance of the various regions is explained in the text. (Reproduced, with permission, from "Pulsars: Structure and Dynamics," by M. Ruderman, *Annual Review of Astronomy and Astrophysics,* volume 10. Copyright © 1972 by Annual Reviews Inc. All rights reserved.)

too big to be accounted for by readjustments in the crust. If these glitches are related to interior events, it is the solid core of the pulsar that is changing its shape as the pulsar slows down. These readjustments presumably occur because the pulsar, in slowing down, must readjust its internal structure to a slower rate of rotation.

X-ray pulsars

Many sources of celestial x rays—of which more will be said in Chapters 5 and 6—have been identified with neutron stars in double-star systems. Thirteen x-ray sources pulsate, some with short periods of a few seconds that are similar to the radio pulsars and some with periods as long as half an hour. Analysis of these objects was an active area of research throughout the 1970s, and some new insights into the nature of neutron stars have come from these observations. It is likely that neutron stars in x-ray pulsars are the same kinds of neutron stars as those that make radio pulsars. Only one radio pulsar, PSR 1913+16, is found in a double-star system; all the x-ray pulsars are in double systems.

The masses of the pulsars in the x-ray systems can be measured. In any double-star system, the two stars are held together by the force of gravity. Measurements of the period and size of the orbit of the two stars around each other provide information on the masses of the stars in the system. (Chapter 5 will explore the way that this measurement is made in more detail.) Two of the x-ray pulsars, SMC X–1 and 3U 0900–40, contain relatively massive neutron stars, with minimum masses of 1.1 and 1.4 solar masses. (The names of x-ray sources are a little confusing. SMC X–1 was the first x-ray source found in the Small Magellanic Cloud, a group of stars about 65,000 parsecs away from us. The "3U" refers to the Third Uhuru catalogue of x-ray sources, produced by the x-ray telescope on the *Uhuru* satellite, and 0900–40 refers to the position of the x-ray source in the sky.) The analysis of x-ray pulsars is consistent with the idea that neutron stars form when the cores of red-giant stars become just a little too massive to be held up by the pressure from degenerate electrons.

X-ray pulsars speed up, rather than slow down. Matter is continually dumped on the x-ray pulsar by the other star in the binary system, and this spinning cloud of matter, when it lands on the star's surface, speeds up the rotation of the neutron star. Theorists hope that models of the way in which this matter falls onto the neutron star can provide some information about the way that the spinning, magnetized, pulsating surface is coupled to the stiff crust, superfluid interior, and hypothetical core of the star. Radio glitches and x-ray observations are two probes of the interiors of neutron stars, and they produce different results. Many models can explain radio glitches, but none is compelling. No one has yet provided a complete explanation of the observed period changes in the x-ray pulsars.

Late stages of stellar evolution

Neutron stars and white-dwarf stars

Two types of stellar corpse have been discovered: the neutron star and the white-dwarf star. Does the third type of stellar corpse, the black hole, really exist? Re-examine stellar evolution to ask whether all stars become white-dwarf stars or neutron stars, or whether some become black holes.

To recapitulate: *White-dwarf stars* are held up by pressure from degenerate electrons. They have been observed, for they are large enough to emit detectable amounts of visible light. They evolve by cooling, remaining constant in size because the degeneracy pressure produces the same force to balance gravity no matter how hot the star is. Several billion years after a white-dwarf star is formed, it becomes cooler than the sun, fades, and is difficult to discover.

Neutron stars are held up by degeneracy pressure from closely packed neutrons. Tens of kilometers across, they are too small to be seen directly. But young, recently formed neutron stars can be seen as *pulsars,* sources of pulsed radio radiation. Neutron stars in binary systems emit x-ray pulses, and two are seen as optical pulsars. A few million years after they are formed, the radio pulses stop, and the neutron star becomes invisible to us.

There is one common feature in the evolution of neutron stars and white-dwarf stars. Neither star changes its structure while it is evolving. It has finished falling down the track of stellar evolution shown in Figure 3-7; the tendency of gravity to make stars contract has been stymied. Degeneracy pressure acts as a wall, a prop, preventing the star from collapsing further. These two states of stellar evolution are the end of the line. Once a star has become a neutron star or white-dwarf star, it will evolve no further, if left to itself.

But degeneracy pressure can oppose gravity only if a neutron star or white-dwarf star is small enough. Electron degeneracy, the force supporting a white-dwarf star, can work only if the mass of the white-dwarf star is less than about 1.4 solar masses (recall Chapter 2). The same phenomenon occurs in the world of the neutron star. Try to make a neutron star larger than some limit, and neutron degeneracy pressure just cannot do the job that it does in smaller stars. Gravity will then overwhelm neutron degeneracy pressure, and the star will collapse forever, becoming a black hole.

How massive can a neutron star be? It depends. Conventional ideas about the behavior of neutron-star matter at these densities indicate that the upper limit to neutron-star mass is somewhere between 1.7 and 2.7 solar masses. The discovery that some of the x-ray pulsars have masses that exceed one solar mass rules out some ideas of the behavior of matter at these very high densities.

But conventional ideas about neutron-star matter could be wrong, and an important and still somewhat open question about the death of stars is how massive neutron stars could be if the stuff of which they are made is as stiff as it could possibly be under the stress of gravity. We ask "How massive could a neutron star possibly be?" rather than "How massive is the largest probable neutron star?" Devil's advocates have come up with models for "obese" or overweight neutron stars, neutron-star models that exceed the mass limits set by conventional ideas. Many of these models for obese neutron stars have been placed on a reducing diet by reassessments of the calculations.

Later on, in Chapter 5, we shall tell the full story of the probable discovery of black holes; the possible existence of very massive neutron stars is a small part of it. We think we have discovered a black hole in a double-star system called Cygnus X–1 because the system contains a very small object that is too massive to be a neutron star. The mass of this small object is at least ten solar masses. It is too massive to fit conventional models of neutron stars. Can it fit any possible models of neutron stars?

First ask what sets the limits to the mass of a neutron star. The maximum conventional mass of 2.7 solar masses is based on a particular idea of how stiff matter is at high densities and on a knowledge of how

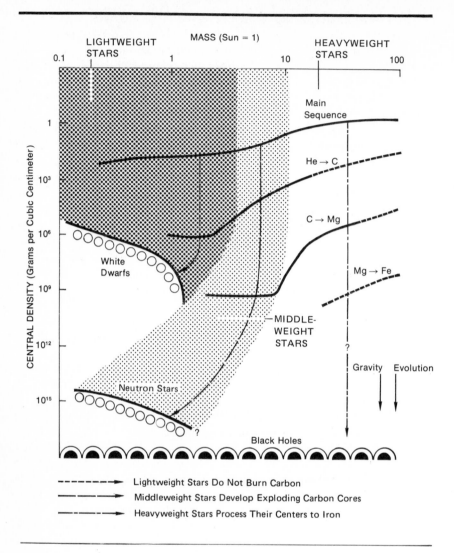

FIGURE 3-7   Overview of stellar evolution, expanding on Figure 2-4. Gravity compresses a star's core, carrying it downward in the diagram. Lines above circles represent the barrier that degeneracy pressure provides. Once a star arrives at this stage it will collapse no further. Three types of stellar corpses are represented at the bottom: white-dwarf stars, neutron stars, and black holes. The working hypothesis discussed in the text argues that lightweight stars (medium shading) make white dwarfs, middleweight stars (light shading) make neutron stars, and heavy-weight stars (no shading) make black holes. Dashed curves are extrapolated from published calculations.

gravity reacts. Take the two key concepts—that of stiffness and that of the proper theory of gravity—and see what can be done to the mass of a neutron star if we push these concepts to their outer limits.

The stiffer matter is, the better it is able to resist pressure and the bigger the maximum mass of neutron stars can be. Yet there is a probable limit on how stiff matter can be, a limit that is set by our acceptance of the notion of causality. The assumption of causality—the notion that a cause must precede the effect that it produces—is one of the foundations of science. Suppose that I, a baseball player, swing a bat and hit a baseball a long distance, far enough to go over the center-field fence of Yankee Stadium (dream!). Every person who analyzed this event would see that the contact between bat and ball preceded the travel of the ball over the fence. Cause must precede effect: The crack of the bat precedes the flight of the ball. For technical reasons that transcend the scope of this book, acceptance of the notion of *causality* corresponds to acceptance of the idea that the velocity of any signal—be it sound, radio waves, or light—cannot exceed the speed of light. This idea is one of the basic precepts of special relativity. (Readers interested in this topic might find one of the many good popular books on special relativity to be interesting and stimulating.[6])

So what does the speed of light have to do with the stiffness of neutron-star matter? To my knowledge, no one has done detailed calculations, but it is generally agreed that this limitation on the speed of light can be related to a similar limitation on the stiffness of matter. Sound travels faster through stiff material than it travels through soft material, and, if causality is to be preserved, a sound signal can travel no faster than a light signal. Using this assumption, several theorists have calculated the maximum mass of neutron stars. Different calculations produce a result of from three to five solar masses. I pick a value of four solar masses for use here. If you abandon causality, but still stick to Einstein's general theory of relativity, you can theorize heavier, obese neutron stars—stars as heavy as eleven solar masses.[7]

Another way to make overweight neutron stars is to presume that Einstein's general theory of relativity, the prescription of the way that gravity acts, is not correct. (We shall discuss other theories of gravity in Chapter 6.) Many, many other theories of gravity have been proposed, theories that preserve the general spirit of Einstein's theory but differ in important details. Virtually all alternative theories fall by the wayside. There are a few that can be barely reconciled with the observations, but these allow for very massive neutron stars, with a limit of 81 solar masses in one theory, and no limit in another. Those of you who seek a definitive statement on obese neutron stars—neutron stars that are more massive than the classical limit of four solar masses set by causality and the general theory of relativity— will not find one in this book. Personally, I don't believe that such objects exist. But they *might* exist—and you and I should keep our minds open to that possibility.

To summarize: Our present understanding of the way that matter behaves indicates that neutron stars can be no more massive than 2.7 solar masses. The notion that, in neutron stars, signals like sound waves travel no faster than light, combined with the theory of general relativity, sets a limit of four solar masses. Other, rather implausible theories of gravitation allow for very large neutron stars. If you are interested in the technical details behind this argument and want some references to the pertinent scientific literature, see note 7.

From red-giant star to the final state

The pieces of the star-death puzzle are all here now: red-giant stars in various evolutionary stages, white-dwarf stars, neutron stars, and perhaps black holes. But how do they fit together? We don't quite know, but we astronomers are working on it. Some evolutionary paths seem well understood, both in theory and observation. Others are less clear. This final section of the chapter tries to put the pieces together as best we can. The puzzle, with tentative connecting links, is depicted graphically in Figure 3-7 for the benefit of those who prefer to see pictures. If you find Figure 3-7 a bit abstract, don't worry; the text includes the story.

At the end of the red-giant stage, a star consists of a small core that has gone as far as it can with nuclear fusion reactions. During this stage, the core has continued to contract, pausing at various way stations when nuclear reactions can temporarily supply pressure to counteract the gravitational squeeze. The main-sequence stage, in fact, is such a way station, but a long-lasting one. But eventually a red-giant star reaches a point at which further contraction does not produce the rise in temperature needed to ignite the next phase of nuclear burning. The distance that any particular star proceeds along the sequence of nuclear reactions depends on the temperature reached in the core, and ultimately on the mass of the star.

Consider lightweight stars first. Their evolution is the simplest and best understood. These stars (at the left of Figure 3-7) have initial masses less than roughly three solar masses. Their cores under the influence of gravity, compress and heat up until they fuse helium nuclei and form carbon and oxygen nuclei. The carbon-oxygen core is the end of the line; this core never ignites. They then shed mass in the form of planetary nebulae and reach their final resting place as white-dwarf stars. (In Figure 3-7, they move downward and to the left in this final gasp of their evolution, moving to the left when they lose mass and sliding downward when they contract. You can think of this picture as a giant pinball machine, with stars sliding downward under the influence of gravity.)

Middleweight stars, too, evolve carbon cores. Yet their interiors are hot enough so that the carbon atoms can fuse and form magnesium atoms, releasing energy and supplying pressure. However, theorists face a problem in determining what happens next. Rotation, magnetic fields, and prescriptions of the way that matter interacts at near-nuclear densities all complicate the calculations. Some theoretical models of middleweight stars produce supernova explosions—that's fine, since supernovae exist—but these model stars blow up completely, leaving no neutron star. We know that some stars blast their outer layers off as supernovae and leave rotating neutron stars, pulsars, as a remnant. Currently the best guess is that it is the middleweight stars that follow this evolutionary sequence.[8]

A hypothetical scenario for the late evolution of middleweight stars, a working model of the late 1970s, is as follows: A middleweight star develops a massive core with a mass that exceeds the limiting mass for white-dwarf stars, the Chandrasekhar limit of 1.4 solar masses. Such a core cannot be supported by the pressure of electron degeneracy and collapses. If the core is not too big, the pressure of neutron degeneracy prevents its final collapse to a black hole, the collapse stops and reverses direction, and the star explodes. The debris of the explosion consists of an expanding shell of gas that we see as a supernova. Once the gas has drifted away, a neutron star is left in the center.

To a great extent, this picture is based on limited models, blind faith, and a few observed facts. It is consistent with the association between supernovae and neutron stars that is demonstrated by the Crab Nebula. One other supernova remnant—a huge gas cloud in the constellation Vela in the Southern hemisphere—also has a pulsar in its center. The neutron stars in x-ray sources are all too massive to be white-dwarf stars.

Heavyweight stars, stars larger than 10–20 solar masses, develop extremely massive cores that can go through a whole series of fusion reactions. Iron is the end of the line of nuclear fusion, for no star can produce energy by fusing iron. Additional energy must be supplied to an iron atom to make it into something else, whether you want to split it or fuse it. Some preliminary attacks on the theoretical problems indicate that there is nothing that can stop the cores of these heavyweight stars from collapsing indefinitely. It is possible that the collapse of these massive cores produces a supernova explosion. Unless overweight neutron stars exist, these cores are far too massive to make anything but a black hole. The cores collapse, collapse, collapse, becoming smaller, smaller, and smaller, until the star disappears, leaving a black hole as the only indication that a star was ever there.

Probably about the only part of the star-death scenario of the last few paragraphs that can be said to be firmly grounded is the idea that low-mass stars make white-dwarf stars. Planetary nebulae (Figure 2-9) have in their centers hot stars that more or less resemble white-dwarf stars. Computation of the number of white-dwarf stars that should be around in the galaxy as corpses of low-mass stars agrees reasonably well with the observed number that are around.[9]

Yet another complication is introduced when we consider that many stars exist in binary systems. One possible scenario for the making of a black hole is to postulate a neutron star in a binary system and then imagining the companion to the neutron star dumping enough mass onto it to make a black hole. Alternatively, a white-dwarf star in a binary system could gain mass by stealing it from its companion and becoming a neutron star — or even a black hole, if it ate enough extra mass.

Thus our only firm conclusion connecting stellar corpses with the red-giant stage is the statement that low-mass stars produce white-dwarf stars. Larger stars may make white-dwarf stars, neutron stars, and black holes in some proportion. The masses of the heaviest known stars exceed 40 solar masses. These stars must lose a great deal of mass to become stable neutron stars. Stars tend to form cores that contain a third of their mass, and it seems reasonable to state that these stars form cores that cannot become neutron stars. They should become black holes. But we can't say that they *must* become black holes.

After all this academic hedging, you are probably thinking, "Get out of the computer center and into the real world. Go find black holes in the sky, not in computer models of the late stages of evolution of massive stars." That is where we are headed next. It is comforting to know that massive stars could make black holes. But first, another chapter from theory. You have to know what a black hole looks like before you try to go out and find one.

White-dwarf stars and neutron stars exist. These stars have managed to win the battle with gravity and stop collapsing, thanks to degeneracy pressure. White-dwarf stars can be seen in the sky; neutron stars have been observed as pulsars. Observations of radio and x-ray pulses from neutron stars have provided some information about the structure of these objects.

But neutron stars must evolve from stars that, at the end of their lives, are fairly small, no more massive than three or four times as massive as the sun. The only firm evolutionary connection between living stars and dead stars is the connection between low-mass stars and white-dwarf stars. One can reasonably suppose that a third type of stellar corpse—the black hole—exists, but we need to know more about stellar evolution before this line of attack can provide us with definitive evidence that black holes do exist.

A better answer to the question, Do black holes exist? would come from the real world with the discovery of one. Let's go and look. First we have to know what we're looking for.

4 JOURNEY INTO A BLACK HOLE

A black hole is the area of space surrounding an object that has shrunk to such small dimensions that its gravity becomes overwhelming. Once anything—even light—is inside the black hole, it cannot escape from the gravitational influence. Hence the name *black hole*. In principle, black holes of any size can exist, but it is difficult to see how a black hole containing less than 1.4 solar masses would form, for such small objects would form white dwarfs or neutron stars when they collapsed. A 1.4-solar-mass black hole would be just five kilometers across.

But a capsule definition of a black hole does not provide enough information to enable us to discover one. We want to bring black holes from the status of theoretical objects in the model world to the status of real objects in the real world. The search for a black hole must begin with a description of what one looks like and how it interacts with the outside world. What black-hole phenomena would render a black hole detectable? Exploration of the properties of a black hole also affects the world of physics, because physicists would like to use black holes as the ultimate testing ground for Einstein's theory of gravitation. Holes are the only places in the universe where gravity is stronger than all other forces. Some black-hole characteristics are quite odd. The theory states that there should be a singularity in the hole's center, a point at which matter is crushed to infinite density and zero volume. Does this singularity really exist? The idea sounds physically absurd. Can a theory that predicts such crazy things be correct?

The structure of black holes can be examined from two points of view. The centerpiece is a thought experiment, in which we follow the adventures of a courageous, suicidal, and indestructible astronaut who undertakes to explore a black hole by falling in to see what is there. One point of view is taken by the outside world, here represented by a rocket ship orbiting the hole at a safe distance. The other point of view is taken by the astronaut. This second approach is purely theoretical, for the astronaut could never return from the hole's interior to tell us whether our ideas are correct or not.

The standard black hole described in this chapter is a product of Einstein's theory of gravitation. This black hole is one that forms after the collapse of a nonrotating star; it obeys Einstein's theory. Adding spin to the black hole complicates the details, but does not affect the essential properties. Extending the black-hole idea to hypothetical holes that do not form from collapsing stars, or trying to modify Einstein's theory so that various uncomfortable aspects of black-hole phenomena go away, opens up new landscapes that lie at the frontiers and fringes of black-hole research (more on this in Chapter 6). The phenomena described in this chapter are based on well-understood theoretical results drawn from Einstein's theory of gravitation.

Black holes are, so far, entirely theoretical objects. Since it is plausible to expect that they exist, their properties are worth exploring. No black hole has yet been found, although there is one object that certainly looks very much like one. Black holes exist primarily in the model world. It is very tempting, especially for people who like science fiction, to succumb to the Pygmalion syndrome and endow these model black holes with a reality that they do not yet possess. Beware of this.

History of the black-hole idea

Pierre Simon, Marquis de Laplace, first thought of black holes in 1796. His initial musings were based on Newtonian gravity and Newton's now discredited corpuscular theory of light. Newton thought of light as little pellets or corpuscles, having properties like very small billiard balls. Laplace realized that such corpuscles could not escape from the surface of a sufficiently massive body. He wondered whether space would be full of these *corps obscurs,* as he called them. Maybe they would be as numerous as stars? However, there was no way to test his idea, and it disappeared into the libraries, never quoted or explored by others.

Shortly after Einstein's theory of gravitation appeared, the German physicist Karl Schwarzschild calculated what space would look like surrounding a point-mass. He thus discovered the standard black hole as an inhabitant of the model world, but he, like Laplace before him, had no idea whether such a body could really exist. This was not determined until 1939, when J. Robert Oppenheimer and a student, Hartland Snyder, showed that a cold and sufficiently massive star must collapse indefinitely, becoming a black hole. The Oppenheimer-Snyder work, appearing about the same time as the Oppenheimer-Volkoff paper on neutron stars, reached much the same conclusion: Black holes could exist. They might be real objects, not just mathematical games that people played with Einstein's theory. In the 1960s, with a revival of interest in Einstein's general theory of relativity, black holes were intensively investigated and their detailed properties were elucidated.

This history bears some resemblance to the early history of neutron stars. Both types of stellar corpse were first known as theoretical objects. Very little research was done on them until the 1960s, when advances in astronomical instrumentation and a revival of interest led them to be investigated more intensively. Unfortunately, black holes are somewhat more difficult to find than pulsars, as will become evident shortly.

Black holes and Einstein's theory of gravitation are very closely tied together. You cannot describe a black hole even remotely well with Newton's theory of gravity because Newton's theory works only when gravity is weak or speeds are small. Newton's theory may work when you try to calculate the trajectory of a thrown baseball, but close to the surface of a black hole, the effects of general relativity are overwhelming.

If black holes are so closely connected to Einstein's theory, you might well ask, What happens to them if Einstein is wrong? But Einstein's theory is almost certainly the correct theory of gravitation; it is accepted by almost all working physicists. Furthermore, most rivals of Einstein's theory are really modifications of it, descriptions of gravity that differ in detail but not in spirit from Einstein's original theory of general relativity. Experiments have ruled out almost all alternative theories of gravitation. (Chapter 6 will describe these experiments.) Those alternative theories that add a small additional effect to the basic framework of Einstein's theory amount to slight variations on a theme, and they produce black holes that are, for all intents and purposes, identical to those described in this chapter. There is one theory that is almost ruled out by the observations: Rosen's "bimetric" theory of gravitation. It is quite different from Einstein's in that it does not produce black holes.

The centerpiece of any of these theories of gravitation is the idea that gravity when it is near a strong source of gravitation, like a black hole, modifies the way that time flows and the way that distances are measured. The next few pages follow the journeys of several hypothetical astronauts who venture forth into these strange parts of the universe.

The view from a distance

One way to approach the understanding of what happens near a black hole is to suppose that you are a spaceship pilot of the future who happens to come upon one. The spaceship scenario is not necessary for our thought experiment, but it is true that you have to be quite near a black hole to really see what is going on. Furthermore, when you want to explore the black hole's immediate surroundings, you have to be close enough to drop probes into it and see what happens to them.

You can only sense the existence of the black hole through its gravity. You cannot see a black hole. No light escapes from it—that is why it is called a black hole. The first noticeable effect sensed by a spaceship would be a weak but relentless gravitational pull. The spaceship would begin to fall toward the black hole. There would be nothing very unusual about this pull, though. Gravity is ubiquitous in the universe, and any massive object would deflect the path of a spaceship.

If you wanted to explore the black hole, you might well choose to go into orbit around it. The spaceship's motion past the hole would prevent it from falling into the hole. The ship would fall around the hole in the same way that the moon falls around the earth, following an orbit. You could measure the mass of the hole by determining exactly how much the hole was pulling you off the straight-line path you were on before you encountered the object. If you were in orbit, you could measure how long it took to complete one circuit of the hole. The more massive the hole, the faster the orbiting. (A bigger hole pulls on you harder, and you have to travel faster to stay in orbit around it.) If, for example, it took 3.7947 months to make a complete circle one astronomical unit away from the hole, you could deduce that the hole had ten times as much pulling power, or ten times as much mass, as the sun. (One astronomical unit is the distance from the earth to the sun, 1.495985×10^8 km.) This is just Kepler's third law, used by astronomers to deduce the masses of double stars (described in more detail in Chapter 5). Nothing new here.

It is only when you look toward the hole that you see something a little odd. Most ten-solar-mass objects in the universe are visible. You would expect to see some sort of star at the center of your orbit; a ten-solar-mass star is generally a bright blue one. You would see nothing. You would be in orbit around an invisible object. The hole itself would be 0.08 second of arc across, or as big as a dime 15 miles away. It would take a 400-inch telescope to see the hole as a disk, even if there were anything there to see.

If you were lucky, you might be able to detect the hole in another way, since you would see light rays from stars on the other side of the hole bend as they passed by the hole (Figure 4-1). You could do this only if there happened to be stars in the right places. According to Einstein, the paths of all particles, even photons, are affected by gravity. Thus the path of a photon or light ray is bent by a gravitational field in the same way that the earth bends the path of a thrown baseball and causes it to fall. Since the paths of light rays from distant stars would be bent as they passed by the hole, these stars would seem to be out of position. The light would have to fall around the hole to reach the spaceship. This bending of light has been observed near the sun as a shift of star positions during a solar eclipse. Thus there is nothing very new about this effect; it is just that near a black hole the effect would be considerably larger.

Unfortunately neither of these two methods that a spaceship pilot near a black hole could use to detect the hole's presence would work from the surface of the earth, far away from the hole. Obviously you need to be near the hole to go into orbit around it. The bending of light by the hole, if seen from the earth, would be minuscule. No equipment now available could detect it. Observations of holes must be based on phenomena arising from the interaction of holes with the material around them.

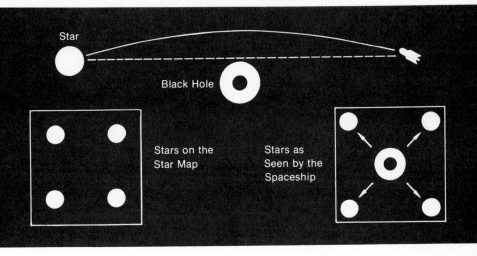

FIGURE 4-1 If you look toward a black hole from a distance, light must fall around the hole to reach your eyes. Stars near the hole would appear to be out of position.

Thus our exploration of the hole will have to be extended to the depths of the hole. Furthermore, it is impossible to find out very much about the nature of a black hole simply by looking at it, since there is nothing much to see. Our spaceship will have to send a probe toward the hole and see what happens to it.

In the next few pages, for the sake of definiteness, I shall present the numerical details as they would apply to an astronaut exploring a hole of ten solar masses. Black holes forming from stellar collapse would be roughly this size. It is unlikely that any significantly smaller holes exist in the real universe; no one has figured out a way to make a hole of less-than stellar-mass. Large holes with upwards of 10^6 solar masses may exist at the center of active galaxies and quasars as the energy source of these objects, according to one idea. Events around a large hole would be qualitatively the same as events around a smaller hole, except for the strength of the tidal forces, which would be the first phenomenon encountered by a probe that was dropped into the hole.

Tides near a black hole

As the probe approaches the black hole, nothing unusual happens for a long time. Since the peculiar effects of a black hole are evident only close to the hole, this is not surprising. The first uncomfortable effect is noticed long before the neighborhood of the hole is reached, but like the bending of light, this effect is only a familiar force amplified to uncomfortable proportions; that is, tides.

Consider the effects of gravity on a person, perhaps our heroic astronaut falling to his doom, as he falls feet first toward the black hole (Figure 4-2). His legs are nearer the hole than his head, and the gravitational force pulling on his legs is stronger than the force on his head. The difference between these two forces is the tidal gravitational force, which, if unopposed, stretches the astronaut out into a long cylinder. The force arises because the closer you are to a massive object, the stronger the gravitational force is. These tides are a common feature of the interaction between two bodies, such as the moon and the earth. This tidal effect produces the ocean tides that are a familiar fact of life along the coast. They act on our bodies all the time, since our feet are nearer the center of the earth than our heads. On the earth, however, they do not present any serious threat, as they are very weak. Near a black hole they are much stronger.

Another tidal force acts as a straitjacket, squeezing the astronaut's shoulders together. All parts of the astronaut fall toward the center of the black hole. In particular his two shoulders fall on converging paths. Gravity draws them together. It is a somewhat gruesome fate for our hero, being stretched out as though on a rack and compressed by this gravitational straitjacket. Bones and muscles must resist these forces if the body is to survive. How close can anyone get to the black hole and put up with this sort of treatment?

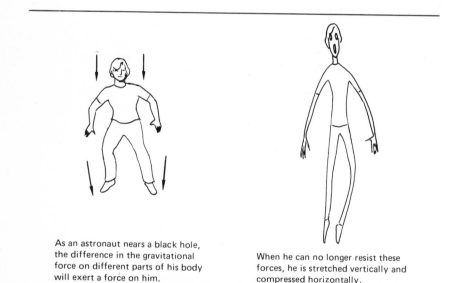

As an astronaut nears a black hole, the difference in the gravitational force on different parts of his body will exert a force on him.

When he can no longer resist these forces, he is stretched vertically and compressed horizontally.

FIGURE 4-2   Tidal forces would distort an astronaut's body near a black hole. (The difference between the forces on different parts of the astronaut's body is exaggerated.)

Optimistically, the human body can withstand a strain of ten times the earth's gravity without breaking. Our astronaut would be 3000 kilometers from the ten-solar-mass black hole when the tidal forces became this strong. He would be killed by them before he ventured any closer. It is not easy for a live astronaut to investigate the properties of a ten-solar-mass hole.

A very large black hole would be a more favorable candidate for investigation, since you can get closer to it before the tidal stresses become severe. If you were investigating a hole larger than 10^4 solar masses, you could reach the inside of the hole before the tidal forces pulled you apart. Holes this big may exist at the centers of galaxies, but as the smaller holes are likely to be more common and are certainly more detectable from the earth, I shall stick with them.

These tidal forces are the black-hole phenomenon that gives us a chance to observe real black holes. As gas falls toward a black hole, it is compressed by these same tidal forces that make life unpleasant for our imaginary astronaut. As this gas is compressed, it heats up. Hot gas emits high-energy radiation like x rays, and it is these rays that are the sign of a black hole. Not all x-ray sources are black holes; you must closely investigate any black-hole candidate to rule out other possible sources for these x rays. (I shall return to this subject in Chapter 5.)

But these horrendous tides are not the phenomenon that makes the black hole one of the strangest concoctions that has been extracted from Einstein's theory of gravitation. The essence of a black hole is the *event horizon*, the point of no return. At the event horizon, you would have to travel at the speed of light to escape from the black hole. Since no material object can travel that fast, nothing can return to the outside world once it has stepped over this invisible boundary. As we watch our probe fall deeper into the hole, we explore the neighborhood of the event horizon.

Approach to the event horizon

The black hole affects space and time around it in two ways. Its gravity distorts and hinders the passage of signals from objects near it as they try to communicate with the outside world, and greatly distorts the passage of time. The event horizon is the edge of a black hole. Once past it, you are inside the hole, caught in its grip forever. You cannot return to the outside world. The horizon is a spherical boundary whose radius depends on the mass of the black hole. Fortunately, this radius is quite small, so the hole is small too. The radius, also called the *Schwarzschild radius* in honor of the discoverer of black holes, is numerically equal to 2.95 kilometers times the mass of the hole in solar masses. Our ten-solar-mass hole is thus 30 kilometers in radius or 60 kilometers across; an object this small is very difficult to see in interstellar space, much less run into. Because black holes are so small, the chances of a collision between the earth and a black hole are extremely remote.

The essence of Einstein's theory of gravitation is that gravity acts on particles by distorting space and time. Thus our exploration of the ten-solar-mass hole will proceed mostly by dropping clocks in the vicinity of the hole and seeing what happens to them. The behavior of falling clocks and signals from them are affected by the motion of the clocks themselves. For instance, the clocks slow down because they are moving. (This is one result from Einstein's *special* relativity theory, which has nothing to do with gravity.)

Thus our thought-experiment scenario for exploring the hole will have to be a bit more complex. Our exploring probe, manned by an inde-structible astronaut, sets forth on its journey into the interior with a large collection of clocks—reliable and accurate clocks, strong enough to with-stand the great tidal forces near the hole. Every once in a while, the astro-naut releases a clock, tossing it into orbit around the hole. When it is in orbit, it is not moving very fast relative to the distant observer, so we can watch it from our distant vantage point and see how the black hole's gravity affects space and time, entirely apart from the way that the motion of the probe hurtling to its doom affects the way that its clocks and meter sticks work. Simultaneously, we ask what our own clocks read during the progress of the journey.

The events we see as we follow the black-hole adventure are sum-marized in Table 4-1.[1] The astronaut puts the first clock into orbit when he is 300 kilometers, or 10 Schwarzschild radii, away from the hole. What odd phenomena do we observe?

The first odd effect is that light from this clock, in orbit around the hole at a distance of 10 Schwarzschild radii, is redshifted. The photons from this clock lose energy as they struggle out of the intense gravitational field near the hole. They are transformed from energetic short-wavelength photons into tired long-wavelength photons, just as a person loses energy climbing a flight of stairs, doing work against the earth's gravity. A loss of energy is a loss of frequency, or an increase in wavelength. Red photons are long-wavelength photons. Therefore this phenomenon is called a *gravita-tional redshift*.

The column labeled Redshift lists the quantity, conventionally noted by the letter z, equal to the fractional change in the wavelength of light emitted by the clocks. (Mathematically, if the wavelength emitted by the clocks is λ and the change is $\Delta\lambda$, then $z = \Delta\lambda/\lambda$.) Thus, if the clocks are illuminated with green light with a wavelength of 5000 angstroms, that light will be shifted by 250 angstroms as it travels to the distant rocket ship (0.05 = 250/5000). Such effects would be observable. A shift of 250 ang-stroms moves green light into the yellow part of the spectrum, toward the red.

TABLE 4-1   THE EVENTS WE SEE AS WE FOLLOW THE BLACK-HOLE ADVENTURE

DISTANCE FROM HOLE CENTER		REDSHIFT	RELATIVE CLOCK RATES	TIME	
In kilometers	In multiples of the Schwarzschild radius			As seen by the rocket ship	As seen by the probe falling in
1 A.U.	4.96×10^6	0	1	0	0
300	10	0.05	1.05	204 hr 33 min 50.1129 sec	204 hr 33 min 49.6681 sec
240	8	.07	1.07	50.1135*	49.6687*
180	6	.10	1.10	50.1141*	49.6692*
120	4	.15	1.15	50.1148*	49.669666*
90	3	.22	1.22	50.1150*	49.669854*
60	2	.41	1.41	50.1153*	49.670012*
45	1.5	.73	1.73	50.1155*	49.670078*
33	1.1	2.32	3.32	50.1157*	49.670091*
30.03	1.001	30.25	31.25	50.1162*	49.670123*
$30 + (3 \times 10^{-8288})$	$1 + 10^{-8289}$	$10^{4144} - 1$	10^{4144}	205 hr	49.670133*
30	1	∞	∞	∞	49.670133*
15	0.5	—	—	—	49.670177*
0	0	—	—	—	49.670200*

* All items marked with the asterisks refer to 204 hr, 33 min plus the tabulated number of seconds.

Along with the gravitational redshift, the clocks close to the hole seem to slow down, as is shown in the column Relative Rate. This column lists the number of seconds ticked off by the distant observer's clock in the time that it takes the clocks near the hole to tick off one second. Thus the rocket ship's clock would tick off 1.05 seconds for every second ticked off by the clock 300 kilometers away from the hole. Clocks near the hole seem to run slowly, and events will take place in slow motion.

Once again, the gravitational redshift and the slowing of clocks are nothing very new, but close to a black hole they become extremely large. The gravitational redshift has been observed elsewhere—in white dwarfs, in the sun, and in photons sent from the basement to the top floor of Jefferson Physics Laboratory at Harvard. These two effects are related; if you run your eye down the two columns you will notice that the relative clock rate is simply $1 + z$, where z is the gravitational redshift. A relation between the rates of two atomic clocks at different altitudes was one effect observed when atomic clocks were flown around the world in commercial

Probe Falling In

The Event Horizon

Time (Seconds) ⟶

FIGURE 4-3   A movie showing how a distant observer would see a space probe approach a black hole. As the space probe falls in, its motion appears to freeze at the event horizon. (The numbers are meant to apply to the fall of a rocket ship toward a ten-solar-mass black hole as described in the text.)

airliners, but the principal effect observed in that experiment was not the effect of gravity. These peculiar effects near a black hole are just more dramatic versions of effects verified experimentally here on earth.

The astronaut's own clock and ours would disagree on how long it took him to reach the 300-kilometer checkpoint. He says that he deposited the first clock in orbit 204 hours, 33 minutes, and 49.6681 seconds after he left orbit. We should imagine his fall to that point to have taken a little longer, 204 hours, 33 minutes, 50.1129 seconds, a difference of 0.4448 second. This half-second difference between our clocks and the astronaut's clocks seems unimportant, for now. But wait. All these strange black-hole effects will become bigger as we probe closer to the event horizon.

The probe continues to fall downward as our astronaut comes closer to the event horizon, the goal of his mission. The gravitational redshifts become larger. At 120 kilometers the redshift is 0.15, and the light illuminating the orbiting clocks looks yellow to our eyes, with a wavelength of 5600 angstroms. The orbiting clocks have slowed down in proportion. They tick every 1.15 seconds according to our clocks, sitting in the orbiting rocket ship a safe distance from the black hole. That annoying half-second difference between our clock and the astronaut's is getting a little larger.

We must look fast at the orbiting clocks 120 kilometers away from the hole. Their orbits are unstable. They may be able to remain in orbit for a while, but any deviation from a circular orbit will cause them to be captured and eventually swallowed by the hole. As the astronaut approaches the hole, putting clocks into orbit as he travels, the redshift of the light from the clocks becomes larger and larger. At 90 kilometers, the

50·1163 51·1163 INFINITY

Time (Seconds) ——▶

effect is truly a redshift as portrayed here. The green light illuminating the clocks will be red, with a wavelength of 6100 angstroms, by the time it reaches our eyes. At 60 kilometers, the clocks will have their light shifted beyond the red to 7000 angstroms, in the infrared part of the spectrum, by the hole's powerful gravity. We should have to look at them with an image tube, a device developed for use in Vietnam, which picks up infrared radiation. (This gadget has had many peaceful applications in astronomy, as it improves the effectiveness of telescopes.)

But the image tubes work well only up to a point, as the redshift becomes larger and larger the closer you get to the hole. When the astronaut is 30.03 kilometers from the hole's center, or 0.03 kilometers (30 meters) from the event horizon, the supposedly green light illuminating the clocks will have a wavelength of 150,000 angstroms, far in the infrared, beyond the range of an image tube. Is there no end? Will this redshift never stop increasing? No, there is not; the redshift of photons increases without limit as you approach the event horizon.

Along with the increase in redshift comes another, more bizarre effect. Clocks near the hole are slowing down along with the redshift, as the relative clock rate is $1 + z$. Events near the hole take much more time to occur. Thirty-three kilometers away from the hole, where the redshift is 2.32, the relative clock rate is 3.32. Events this close to the hole pass at roughly one-third their normal rate.

Nearer and nearer the horizon, the slowdown of clocks would increase. The orbiting clocks would tick slowly, slowly, more and more slowly—tick, tick, ... At the event horizon, what would happen? It would take an infinite time until the next clock tick. Events would be frozen. Time comes to a stop at the event horizon.

What about the falling astronaut? His clock and ours are shown in Figure 4-3, which depicts graphically the events shown in the table. That half-second difference between his clock and ours would become larger and larger as he approached the event horizon. If we were monitoring his heartbeat, it would be recorded as slowing down too, along with his clocks. He would seem to stop falling, as his fall would be frozen at the event horizon. It would be like watching a movie with someone slowing down the rate of the projector. The slowdown occurs very abruptly at the edge of the black hole. Our clock would not advance to 205 hours until the astronaut was within 3×10^{-8288} km of the event horizon. (To write out 3×10^{-8288}, I would have to put 8287 zeros between the decimal point and the 3. It would take three pages in this book to write out that small a number.) His clock would be frozen, and ours would tick on. We would never see his clock go beyond 204 hours, 33 minutes, 49.670133 seconds. We would never see him fall through the event horizon. He would inch closer and closer to it, ever more slowly, but he would never pass through.

The frozen star

If you happened to watch the formation of a black hole, you would see something similar (Figure 4-4). The collapse would proceed quite rapidly at first. The light from the star would be redshifted more and more as the star came closer and closer to the horizon. (Here the redshift of the star is rendered as a darkening.) Just short of the horizon, the collapse would slow down abruptly, because the star's own gravity would cause everything to seem to happen in slow motion to a distant observer. The collapse would be effectively frozen just short of the event horizon.

Remember, it is only exceedingly close to the event horizon that the collapse appears to freeze. The large redshifts at this point would cause the star to appear to be black. The freezing of the collapse occurs only when the redshift is extremely high. After 4.6×10^{-5} sec the redshift is 10, if you start counting from the time the star has a radius of 1.5 times the Schwarzschild radius. After another 4.6×10^{-5} sec, the redshift has increased tenfold again, to 100. Because the star is emitting its light in discrete photons, there is a time at which the star sends its last photon out to the outside world. Detailed calculations indicate that the last photon from a ten-solar-mass star would emerge less than 0.01 second after the star's surface passed the 1.5-Schwarzschild-radius, or 45-kilometer, point. The collapsed star would be black, and its collapse would be frozen. Hence another name for black holes: frozen stars.

The event horizon as a limit

The preceding section points out that the event horizon is a limit. You cannot see anything happen *at* the event horizon, as no photons can reach you from there. As you look closer and closer to the event horizon, time slows down without bound. Closer, closer, closer the clocks go slowly, more and more slowly. Paradoxical place, the event horizon.

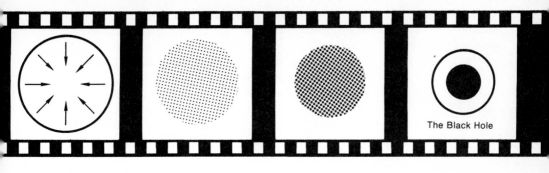

Time ⟶

FIGURE 4-4   A movie showing the collapse of a star. As a star collapses to form a black hole, it dims very rapidly. It emits its last photon less than 0.01 second after it becomes smaller than 1.5 Schwarzschild radii. Compare with Figure 4-3.

The concept of the event horizon as a limit can perhaps be better illustrated by one of Zeno's paradoxes. Suppose you want to go through a door, and you are six feet away. For reasons best known to yourself, you decide to approach the door slowly, covering half of the remaining distance with each step. At first, this seems like a reasonable approach; your first step takes you three feet toward the door, and you have made progress. But you will never get through the door if you play the game according to the rules. The second step leaves you 1.5 feet away, the third 9 inches, the fourth 4.5 inches, the fifth 2.25 inches, and so on. No step will ever take you through the door, as you can only approach it with each step. The same thing happens as you look toward a black hole. If you try to watch someone enter the interior, it seems to take the person longer and longer to get there, as he or she travels more and more slowly.

Looked at from the outside, the event horizon seems to be a very strange place. Somehow the idea of time coming to a stop at the event horizon doesn't quite jibe with the way that the world is supposed to work. What sort of place is the event horizon, anyway? To explore the nature of the event horizon and the world inside it, the interior of the black hole, we shall have to succumb to the Pygmalion syndrome (recall Chapter 1) and leave the realm of the real world. Anyone who fell through the event horizon in an attempt to verify experimentally the theoretical results about to be presented could never return to tell us that we were right.

Yet there are good reasons for indulging in this theoretical exercise of imagining what a trip beyond the event horizon would be like. The idea that time comes to a stop there makes you think that maybe Einstein's theory breaks down at the event horizon. If this is true, then the very existence of black holes is open to question and the validity of Einstein's theory elsewhere in the universe is questionable. The theory is supposed to be valid anywhere in the universe, including in the vicinity of the event horizon.

It turns out that the peculiarity of space-time near the event horizon—the idea of time coming to a stop—is just a consequence of our point of view. If we follow our courageous astronaut through the event horizon, we find that the horizon is not such a strange place after all. Yet I repeat that what follows is theoretical only, as no one who fell into a black hole could come out again to tell us what was really there. (The speculation that in fact you *could* emerge from a black hole is dealt with in Chapter 6.)

Through the event horizon

Look at Figure 4-3 again, and at Table 4-1, this time paying attention to the astronaut's clocks. Unlike the external observer, he will not see his fall toward the hole freeze at the event horizon. 204 hours, 33 minutes, and 49.6681 seconds after he left the rocket ship, he would be 300 kilometers away from the hole, and his clocks would be in general agreement with the clocks back on the spaceship. Only a split second later, at 49.670133 seconds, he would fall through the event horizon. As he approached the hole, he would not notice any slowing down of clocks. He would not be able to see the surface of the frozen star, for it would be black. It would look like a hole. As he fell, he would see events around him (if there were anything happening) escape from their slow-motion mode as seen from the outside and proceed normally. When he fell through the horizon, he might have to endure some discomfort from the ever-increasing tidal forces. But the tidal forces would stay within bounds at the horizon, and a suitably built astronaut or probe would survive.

I cannot emphasize too strongly that at the event horizon, someone falling through would not experience any impossibly odd physical effects. There is no sign at the event horizon warning of the danger inside, no infinite tidal gravitational forces that would pull one apart before one got in. Look at the way the rocket ship sees the black-hole adventure in Table 4-1. See? The probe falls through the event horizon perfectly happily, in a reasonable amount of time, from its point of view.

The absence of any pathological effects at the event horizon means that Einstein's theory does not break down there. It is only our point of view from the outside that produces the odd effect of time's coming to a stop. If you adopt another point of view, the infinite redshifts and frozen clocks disappear. They are only ephemeral, a consequence of our point of view from the outside. Einstein's theory is still valid, and it is all right to use it to predict what happens up to and through the event horizon.

Why a black hole?

For a long time, the term used to describe the subject of our investigation was not *black hole* but *frozen star*, or *collapsar*. We see now that the term *black hole* is quite appropriate. The appropriateness of the term *black* was discussed before. From the outside, the star would fade to invisibility in one-hundredth of a second or less if you were watching it collapse. You could not even see its surface by shining a flashlight on it, as the light from the flashlight would catch up with the collapse on its way in—the collapse would be unfrozen from the point of view of the ingoing light. It would be swallowed up by the hole. No, you cannot see the frozen star. It is black.

A similar fate would befall any foolish astronaut who sought to prove that it was not a black hole, that it was only a frozen star there in space. If he tried to scoop up a piece of the frozen star, the collapse would always unfreeze just enough to stay ahead of the scoop. If he tried too hard to pick up a piece of the star, he would pass beyond the event horizon, and disappear from his friends in the world outside. Yes, this object is truly a hole, too. The accepted term now is *black hole*. If you fall into it, you fall, and you fall, and you fall, until . . . .

The interior of a black hole

There is one very serious problem with being inside the event horizon. You can never get out again once you are there. The event horizon is a cosmic turnstile. One way only. Anything, whether it is light, space probe, TV set, rock, rocking chair, or unfortunate astronaut, can go only in one direction: *inward* (Figure 4-5). Some serious frontiers or speculative fringes of black-hole research suggest that there may be possible exceptions to the one-way behavior of the event horizon. Work by Stephen Hawking indicates that any black hole will eventually evaporate—but, for a ten-solar-mass hole, 10^{66} years will pass before this happens. The fringe areas of black-hole research involve discussions of mirror images of black holes—white holes—in which material erupts. These white holes probably don't exist, however. See Chapter 6.

The one-way nature of the event horizon forces us to rely on theory in order to probe these lower depths of black holes. These theoretical calculations, although detached from the real world, are not totally meaningless, for they provide insight into the nature of the theoretical picture of a black hole and provide guides to the situations that Einstein's equations can be applied to. With these reservations, let us follow the probe as theory carries it toward the black hole's center. (I'll call it a probe now; the idea of subjecting a person to these hypothetical but nevertheless horrifying experiences is too lugubrious.)

The probe is pulled relentlessly toward the center of the black hole. As the probe approaches the center, the tidal forces become stronger and

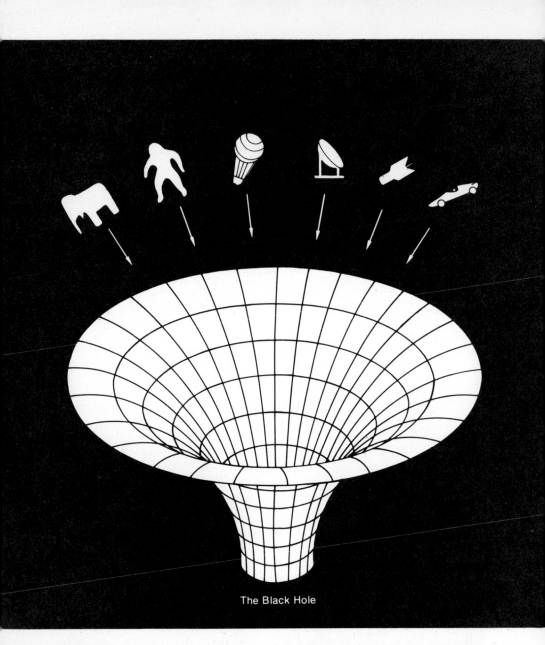

The Black Hole

FIGURE 4-5   Everything falling into a black hole loses its identity. You do not know whether it was a space probe or a TV set that fell in. (Adapted from Remo Ruffini and John A. Wheeler, "Introducing the Black Hole," *Physics Today,* January 1970, p. 31, © American Institute of Physics.)

stronger. They increase indefinitely, so the probe will be destroyed by them before it actually reaches the center. The probe could struggle against gravity, trying to escape this fate by turning on its rocket engine and darting here and there, but it could only postpone the inevitable for a very short time. The tentacles of gravity have caught it, and it must fall to destruction at the center of the hole. In a ten-solar-mass hole, it would fall quite fast. If it did not start its engine in an attempt to escape, it would reach the center 67 millionths of a second after it passed the horizon.

What is at the center? Einstein's theory does in fact break down here. The theory presents us with a very bizarre object, a singularity. A singularity is an absurdity. It is a point containing all the mass of the hole. The singularity has zero volume, and the density of matter is infinite. The tidal forces are infinite. So the theory says, anyway.

The idea that there is a singularity at the center of a black hole makes many physicists feel uncomfortable. When a theory starts producing infinities in models derived from it, a reasonable feeling develops that the theory is wrong. Standard black-hole theory (the subject of this chapter) is based on Einstein's theory of gravitation, so a standard black hole has a singularity in the middle. Numerous people have tried to modify Einstein's theory of gravitation so that the singularity goes away.

To a certain extent, though, such modifications are beside the point. Remember the Pygmalion syndrome again. The whole purpose of this exercise of following an astronaut or probe into a hole was to see where inside the hole Einstein's theory breaks down, and in particular, whether it breaks down at the event horizon. What happens inside the event horizon has no effect on the outside world, since anything that falls in can never get out again. The interior of a black hole is cut off from our universe by the event horizon, so what happens there does not affect us.

Types of black holes

Rotating black holes

Real stars rotate, but the standard black hole described in the last few pages does not. One would expect a spinning star to produce a spinning black hole, but how would the spin of a black hole affect the hole's properties? Roy Kerr developed a mathematical description of a spinning black hole in 1963, and so spinning black holes are sometimes called "Kerr black holes," in contrast to the nonrotating Schwarzschild black holes, named for black-hole pioneer Karl Schwarzschild.

The event horizon of a rotating black hole would be found in the same place as the event horizon of a nonrotating black hole of similar mass.

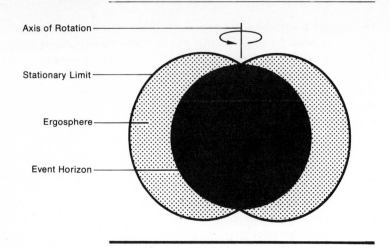

Axis of Rotation

Stationary Limit

Ergosphere

Event Horizon

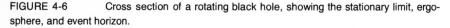

FIGURE 4-6   Cross section of a rotating black hole, showing the stationary limit, ergosphere, and event horizon.

The phenomena that occur near the event horizon, though, are a little different. Suppose you approached a rotating hole from the top, falling down the pole of rotation. The story described earlier in this chapter would be repeated in every detail: You would fall through the event horizon, never to be seen again. You would notice nothing peculiar near the horizon, given that you could survive the tidal gravitational forces and still see anything. The friends you left behind would see you fall more, more, and more slowly as you neared the limit of the event horizon.

But suppose you fell toward the black hole in a different direction, approaching it toward the equator, for example. Far away from the hole, the situation would be the same. Close to the hole, the hole's rotation tends to sweep objects around it, in the same direction that it spins. There are two consequences of this tendency. One—which might be able to be observed in real situations—is that small particles orbiting in the direction of the hole's spin could be closer to the hole and still stay in orbit than they could in the case of a nonrotating hole. The other difference involves a second boundary—the static limit. Before meeting the event horizon, an intrepid black-hole investigator would encounter this invisible boundary. Inside the static limit, it would be impossible to remain at rest, since the spin of the hole would carry you madly around in the direction that the hole is rotating. Yet you could still send signals to the outside world, and even escape from the hole's influence by turning on your rocket engine. This all presumes that you still *have* a rocket engine—and that it and you have not been destroyed by the tidal gravitational forces.

Investigators have explored various methods of extracting energy from a rotating black hole. Such explorations are motivated partly by curiosity and partly by the need to find sources of enormous energy to explain the powerful emissions of quasars and active galaxies, the subjects of Part 2 of this book. One proposal suggests that you drop a rocket engine inside the static limit, turn on the rocket, and watch the black hole's rotation shoot out the rocket at a high speed. Detailed calculations of this process show that the rocket would accelerate, but no more than it would if you did the same thing far from a black hole.

Another proposal deals with light waves focused into the space surrounding a rotating black hole. Given exactly the right conditions, these light waves could be amplified. The article announcing this mechanism is titled "The Black Hole Bomb." It talks about the possibility that a suitable mirror could be placed outside a rotating black hole to focus the light, and that eventually the amplified light waves would break the mirror. Realistically, more light would be lost down the hole than would be amplified, and this particular black-hole bomb would be a fizzle.

These investigations illustrate the strange phenomena that might occur near a rotating black hole. So far, many theories about black-hole powerhouses have turned out to be weak, but proposals for extracting energy from rotating black holes continue to appear. One of them may turn out to provide the powerhouse for the quasars.

In principle, electrically charged black holes could exist. Yet such objects are unlikely to form in the real world, for a charged black hole would repel objects of similar charge that tried to fall down it, and would attract oppositely charged objects, thus eventually neutralizing itself. Electrically charged black holes are, for the most part, similar to the neutral, uncharged ones.

Black holes have no hair

Mass, spin, and electrical charge—these are the three properties that we have considered so far. It is mass that governs the size of the event horizon and the scale of the black-hole phenomena that I have discussed. Spin and charge slightly modify the properties of the black hole. What other properties might a black hole have? Does it make any difference whether you make a black hole out of an iron stellar core or out of a carbon stellar core? Can you tell whether someone threw a cupful of white-dwarf stuff, a pinhead of neutron-star stuff, or 24 elephants down the hole? The somewhat surprising answer is no.

During the early 1970s, most theorists believed that mass, charge, and angular momentum were the only properties that a black hole could have, but no one had proved that this must be so. Some other property could perhaps remain after a black hole formed. Black-hole theorists Kip Thorne and John Wheeler referred to this other property as "hair," an attribute that could distinguish black holes of similar mass and spin from each other in the same way that the color, length, or style of hair distinguishes one human being from another. But in 1975, investigators proved the last of a series of theorems, showing that mass, spin, and possibly electrical charge are the only black-hole properties allowed by Einstein's theory of gravitation. These theorems are collectively (if informally) described by the short phrase "Black holes have no hair." Black holes can differ greatly in the amount of mass that they can have, and can differ in spin, but differ in no other ways—except possibly electrical charge. This completes our description of isolated black holes.

Isolated black holes are fascinating in many ways. The strange behavior of space and time near the event horizon is mind-expanding. The very existence of an event horizon, a point of no return, has stimulated many a science-fiction writer. Yet, in some other ways, isolated black holes are quite uninteresting. Black holes are black and the sky is black; black on black does not make for good observing. Present-day astrophysicists have probably discovered black holes by observing the interaction of holes with the world around them. We shall explore these interactions in Chapter 5, which deals with the search for black holes that form from stars.

You can look at a black hole from two points of view—the outside and the inside. The view from the outside is the only one that can be experimentally verified. Looking at a black hole, you see an event horizon, where time has come to a stop. Surrounding the event horizon, barely outside it, is the surface of the collapsing star that formed the black hole, from the collapse of this star that has been frozen. The frozen star emits no light, so it is black.

But time has not really come to a stop, since it is only your point of view, from the outside looking in, that makes you think it has. If you follow, theoretically, the adventures of a space probe dropped toward the hole, you will find out that the probe falls into the hole. Time has not stopped at the event horizon from the point of view of someone falling in. Outsiders cannot see the collapse go to completion, but someone falling in will pass straight through the event horizon, enduring discomfort only from the tidal forces. (A person would have to be pretty strong to endure those forces.) The person falls through the event horizon, into the speculative arena known as the interior of the black hole. No probe, rocket, astronaut, or anything else, once inside, can escape a standard black hole. The object is caught and pulled inexorably toward the central singularity, where Einstein's theory of gravitation breaks down as the forces of gravity take off toward infinity.

The inclusion of spin and electrical charge as black-hole properties does not fundamentally affect the nature of the event horizon or the singularity in black holes, although some details regarding the behavior of orbiting bodies near the event horizon are slightly modified by rotation. Recent work has shown that mass, spin, and electrical charge are the only properties that a black hole can have. Thus the description of a black hole in this chapter, suitably scaled to different masses, can be applied to any black hole in the universe.

5 THE SEARCH FOR BLACK HOLES

Do black holes really exist? So far, we have examined a theoretical object, a component of the model world. Black holes are fascinating objects, but before you let yourself become too intrigued by the mysteries of the event horizon, you should try to ascertain the existence of some of these things in the real world. The search for black holes is somewhat more focused than the search for neutron stars because we have some ideas of what we should look for. In some respects, however, the stories are similar, in that it was first shown that black holes might exist (Schwarzschild, in 1916) and then it was shown that massive stars might actually become black holes (Oppenheimer and Snyder, in 1939). Now there is a good possibility that they might be observable.

Chapters 2 and 3, which provided a sketchy overview of the life cycles of stars, reached the conclusion that massive stars may end their lives as black holes. In summary, at the end of its life, a star has burned all of its nuclear fuel. It can no longer keep its inside hot enough to hold up its surface layers, so it shrinks. If a star leaves a small corpse, the dead star can prevent itself from totally collapsing because degeneracy pressure can hold the star up. Since degeneracy pressure does not arise from heat, it continues to operate even as the star cools. Stellar corpses with low masses thus end up as white dwarfs or neturon stars.

But if a dead star is more massive than some limiting, critical mass, degeneracy pressure is overpowered by the weight of the star. Degenerate matter is simply not strong enough to hold up a cold, dead star with a mass greater than a certain limit. The exact value of the limiting mass is a matter of some debate. As long as general relativity is valid, this limiting mass cannot exceed four solar masses, and is probably about two and a half solar masses. Many stars start their lives with more material than this limiting mass. To avoid becoming black holes they must shed mass in the red-giant stage. Although stars are known to shed mass as red giants, they probably don't shed enough to enable the largest stars to avoid the ultimate fate of the cosmic collapse—the black-hole stage. We astronomers cannot firmly argue that massive stars must become black holes, but they probably do.

Where are the black holes hiding?

Black holes come in all sizes, but in a limited variety of shapes. The properties of a black hole are mass, spin, and charge; we can therefore seek black holes of any mass. Yet the search for black holes is easier if we pick black holes that evolve from massive stars as the target. They are the ones whose origin we can understand reasonably well. Further, in double-star systems, stellar black holes can suck matter away from the companion star and emit x rays that we can see and analyze. It is the analysis of double-star systems that provides us with the best evidence that black holes exist.

Different sizes of black holes can also be looked for. Mini-black holes—black holes far smaller than stars—might have formed in the gravitational tangles of the Big Bang, the explosion that marked the beginning of the universe. A remarkable result of black-hole research in the 1970s is the theory that these mini-black holes evaporate eventually. But no one has yet found an evaporating mini-black hole. (Consideration on this exciting frontier area is deferred until Chapter 6.) Black holes larger than stars may also exist. These giant black holes may supply power to the x-ray sources seen in large star clusters. They may be the powerhouse that supplies energy to the quasars. If the expansion of the entire universe should ever come to a stop, then the universe would be one cosmic black hole.

But these Lilliputians and Gargantuans of the black-hole cast are hard to identify unambiguously. We understand stars and double-star systems far better than we understand mini-black holes, supermassive objects in the center of star clusters or galaxies, or the universe itself. We shall reserve mini-black holes and black holes in star clusters for the frontier areas treated in Chapter 6. The enigmatic quasars are the subject of Part 2, and I'll deal with the universe in Part 3. So, postponing these other topics until later, let us turn to the most promising place to search for black holes: double stars.

Spectroscopic double stars

If black holes are invisible, how can we find them? The simplest effect that a hole has on its surroundings is that it causes rockets, planets, other stars, or anything nearby to fall toward it or to orbit it. If the earlier illustration of a rocket ship orbiting a black hole is changed slightly, and we let a star be substituted for the rocket ship, then we have a black hole that can be detected. You can look and see the star orbiting something, and you cannot see the object it is orbiting. Therefore it is orbiting a black hole.

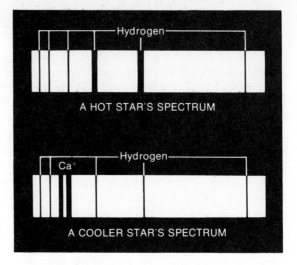

FIGURE 5-1
Stellar spectra
(schematic)

This picture is nice and simple when you have a close view. Unfortunately the stars are far away, and from a distance the situation is proportionately more ambiguous. How can you detect that a star is orbiting some invisible companion if the star is tens or hundreds of parsecs away? Close analysis of the message of starlight is required. There are two main questions: How do we know that the star is orbiting in the first place? and, How do we know that the companion is truly invisible and not just a dim star hidden by the light of the star we can see?

To answer both questions, we must analyze the star's spectrum. Starlight, like sunlight, is a mixture of all colors of the rainbow. To photograph a star's spectrum, we disperse the starlight into its different colors or wavelengths and photograph the result. Sketches of stellar spectra are shown in Figure 5-1. One can also, using varying techniques, determine the color of the star. All that we know about a star must be determined from its spectrum and its color, for the only messages that we receive from the star are the messages of radiation—light, radio waves, x rays, and other forms of radiation.

The most striking feature of the stellar spectra shown in Figure 5-1 are the dark lines crossing them. These dark lines indicate that less light is being emitted at that particular color, since in the spectrum the color of the light is changing from blue on the left to red on the right. Dark lines exist because particular atoms in the surface layers of the star are absorbing light and preventing it from reaching us. (The Preliminary section explains this phenomenon in more detail.)

What can we learn from these dark lines? From their pattern, we can learn what type of star is emitting light. We can classify the spectrum, and from this classification we can derive the star's temperature, size, and luminosity.

A star's temperature is the principal factor governing the appearance of its spectrum. Temperature classifications of stellar spectra are designated by letters, but the order is somewhat jumbled for historical reasons: O, B, A, F, G, K, M. For example, the bottom spectrum in Figure 5-1 is a G-type spectrum; the two very dark lines at the left come from ionized calcium atoms that are found mostly in stars with the same temperature as the sun, a G-type star. The top spectrum is that of a hotter A-type star; the calcium lines are absent.

Yet spectroscopists can tell more than just a star's temperature. The art of spectral classification has advanced to such a point that by examining a star's spectrum they can determine its size: supergiant, giant, or dwarf. If we can determine a star's temperature and size, we can also determine its luminosity.

We now have an idea of how to answer our second question about a star with an invisible companion. How do we know whether the invisible star is a dim companion star or a black hole—not a star at all? We classify the spectrum of the star that we see. From its spectral class, we can determine its luminosity. We then determine how faint its companion must be to remain invisible. We can then ask, Is it reasonable that the companion is so faint?

But how do we know that the companion is even there? We are not close enough to the star–black hole pair to see the star executing a beautiful elliptical orbit as it dances around the black hole. For some pairs of stars, you can actually see the stars move around each other. But none of the black-hole candidates is amenable to such an analysis. Once again, we must count on the spectrum to provide help. The exact color or wavelength of the dark lines tells us the answer. If the wavelength of these lines changes in time—if the lines appear to shift in wavelength and then shift back again— we know that the star is alternately moving toward and away from us. If it is moving in such a fashion, it is orbiting something. Figure 5-2 shows what the observing astronomer would see.

Doppler shifts

How does an observation like the one in Figure 5-2 tell us that there is a companion star? It is the changing color or wavelength of the dark lines in the star's spectrum that tells us that the visible star is moving in an orbit. This changing color arises from the Doppler effect. Light from a star moving toward the observer looks bluer than it did when it left the star, and light moving away from the observer looks redder. For example, if two astronomers on two planets watch a star move toward one of them and

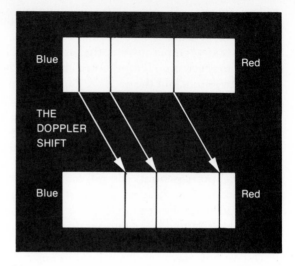

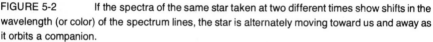

FIGURE 5-2   If the spectra of the same star taken at two different times show shifts in the wavelength (or color) of the spectrum lines, the star is alternately moving toward us and away as it orbits a companion.

away from the other, each will see a different spectrum (Figure 5-3). The astronomer on the right, seeing the star approach, will see a blueshifted spectrum. The one on the left will see a redshift. This shift connects the motion of a star to the colors or wavelengths of light in its spectrum.

What causes this Doppler shift? Consider a terrestrial analogy, shown in Figure 5-4. A police car moves down the street, like the star moving through space. This car is equipped with a siren tuned to middle C. Consider what two people on the sidewalk observe, as opposed to what a person inside the police car hears. As the car approaches the observer on the right, sound waves from the siren tend to pile up. The car is catching up with the sound waves it emits. As a result, the observer on the right senses that the length of the sound waves is shrinking, and hears a pitch higher than middle C. Similarly, the observer on the left, with the police car speeding away from him, experiences sound waves that appear stretched out, and perceives a wavelength that is longer and a sound that is lower-pitched. The person inside the car does not sense that the sound waves are either stretching out or shrinking, but simply hears middle C. (The pitch of a musical sound is related to its wavelength. Short-wavelength sounds are high-pitched; long-wavelength sounds are low-pitched.)

The extent of the Doppler shift is governed by the speed of the moving object, in this case the police car. The faster the car moves, the more the sound waves pile up (or stretch out), with consequent greater shift. In principle, a blind person with perfect pitch could determine how fast the police car was traveling. The blind person could recognize the pitch, and knowing what pitch the siren was tuned to, could thus determine

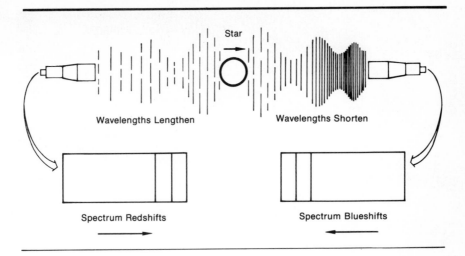

Wavelengths Lengthen   Wavelengths Shorten

Spectrum Redshifts   Spectrum Blueshifts

FIGURE 5-3   Astronomers see Doppler shifts in stellar spectra, as the colors (or wavelengths) of the dark lines change depending on the motion of the star. Compare with Figure 5-4.

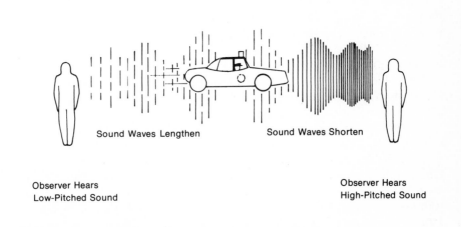

Sound Waves Lengthen   Sound Waves Shorten

Observer Hears
Low-Pitched Sound

Observer Hears
High-Pitched Sound

FIGURE 5-4   Pedestrians on a sidewalk can hear Doppler shifts from the siren of a passing police car.

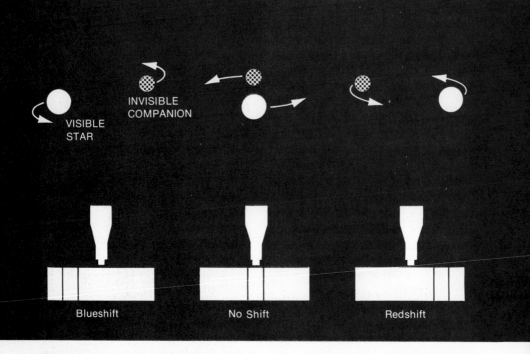

VISIBLE STAR

INVISIBLE COMPANION

Blueshift

No Shift

Redshift

FIGURE 5-5   A single-lined spectroscopic binary, in which a star orbits a dark companion, will show a shift in its spectrum as the visible star moves toward and away from the observer in its orbit.

the car's velocity. You can hear the shift in pitch of a police car siren as one travels by you on the roadway. Try it sometime.

(To obtain a mathematical expression for the Doppler shift, express the shift in wavelength as $\Delta\lambda$, the wavelength as λ, the wave velocity as c, and the object velocity as v. The relation between these quantities is wavelength shift/wavelength = object velocity/wave velocity, or $\Delta\lambda/\lambda = v/c$. This relation holds only if v/c is small.)

In practice, the astronomer examines spectra of a star taken at various times. When the star is moving toward the earth as it orbits its companion, the lines in its spectrum are shifted toward the blue end of the spectrum, as the light waves are compressed by the star's motion. When the star passes between the earth and the companion, you observe no shift, and when the star is receding from the earth, you observe a redshift. The whole scheme is shown in Figure 5-5. (When you do work of this kind, you need to be able to measure the exact wavelength of the lines in the spectrum. You can do this by exposing the spectrum of a fixed source on the same photographic plate.) A star that shows changing shifts in its spectrum like those in Figure 5-5 is known as a single-lined spectroscopic binary: *single-lined,* because you observe only one spectrum, *spectroscopic,* because it takes a spectroscope to determine what is going on; and a *binary,* because the shifting pattern of spectrum lines indicates that there are two objects in the system, that the star whose spectrum you see is orbiting something.

A case history: Epsilon Aurigae

From this discussion, you might be tempted to conclude that all single-lined spectroscopic binaries are black holes. Such a conclusion would be premature. All that you know about a single-lined spectroscopic binary is that the system encompasses another object or other objects. Whether that object is in fact a black hole is a further question requiring thoughtful analysis. Each case must be considered individually. The investigator has to ask, Is there any way to explain what is observed in this system without involving a black hole? For each individual star, the arguments are slightly different.

The search for black holes in binary systems begins with one particular candidate, Epsilon Aurigae. I start here not because Epsilon Aurigae is the best black-hole candidate, for it is not the best. But it does illustrate some of the basic physical laws that underlie the quest. These basic laws are not just set forth abstractly, but are applied to a particular system. Further, this system may contain a black hole. However, the evidence is not so strong that astronomers believe that it *must* contain one.

Auriga is a constellation that is fairly easy to recognize. It is a pentagonal group of bright stars, north of Orion, and it passes almost directly overhead in the after-dinner sky of January (see Figure 5-6). Auriga contains the bright star Capella. Just south of Capella lies a group of three stars, and Epsilon is the northernmost, brightest star of the group. It is a very peculiar spectroscopic eclipsing binary. The lines of its spectrum shift in much the same way as described above. But in addition, every 27 years the bright star becomes dimmer as the *secondary* (as the companion is called) passed in front of the bright primary. All investigators agree that the secondary contains a cloud of dust that blocks out some of the light of the primary as the secondary passes between us and the primary, thus causing the eclipse. During the eclipse—and only during the eclipse—you see a second spectrum, quite similar to the spectrum of the primary star. Whether this second spectrum means that the secondary is actually emitting light of its own is unclear at present. The secondary is completely invisible outside of eclipse. Now one must attack the question, Is this invisible secondary, at the center of this immense dust cloud, a black hole?

To analyze this question, we must determine the properties of the secondary as best as we can from the information available. In particular,

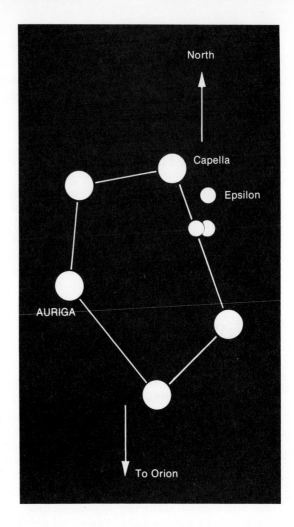

FIGURE 5-6
The constellation of
Auriga and the location
of the peculiar binary star
Epsilon Aurigae

we need to know its mass. What do we know? We know that the primary, the star whose spectrum we see, is being pulled around by the secondary, for there are Doppler shifts in the spectrum of the primary. We can estimate the strength of the pull that the secondary is exerting. To go from the strength of this pull to the mass of the secondary, we need to know a little bit more about how double stars work.

Two stars orbiting each other are held together by gravity. The strength of the pull necessary to keep the stars from flying off in opposite directions depends on the speed of the stars in their orbit. Fast-moving stars need a strong gravitational force to hold them together, while slower-moving stars need less force. The strength of the force depends on the size of the orbit and the masses of the two stars, according to the laws of gravitation. The more massive the stars, the stronger the force; and the larger the separation between the stars, the weaker the gravitational force. Thus, if a pair of stars orbit each other rapidly at a great distance, we can deduce that they must be massive, since only massive stars can exert a strong gravitational force over a great distance.

(Mathematically, the relation described in the last paragraph is known as Kepler's Third Law, relating orbital period P in years, the distance between stars R in astronomical units, and the total mass M of the system in solar masses. This relation is $P^2M = R^3$. You need not understand the mathematical relation in order to understand what follows.)

Thus you can measure the mass of a double-star system if you know its orbital size and orbital period. Measuring the period is easy. Keep obtaining spectra and watch the lines shift back and forth. The amount of time it takes the lines to go through one full cycle is the period. The extent of the shifts tells you how fast the stars are moving toward and away from you. If you know whether the stars are moving directly at you or are moving on an angle, you can determine the orbital speed. You then know how fast the star travels in its orbit and how long it takes to get around, and you can calculate how big the orbit is.

Now you know the combined mass of the two stars. Sometimes you see both stars moving around each other, and you can tell which one has the greater mass, the greater inertia. But if you could *see* both stars in a double-star system, neither one would be a black hole. In a single-lined spectroscopic binary like Epsilon Aurigae or any other black-hole candidate, you need to estimate the mass of the visible star. You know the total mass of the system, and what's left over that you can't see is the mass of the invisible star, the black-hole candidate.

Because the analysis involves making some assumptions, different investigators obtain different models for the Epsilon Aurigae system. But the smallest mass anyone has found for the invisible companion is eight solar masses. An eight-solar-mass main-sequence star would be seen in mid-eclipse, so the companion isn't a normal main-sequence star. White-dwarf stars or neutron stars would be too dim to see in mid-eclipse, but neither can make up the needed mass of eight solar masses. At first glance, the only thing that would be massive enough to be the companion and still remain invisible is an eight-solar-mass black hole.

But wait. The eclipses of the visible star last two years, and eight-solar-mass black holes are too tiny to produce any visible eclipse—and certainly can't produce an eclipse that lasts that long. The companion is, presumably, surrounded by a cloud of dust that produces the eclipse. Models differ in detail. Is this cloud of dust a sphere? a disk? a ring? a rectangle? No one knows. But whatever it is, it could hide an eight-solar-mass main-sequence star from our eyes. The object at the center of this dust cloud could be a black hole, but it doesn't have to be. Is there any other, less ambiguous, fingerprint that we can use to identify a black hole in a double-star system? X-ray astronomy may provide the missing clue.

X-ray sources and black holes

The basic problem that Epsilon Aurigae presents as a black-hole candidate is the difficulty of distinguishing between a black hole and a normal star embedded in a dust cloud. We could tell, from the shift of lines in the spectrum of the visible primary star, that there was another star pushing the primary around. An estimate of the strength of this pull indicated that the companion was sufficiently massive to be a black hole rather than a white dwarf or neutron star. But is the companion an evolved star, so massive that it is certainly a black hole, or is it just a normal star hidden from our eyes by this vast dust cloud? Epsilon Aurigae does not provide any evidence bearing directly on this question. The answer may emerge eventually, but only after a series of detailed models have been made. In the search for black holes, attention now focuses on evidence that can prove that the star system of interest contains an evolved star. Neutron stars and black holes are evolved stars that are so small that in the literature they are often referred to as collapsed objects, although only a black hole is, strictly speaking, collapsed.

What does a collapsed object do that stars don't do? A collapsed object is very small—some tens of kilometers across, or maybe hundreds, at the utmost—while a massive star is quite large, millions of kilometers across. Remember the gruesome fate of an astronaut who ventures too close to a black hole. As the astronaut falls toward the hole, all body parts try to fall toward the same point. The astronaut is squashed by the strait-jacket of gravity, as shoulders, head, feet, and the entire body fall inward toward the center. The whole astronaut is compressed. These compression forces are so strong that they would destroy any realistic space probe constructed with present-day technology that would explore a stellar-mass-sized black hole.

As gas and dust swirls toward a black hole they will be squashed in the same way that the astronaut would be. When gas is squeezed, it heats up. The more it is compressed, the hotter it gets, and the more rapid the swirling motion is as the gas falls toward the hole. This hot vortex of infalling gas eventually gets hot and dense enough that it emits x rays as it nears the collapsed object. Detailed theoretical models confirm this picture. If there is a black hole somewhere in space with a continuous supply of gas falling in on it, x rays come from the compressed, swirling gas just before it reaches the event horizon. X-ray astronomers can detect these high-energy signals that the black hole creates as it compresses the infalling gas stream.

X-ray astronomy

X-ray astronomy, which can provide a key step in the search for black holes, is now growing up. In the 1960s, all x-ray astronomy was done with instruments sent above the atmosphere in rockets or sometimes balloons. X rays cannot penetrate the earth's atmosphere, so it is necessary to go above the atmosphere with a rocket or satellite.

Rockets are advantageous instruments to use when you are just opening up a field and do not know whether you will find anything interesting or not. They are cheaper than satellites, both in dollars and in the amount of time it takes to put an experiment together. It takes only six months to set up a rocket experiment, if you have the instrumentation built, whereas five years often passes between the initial proposal and the launch of a satellite. However, rockets have a great disadvantage in that the observing time on a typical rocket flight is only about five minutes.

The rocket era of x-ray astronomy was extremely productive, considering that the total observing time amassed by various rocket flights was little over an hour. Think how little you would know about the sky if you had just an hour to look at it. In 1969, the rocket era came to a close with the launching of the *Uhuru* satellite. This first x-ray satellite observatory has probably been one of the most productive scientific ventures of all time. (Its name comes from the time and place of its launch: off the Kenyan coast on the fifth anniversary of Kenyan independence. *Uhuru* is the Swahili word for freedom.) Scanning the sky in the x-ray region, *Uhuru* has provided a comprehensive list of about 200 x-ray sources, half of which are located within the Milky Way galaxy. High Energy Astronomical Observatory satellites, along with others launched in the late 1970s, discovered many more x-ray sources, and provided much more information on those first seen by *Uhuru*. Yet it was *Uhuru* that first uncovered the wide variety of x-ray sources in the galaxy. Some of these x-ray sources may be black holes.

A few x-ray sources in the Milky Way galaxy are associated with other forms of dead stars. Some x-ray sources are gas clouds that are the expanding wisps of a star's outer envelope—the debris of a supernova. A few are very hot white-dwarf stars. Most are associated with binary stars, systems in which two stars orbit each other. Many of these binary systems contain neutron stars; the x-ray pulsars (see Chapter 3) are examples of these. A few of these binary systems may contain black holes, and these sources are worth examining in detail.

Flickering x-ray sources

A few x-ray sources are particularly interesting as far as the search for black holes is concerned. These x-ray sources flicker quite rapidly, varying in intensity within thousandths of a second. Of all the galactic x-ray sources, only a few fall into this category: Cygnus X-1 and Circinus X-1 are two of the best-investigated ones. The flickering nature of these sources provides evidence that the x-ray-emitting regions are very small, as small as the inner edge of a whirling vortex of gas surrounding a black hole. (I shall discuss the basis for this conclusion presently.) Recall that Epsilon Aurigae became an unpromising black-hole candidate because there was no reason that it couldn't be a large, evolving star rather than a tiny black hole.

Why does the flickering of the x rays provide information on the size of the x-ray-emitting region? Think about what happens when an object doubles in brightness. In order to double in brightness, a cloud of emitting material must either become twice as bright as it was, or another cloud of emitting material must start to radiate, in a short time interval—let us take 0.05 seconds as an example. Whatever caused the emitting region to suddenly turn on, information of some sort must travel from the stimulus to the responding gas, from the source of the energy to the gas that radiates it away as x rays. If the x-ray source is a little glob on the inner edge of the accretion disk, it must compress or heat up in that short time interval, and the sound waves that cause it to heat up or compress must cross the hunk of gas in that time of 0.05 second. These sound waves travel no faster than light; in fact, they probably travel much slower than light. Thus the gas cloud must be small enough so that light can cross it in 0.05 second—it must be less than 15,000 km across. (In fact, it is probably far smaller than that, but you would have to know how fast sound waves travel through this particular gas cloud to determine the size precisely.) Were this gas cloud larger than 15,000 km, it would be like the proverbial brontosaurus, which took several minutes to realize that its tail had been stepped on—it just couldn't respond in time to flare up quickly. Such a gas cloud is too small to be part of a main-sequence star like the sun, a star whose radius is 695,000 km. These x-ray sources are associated with tiny stellar objects: white-dwarf stars, neutron stars, or black holes. One particular star has been analyzed thoroughly enough so that the evidence favoring the black-hole interpretation is very strong. This object is Cygnus X-1.

Cygnus X-1: the first black hole

Discovery

Although it is not completely certain that Cygnus X-1 is a black hole, there is good evidence that it is. What follows is a cosmic detective story. The sky has presented astronomers with a puzzle. Where does the Cygnus X-1 piece fit in? One hesitates at first to put Cygnus X-1 in the black-hole slot, since invoking black holes to explain anything strange is not proving that they really exist. Although there is still some debate, it is difficult to explain Cygnus X-1 as anything but a black hole.

The story begins in 1965, when Cygnus X-1 was first discovered during a rocket flight. As its name indicates, it was one of the first x-ray sources to be discovered. Its nature was uncertain, since it could not be identified with any obviously peculiar optically observed object. It is located in the Milky Way (see Figures 5-7 and 5-8), and the first x-ray positions

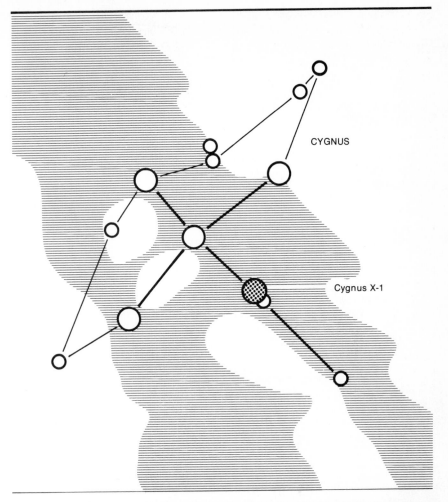

CYGNUS

Cygnus X-1

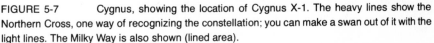

FIGURE 5-7   Cygnus, showing the location of Cygnus X-1. The heavy lines show the Northern Cross, one way of recognizing the constellation; you can make a swan out of it with the light lines. The Milky Way is also shown (lined area).

were not precisely determined. From the initial observations, all you could say was that it was somewhere in the region shown in Figure 5-8. There are a lot of stars in that picture, and it is impossible to check all of them for peculiarities, especially when you do not know exactly how the optical counterpart to an x-ray source will be peculiar. It might even be totally invisible optically, as some x-ray sources are. In the late 1960s, more data were accumulated. The x-ray intensity of the source varied. The nature of the source was still unknown.

In 1971 and 1972, the breakthrough occurred. In March and April 1971, the *Uhuru* satellite indicated a marked change in the x radiation from

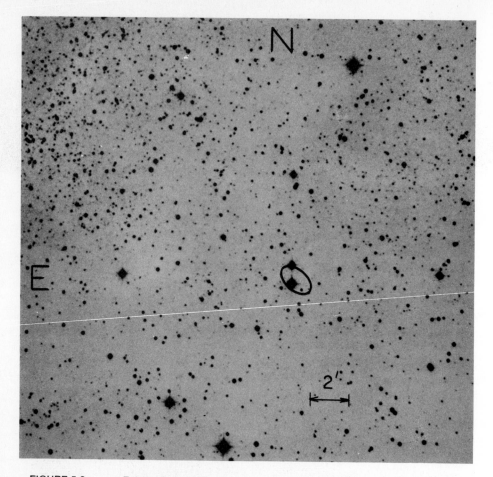

FIGURE 5-8   Enlargement of a Palomar Sky Survey print showing the location of Cygnus X-1, which is the bright star at the lower left of the ellipse. (From S. Rapaport et al., *Astrophysical Journal Letters,* vol. 168, plate L6, 1971, published by the University of Chicago Press. Copyright © by the University of Chicago. All rights reserved. The Palomar Sky Survey is copyright © by the National Geographic Society– Palomar Observatory Sky Survey.)

this source. Lower-energy x rays decreased in intensity to one-quarter of their former value, and high-energy x rays increased in intensity. But most important, a radio source appeared suddenly in the same part of the sky as the x-ray source. The 140-foot telescope of the National Radio Astronomy Observatory had been used to search for a radio counterpart to Cygnus X-1, but had not found one until this change occurred.

It seemed fairly clear that the radio source and the x-ray source were the same object. The importance of the discovery of the radio source was not that the radio noise provided much information about the nature of the source, but that the radio position could be measured much more accurately. At the same time, the x-ray position became more precise. (The ellipse in Figure 5-8 is the measured position of the source from an MIT rocket experiment; the experimenters said that the source was somewhere in that ellipse.) The bright star in that ellipse began to look much more intriguing, especially since the radio position was centered exactly on that star. The uncertainties in the radio position were about the same size as the image of the star in the photograph.

Thus the identification of Cygnus X-1 with a star was confirmed. Optical identification of x-ray sources is a critical step in determining their nature, as spectroscopy can tell you a lot about a star: how big it is, how hot it is, whether it is a double, and so forth. This star is called HDE 226868, Star No. 226,868 in the extension of the Henry Draper catalogue of spectral classifications. Its spectrum shows that it is a B-type supergiant, a large, hot, blue star.

A further discovery in 1971 was the finding, by a Japanese group, that the x-rays flickered very rapidly. This flickering is significant, you recall, because it shows that the x-ray source must be very compact. Cygnus X-1 began to look more and more like a black hole.

Now the optical astronomers started working. They looked for variable Doppler shifts in the star's spectrum. These would indicate that this massive B supergiant star, HDE 226868, was being pushed around by the gravitational forces of an invisible companion. It was crucial to measure the amplitude of these Doppler shifts, and it was also necessary to determine the properties of the visible star. During the 1972 observing season (an optical astronomer can observe stars only when they are in the night sky, and Cygnus is in the night sky only in the spring and summer), Cygnus X-1 was studied intensively. From these observations, a fairly good model emerged that several investigators agreed on. Although each individual adopted slightly different numbers, the basic model was the same. See Figure 5-9.

The star system is a double one, containing a B-type supergiant star and a black-hole companion. Mass flows away from the supergiant star in a stellar wind, produced by the high temperatures in the outer layer of the star. Were the supergiant star isolated in space, the stellar wind would just flow out into the depths of interstellar space, becoming part of the wispy gas between the stars. But the nearby companion gobbles up some of the outflowing mass. This gas swirls around the companion, forming a circular disk of matter around it. Gas in this accretion disk swirls down toward the black hole, is compressed, and emits x rays before it is swallowed up by the black hole. These x rays are one of the key links in the logical chain of reasoning that makes the black-hole interpretation of this stellar system the most favored one.

This gas stream shows up in other ways, too. Characteristic emission lines of hydrogen and helium come from parts of this stream. It is certain that these emission lines do not come from the star, as the Doppler shifts of these lines indicate that the gas producing them is not moving with either star.

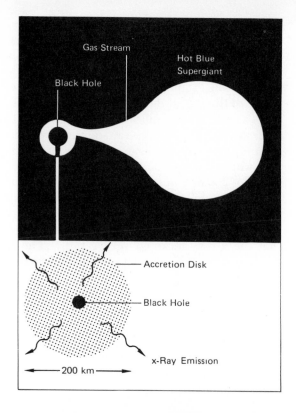

FIGURE 5-9
Model of Cygnus X-1

The model for Cygnus X-1 described above and depicted in Figure 5-9 manages to explain all the phenomena. But an astronomer must be skeptical and ask, "Is there any other believable model that can explain all the observations?" The most important observation here is the existence of the x rays, for the only reasonable way that a star can produce lots of x rays is for the star system to contain a neutron star or a black hole. Which is it? Neutron star or black hole? We have managed to find a source for the x rays, but have we really found a black hole? Remember that neutron stars can have at most three solar masses. To settle the question, we must determine the mass of the companion. The binary-star techniques described earlier were used to answer this question.

The general consensus at the end of 1972 was that the companion was so massive that it must be a black hole. Hot supergiants are generally massive stars, since stars that are not massive, it is believed are cooler and never pass through the blue supergiant stage, at least in normal stellar evolution. Typically, B-type supergiants like HDE 226868, the visible star, have masses of about 30 solar masses. With this large mass assigned to the primary, analysis indicates that the secondary must have a mass of at least five, and probably about eight, solar masses. Neutron stars certainly contain no more than three solar masses, and probably considerably less than that. Therefore the companion is a black hole.

Current research

The black-hole model for Cygnus X-1, illustrated in Figure 5-9, is quite a persuasive one. When all the pieces were more or less in place, in mid-1973, it was definitely time to take this news out of the scientific journals and tell the world that a black hole had been discovered. Enough evidence was in hand that this conclusion was more than mere conjecture or speculation.

But did the completion of the initial stage of analysis of Cygnus X-1 mean that it was time to stop, remain content that we had discovered a black hole, and work on other research problems? Science does not work that way. The model of Cygnus X-1 outlined so far, is deceptively complete. All the observations fit into place, but the explanations are quite generalized. Details remain to be worked out. Observations need to be verified. The analysis of this system continues. The next few pages describe the major topics of concern at the beginning of the 1980s.

How did the Cygnus X-1 system come to be the way it is? Astronomers hardly ever see stars evolve before their very eyes, so they need to use the tools of inference. Is it reasonable that the processes of stellar evolution could produce a system like Cygnus X-1? The discovery of Cygnus X-1 and other systems like it has spurred a revival of interest in theoretical calculations of the late stages of evolution of binary stars.

Right now, there are two stars in the system: a 25-solar-mass star that we see, and another object, the black hole, with a smaller mass—let's call it eight solar masses. In the beginning, the star that is now the black hole had most of the mass of the system. It became a red-giant star, and its envelope expanded just the way that red-giant envelopes normally do, to a point. But eventually the envelope got so big that the second star in the system, the one that we now see, started capturing some of the mass in the system. Streams of gas flowed from the bloated, distorted red-giant star toward the companion. Eventually all the envelope was dumped on the other star, and the core of this first red giant in the system collapsed to become the black hole that we see now. Subsequently this second star started dumping mass on the black hole, producing an accretion disk and the x rays that we see now.

Calculations that provide numbers describing the scenario of the last paragraph[1] indicate that systems like Cygnus X-1 don't last very long. In about 10,000 years, the rate of mass flow will increase because the 25-solar-mass star will get too big. This increased flow will probably choke off the x rays from the accretion disk. What happens then? It's anybody's guess.

So we can understand how Cygnus X-1 came to be the way that it is now. The short duration of the present phase, 10,000 years (short compared to the millions and billions of years it normally takes stars to evolve) explains why we don't see many more of these objects. When we observe the sky, we pick up only a few stars passing through this fleeting evolutionary phase. Most such systems are not yet x-ray sources, or were x-ray sources in the past.

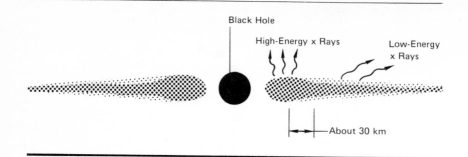

Black Hole

High-Energy x Rays

Low-Energy
x Rays

About 30 km

FIGURE 5-10   Sketch of the inner part of the accretion disk of Cygnus X-1. The two shaded regions show the shape of the accretion disk at two stages. In one state (dark shading), the disk is very flat, except in the extreme inner regions, which emit the high-energy x rays. In the other state, the thick part of the disk is larger. It is the transition from one state to the other that is responsible for the changing character of the x-ray spectrum. [Freely sketched from the models of Thorne and Price, *Astrophysical Journal Letters* 195 (1975), L101-L106, and Eardley, Lightman, and Shapiro, *Astrophysical Journal Letters* 199 (1975), L153-L156.]

Another active area of research involves modeling the accretion disk of the system, trying to understand exactly where the x rays come from. Theorists here seek to produce a model for the accretion disk that produces an x-ray system that resembles the real x-ray system. Some generalized understanding has emerged, but detailed agreement of theory and observation is still in the future. As an example, consider the changes in the x-ray behavior of the system. Sometimes the system emits radio waves and a few low-energy x rays. At other times, after an abrupt change, it emits no radio waves and larger quantitites of low-energy x rays. Alan Lightman and Douglas Eardley have shown that accretion disks are unstable. An accretion disk can be either thick or thin, and the transition from a thick disk to a thin disk is quite abrupt. Lightman, Eardley, and Stuart Shapiro showed that the thick-disk spectrum does resemble the observed x-ray spectrum in a general way, but Shapiro notes that "no details exist."[2] A rough sketch of this model is shown in Figure 5-10.

In the United States in the late 1970s, a new generation of x-ray telescopes called HEAO's (High Energy Astronomical Observatories) was launched. These telescopes provide additional information on the time variation of the x-ray emission. These x rays flicker, varying in intensity in milliseconds. These variations probably arise from the motion of lumps of gas in the accretion disk. A hot spot, a slightly denser gaseous glob, orbits the hole a few times before being sucked away from the inner edge of the disk into the cosmic garbage disposal—the black hole itself..

Suppose that the radiation from the hot spot was more or less beamed in a particular direction. This beam would circle the sky once for each orbit of the black hole, just as a pulsar beam sweeps around the sky once for each revolution of the spinning neutron star. X-ray observatories would see a short, quasiperiodic sequence of pulses of x-ray emission from this swirling hot spot. Observers have claimed to detect these quasiperiodic pulses with the *Uhuru* generation of x-ray satellite observatories, but none of these claims has been unambiguous. If such pulses are ever seen, analysis of them might provide further information about the structure of the accretion disk.

The last few paragraphs have provided a quick overview of one major area of current research on black holes and accretion disks. These projects are the work of "establishment" theoretical astrophysicists, people who subscribe to the majority view that the Cygnus X-1 system contains a black hole in the middle of the accretion disk. How are the black holes formed? How do the accretion disks produce x rays? Are hot spots in accretion disks observable? To progress, scientists must base their work on some paradigm, some underlying model of a particular star system. The black-hole paradigm for Cygnus X-1 seems to be a reasonable one.

Yet this conservative, establishment viewpoint is not the only one that can be taken. Another frontier of research in the black-hole business is pursued by devil's advocates. These people feel uncomfortable with the idea of a black hole as the central object of the accretion disk in the Cygnus X-1 system. Even if they are wrong, their research is useful, for it forces the establishment theorists to reexamine the foundations of the Cygnus X-1 model described above. And if the critics are right, the best candidate for a black hole vanishes, and we come back to square one: Do black holes really exist?

Is Cygnus X-1 really a black hole?

Alternative models

The devil's advocates have asked some probing, pertinent questions about the establishment model for Cygnus X-1. They have not yet engineered a coup, dethroning the black-hole model and consigning it to that remote prison that discredited scientific theories end up in—the prison of obscurity in the annals of science. In fact, at this time, the establishment model still prevails in the view of most astronomers. But even if the establishment model ultimately prevails, the questions asked by devil's advocates have deepened our understanding of double-star systems that contain accretion disks around black holes.

Must the x rays in the Cygnus X-1 system come from an accretion disk? The accretion disk surrounding the black hole is the centerpiece of the black-hole model for Cygnus X-1, for the accretion disk is the source of the x rays. One early proposal for avoiding the presence of a black hole in Cygnus X-1 argued that the massive companion in the system was a normal, massive star, just as the massive companion to Epsilon Aurigae may be a normal star. In this model, the two stars were connected by tangled magnetic fields, which accelerated electrons to high speeds and thus produced x rays.

There are two principal problems with the model of Cygnus X-1 as two magnetized stars orbiting each other, twisting up their magnetic fields, and emitting x rays. One is that the proposed companion star has never been seen—a problem that also arises with some of the other alternative models we shall consider later. A second, more serious problem is that this model makes no specific statements regarding what should be observed in order to confirm it. Since no one has observed any other star system in which two normal stars orbiting at similar distances produce x rays, how is this model to be confirmed? Actually, the model has not been taken too seriously because the accretion-disk models provide a far more natural explanation for the x-ray emission. But who knows? Perhaps this model has been dismissed too easily.

A question that has received more attention recently is, Must the secondary be indeed as massive as the observations say it is? The Cygnus X-1 binary system is far, far away; we must use indirect arguments to find the masses of the two stars in it. The Doppler shifts observed from the visible companion show how much that star is being pulled around by the other object in the system. If the visible star were a lightweight star, a low-mass, invisible companion could pull it around and produce the Doppler shifts that are observed.

A serious competitor to the black-hole model postulated that the visible, blue star is a peculiar, low-mass blue star that is in a strange stage of evolution. A low-mass blue star that could be in a similar stage of evolution has been found elsewhere; its catalogue name was HZ (= Humason-Zwicky) 22.

Yet a star like HZ 22 would have to be very close to us in order to be as bright as the visible star in the Cygnus X-1 system. In 1973, two teams of astronomers indirectly measured the distance to the Cygnus X-1 system, finding it to be about 2.5 kiloparsecs (8,000 light years) away. At this distance, a low-mass blue star with a low-mass companion would be invisible. That particular model for Cygnus X-1, a model that did not involve a black hole, bit the dust. By the end of 1975, a number of investigators, pursuing the problem from many different angles, agreed that the mass of the companion is between 10 and 15 solar masses.

The next question asked by devil's advocates was, Must the accretion disk be around the 10-to-15-solar-mass companion, or could it be somewhere else in the system? Two independent groups of astronomers suggested that the accretion disk could surround a neutron star. A main-sequence star could be the 10-to-15-solar-mass companion that produced the Doppler shifts in the star that we see, and a lower-mass neutron star could be responsible for the x-ray emission.

These models are a little difficult to believe in, for two reasons. They both postulate a massive, 10-to-15-solar-mass star in the system. Such a star would normally be quite luminous, and no other star has been seen in the system. The observations are almost, but not quite, sensitive enough to allow us to conclude that a 10-to-15-solar-mass star in the system would have been detected and that therefore, since it has not been detected, it isn't there. Further, a slight decrease in the x-ray emission from the system has been observed when the supergiant star is between us and the massive companion. A natural way to produce these reductions in x-ray intensity is to have the x rays absorbed by the supergiant star or by the gas stream flowing from the star to the accretion disk. In addition, it is a little difficult to understand how a triple system with a supergiant, a massive main-sequence star, and a neutron star might have evolved. This triple-star alternative to the black-hole model is almost, but not quite, shot down.

Another question about the black-hole model has been asked before, in considering what types of stars would form black holes. "Must a 10-to-15-solar-mass object at the center of an accretion disk be a black hole? Could it be an overweight neutron star?" With a limiting mass of 10-to-15 solar masses, such an object could not exist on the basis of conventional ideas of how matter behaves at high densities and on the basis of Einstein's theory of general relativity. You have to abandon general relativity to make overweight (or "obese") neutron stars that are as massive as Cygnus X-1. (This issue was discussed at the end of Chapter 3.)

Cygnus X-1 as a black hole

The story is not yet finished. More challenges to the establishment black-hole model for Cygnus X-1 could be\mounted. Yet the established model has so far stood the test of time. Some pointed questions have been asked, some viable alternatives proposed, and many of the questions answered. One cannot avoid the presence of a black hole in Cygnus X-1 without invoking some fairly complicated models. Cygnus X-1 is *probably* a black hole; astronomers and physicists have not proved that it *must* be one. It is worth reviewing the steps in this somewhat complicated logical chain:

1. HDE 226868, the blue supergiant star that we see, is part of a stellar system containing the x-ray source Cygnus X-1.

2. Analysis of the Doppler shifts in the spectrum of this single-lined spectroscopic binary indicate that there is a companion in the system with a mass of 10-to-15 solar masses.

3. The x rays come from an accretion disk surrounding a compact stellar object (neutron star or black hole).

4. The massive companion is at the center of the accretion disk; the massive companion is the compact stellar object.

5. A compact stellar object with a mass of 10-to-15 solar masses is a black hole.

If Cygnus X-1 is to be proved a stellar system containing a black hole, each link in this chain of reasoning must be sound. So far, the links have been tested and they ring true. The questioning is an important part of the scientific process. Later on you will see similar questions posed with regard to various models of various astronomical phenomena. Up to now, the black-hole model of Cygnus X-1 has withstood the test of being assaulted by alternative models constucted by devil's advocates. The black-hole model is the simplest way to explain the observations. Therefore Cygnus X-1 probably is a double-star system with a black hole in it.

Other black-hole candidates

Lurking behind the lingering doubts that Cygnus X-1 is indeed a black hole is a hesitation to place too much reliance on one particular object in concluding that black holes really exist. Cygnus X-1 may be a strange, peculiar system. Nature may be playing tricks on us. Cygnus X-1 could be something else disguised as a black hole. But it is unlikely that the same peculiar disguises, the same tricks of Nature, would be at work in two different stellar systems. One system, one object, can always eventually be dismissed as "peculiar." But several objects, all of which act in the same way, cannot be passed off so easily. Complex models may be able to avoid the need for a black hole in the Cygnus X-1 system. Yet they could not, in all probability, apply to several x-ray-emitting double stars. The discussion so far has concentrated on Cygnus X-1, since it is the best-studied of the massive x-ray binaries. Some star systems exist in which the compact object is massive enough so that it *might* be a black hole, but not so massive that it *must* be a black hole. Two stars in particular, Circinus X-1 and V 861 Scorpii, are the best black-hole candidates (other than Cygnus X-1) at the present time.

Circinus X-1

Circinus X-1 is an intriguing x-ray source, but at this time it is not the best black-hole candidate. It is in a binary-star system, for the x rays turn off at 16-day intervals, indicating that orbital motion is carrying the x-ray source behind some other star from our viewpoint. But, for a long time, no optical star could be identified with this system. It was only in late 1976 that a team of astronomers working at the Anglo-Australian telescope (located in the mountains in Australia) identified Circinus X-1 with a very red, very faint star. This star looks red because its light has passed through a forest of interstellar dust clouds that have absorbed most of the blue light from the system. It is barely visible in blue light, even in photographs taken with the most powerful optical telescopes. Yet it is intriguing because its x-ray intensity fluctuates rapidly, in the same way that the x-ray intensity of Cygnus X-1 flickers in time scales of milliseconds. Since the optical star is so faint, it is not yet known whether it is a single-lined spectroscopic binary star, or whether the star that we see has an invisible companion that could be a black hole. The status of Circinus X-1 as a black-hole candidate stems from its identification as a binary star and the flickering nature of the x-ray source.

V 861 Scorpii

In the summer of 1978, another—potentially more promising—black-hole candidate was discovered by a group of astronomers using the Copernicus satellite. This star had been discovered to be a single-lined spectroscopic binary. Analyses of its orbit completed in the early 1970s by E. N. Walker showed that the invisible companion had a mass between 7 and 11 solar masses, too much for a neutron star. R. S. Polidan, G. S. G. Pollard, P. W. Sanford, and M. C. Locke used the satellite to study mass transfer in binary-star systems. They observed V 861 Scorpii in April 1978 with the ultraviolet and x-ray telescopes on the satellite. They discovered that this star is also an x-ray source, and that the x rays are eclipsed once in each orbital period when the supergiant star or the gas stream prevents the x rays from reaching us.

V 861 Scorpii, then, seems to be an analogue of Cygnus X-1. All the ingredients for a black hole are there: a massive invisible companion, x rays that can come from an accretion disk, and a visible star to supply mass to the accretion disk. The observations so far lack the wealth of detail that is available about Cygnus X-1. Yet what we *do* know about the system certainly makes the case for a black hole in V 861 Scorpii a promising one.

Finding black holes is a tricky business. Since you cannot see a black hole, you can only hope to detect it through its gravitational effect on another star. Such a gravitational effect appears in the changing Doppler shifts in a visible star's spectrum, a phenomenon that indicates that the star is orbiting an invisible companion. Mass transferred from the visible star to an accretion disk around the companion eventually is heated by the swirling motions in the disk, and the accretion disk emits x rays. The case for a black hole in any particular system is built up on inference.

We discussed several black-hole candidates in this chapter. Two equivocal candidates are Epsilon Aurigae and Circinus X-1. Epsilon Aurigae has a massive, invisible companion, but the x rays that would demonstrate the presence of an accretion disk around a small object in the system are not there. Circinus X-1 has the x rays, but the star is so faint that spectroscopic evidence for a massive companion is very difficult to obtain. More promising candidates are Cygnus X-1 and the recently discovered V

861 Scorpii. Of these, far more evidence is available for Cygnus X-1. All the elements of the black-hole model come together in Cygnus X-1. There is a visible star that orbits an invisible companion, feeding matter to it. There is an accretion disk around the companion that emits x rays. And, apparently, the companion is too massive to be a neutron star. Models that attempt to explain the observations without involving a black hole run into difficulties which almost, but not quite, make them untenable. It is not easy to fit the pieces of this cosmic puzzle together unless you make the companion a black hole.

6   FRONTIERS AND FRINGES

The black hole described in Chapter 4 is a well-understood one, to take the viewpoint of the theoretician. As long as you go along with Einstein's theory of gravitation, that kind of black hole is the only kind of black hole there is in the universe. Yet our understanding of the phenomena described in Chapters 4 and 5 is not complete. We now know what black holes look like (for example, that black holes have no hair) but we have yet to figure out how they are produced in Nature. We think that Cygnus X-1 is a black hole, but we aren't sure; so observers and theorists obtain and analyze more data, and perform more calculations in order to better understand this enigmatic stellar system and other systems like it.

There is a component of current research on black holes and related phenomena that goes beyond re-exploration of the familiar territory opened up in Chapters 4 and 5. X-ray astronomers have discovered a veritable zoo of x-ray sources. The x-ray sources located within the Milky Way galaxy contain stellar corpses, neutron stars, possibly white-dwarf stars, and possibly black holes. Einstein's theory of general relativity—the basis for the black-hole picture developed so far—is being tested, and some of its consequences are being explored. The violent end to the lives of massive stars produces gravity waves that ripple across the cosmos, and these waves are currently being searched for.

There are a few frontier areas of black-hole studies that are properly called fringes, since they represent speculative ventures far beyond the boundaries of experimentally tested or even testable theory. These fringe areas are widely publicized. You see reports that black holes are space warps: You can fall into one and come out somewhere else in this universe or in another universe. Although these ideas *could* be true, they are, at our present level of sophistication, flights of fancy into the never-never land inside the event horizon. It is very easy to believe that black holes are such strange objects that, if you accept their existence, then anything weird, even space-warp stories, that is said about them is true. Do not fall into this trap. Black-hole research, like most of science, contains some results that are true, some that are probably true, and some that are speculation— published because they are interesting if fanciful ideas and just *might* be true. I have gathered all these ideas and put them in the latter part of this chapter so that you, the reader, will know what is fact and what is not.

Frontiers: X-ray astronomy

One of the first x-ray telescopes, hurled above the atmosphere by a rocket, discovered the most famous of all the x-ray sources: Cygnus X-1, the best black-hole candidate. The *Uhuru* satellite of the early 1970s discovered many more x-ray sources in the Milky Way galaxy. In the late 1970s, a series of far more powerful x-ray telescopes, the HEAO's, was launched, and we are slowly developing an understanding of the types of objects that can produce large quantities of x rays. The x rays in most galactic x-ray sources come from accretion disks around small objects like neutron stars of black holes. Interpretation of the x-ray observations can produce a deeper understanding of black-hole candidates like Cygnus X-1, and perhaps can lead to the discovery of other strange ways that stars can end their lives.

X- and gamma-ray bursts

In late November 1975, Jonathan Grindlay, a research associate at the Harvard Observatory, noticed some strange numbers in a page of computer printout of data. These numbers came from x-ray telescopes on board the Astronomical Netherlands Satellite (ANS). In one particular one-minute interval, the telescope picked up twice as many x rays as it did at other times. This blast of x rays was not merely a celestial hiccup; at its peak, an x-ray burster like this one is 100,000 times as powerful as the sun.

Grindlay's reaction to his discovery of these numbers was not "Eureka!" but surprise and disbelief.[1] Had these numbers been garbled as many, many millions of bits of data were fed through the tortuous path that information from a satellite telescope must take? Data from a satellite are transmitted to earth as a series of pulses, the zeroes and ones of the binary-number system. These millions of bits are radioed from the satellite, picked up by a NASA ground station, in this case a radio telescope in Santiago, Chile, and sent over telephone lines through various computers before they reached Harvard. Checking revealed that the numbers were real, and that most of the powerful blast of x rays was produced in the first few seconds of that one-minute period. Subsequent checking revealed more bursting x-ray sources. Grindlay and John Heise of Utrecht in the Netherlands announced their discovery. Shortly thereafter George Clark and Garrett Jernigan of MIT confirmed the existence of these bursts with another satellite. And at least one other group had independently discovered this phenomenon.

Satellite astronomy has also revealed other sudden blasts of high-energy radiation from the cosmos. Bursts of gamma rays, different from the x-ray bursts, were discovered by a satellite that was not launched with astronomy in mind. After a nuclear-test-ban treaty was signed in 1963, the United States launched satellites to monitor the extent to which the treaty was being observed. An exploding nuclear bomb emits a blast of gamma rays when nuclear particles rearrange themselves and release the energy that makes the bomb work. An orbiting satellite can detect this sudden eruption of gamma rays, telling us that someone has set off an atomic bomb.

The American monitoring satellites, called Velas, worked very well. Mountains of data, enormous collections of numbers, piled up at Los Alamos. Trained investigators soon fell behind in their efforts to make sense of it all. In 1969, scientists poring over the 1967 data detected a burst of gamma rays. Presumably the characteristics of this burst differed from the bursts expected from nuclear bombs, so it had not been noticed earlier when the data were analyzed for evidence of nuclear explosions. Later generations of satellites were modified so that approximate locations for the gamma-ray bursts could be determined, and it turned out that the gamma rays were coming from the sky.[2] No one knows just how far away the sources of the gamma-ray bursts are, so we don't know if they are feeble phenomena that we see because they are close by or powerful blasts coming from a great distance. But even if the gamma-ray bursters are no farther away than the nearest stars, at their peak they are ten times as luminous as the sun. Since they are probably much farther away, they are likely to be considerably more powerful.

The x- and gamma-ray bursts involve sudden surges of radiation from a particular object. There is another type of variable x-ray source called a *transient* x-ray source. The most dramatic of these appeared in August 1975, when a group from the University of Leicester, in England, using a satellite called Ariel 5, noticed the sudden appearance of an x-ray source in the constellation of Monoceros (the Unicorn). Using an initial letter A for Ariel, this source was dubbed A 0620−00, where, as is often the case, the position of the object in the sky becomes part of its name. By mid-August this source was the brightest x-ray source in the sky, and it then declined. This behavior is reminiscent of the supernova phenomenon, described in Chapter 3. Other, less spectacular x-ray transients have appeared in recent years. What are these x-ray sources?

Globular-cluster x-ray sources

A type of x-ray source that might be an intriguing black-hole candidate is the globular-cluster x-ray source, first discovered in 1974. Globular star clusters are large collections of between 10^4 and 10^6 stars (see Figure 6-1). It is possible that the x rays from these star clusters come from an accretion disk surrounding a 1000-solar-mass black hole located in the center of the cluster. When the globular-cluster x-ray sources were first

FIGURE 6-1   The globular cluster Messier 3. These star clusters contain more than 100,000 stars. Seven globular clusters are x-ray sources. Some theorists suspect that the x rays may come from an accretion disk surrounding a massive black hole in the center of the cluster. (Hale Observatories photograph)

discovered, this seemed like a very good possibility, but now there are other models, not involving giant black holes, that also explain the observations.

Analysis of the *Uhuru* catalog of x-ray sources indicates that seven of the hundred or so globular clusters in the Milky War galaxy are x-ray sources. This number is a bit high; if these clusters contained the same proportion of x-ray-emitting double stars as the Galaxy does, you would expect only one x-ray source, at most, in one of these star clusters. Clearly something about globular clusters causes them to produce more than the average number of x-ray sources. But what is it? Where are the x-ray sources?

Turning to possible theoretical interpretations, let us take the most obvious one first. At the center of a globular cluster, stars can no longer be considered as isolated, tiny gas spheres roaming the vast deeps of interstellar space. Rather, they would be very close to each other. When they go through their red-giant stages, globular-clusters, stars shed mass in the way that all stars do. This mass swirls in toward the cluster center, forms an accretion disk around the massive, central black hole, and emits x rays just before it takes the final plunge into the event horizon. A black hole of a few hundred solar masses would be big enough to do the job and produce the required x-ray luminosity. The idea that globular clusters contain such maxi-black holes seemed at first to be quite attractive, and it may still be the right interpretation.

However, another interpretation is possible. The globular-cluster x-ray sources are not drastically different from other x-ray sources in the Milky Way galaxy. Their main peculiarity is that there are so many of them in these star clusters. George Clark of MIT has proposed that there are special conditions in globular clusters that favor the formation of the type of binary system that tends to produce x rays when one star dumps mass onto a neutron star or black-hole companion.

How do we decide which interpretation is correct, or if some other interpretation is needed? Several questions can be asked, and the next generation of x-ray satellites, the HEAO's, can answer them. Are the x-ray sources at the cluster centers? They would have to be, if the black-hole model were to be the right one. The HEAO's can find out whether these x-ray sources are all, in fact, at the cluster centers.

There are other ways that a 1000-solar-mass black hole at the cluster center would show up, for it would act as a strong gravitational influence and a larger-than-normal number of stars would collect around it. Such a concentration of stars would appear as a bright peak of light in the cluster center. This peak of light has, in fact, been seen in one of the x-ray-emitting globular clusters, but it can be interpreted in other ways.[3] We can also ask whether the appearance of 1000-solar-mass black holes at the center of globular clusters is reasonable, given what we know about these star clusters. Calculations show that a massive black hole at the cluster center would not grow fast enough to reach its present size by swallowing stars in the 15 billion years since the cluster formed.[4]

Thus the interpretation of globular-cluster x-ray sources is ambiguous. As I read the technical literature, I have the impression that the majority view favors the binary-star interpretation of these x-ray sources. The case for maxi-black holes of 1000 solar masses in the center of these clusters is not compelling.

A tour of the x-ray zoo

X-ray telescopes in satellite observatories have discovered an enormous variety of x-ray sources. Some of these sources are related to normal processes of quiescent stellar evolution. For example, the surfaces of some very hot white-dwarf stars emit x rays just because they are hot. But the most powerful x-ray sources involve other types of phenomena, phenomena that play key roles in the search for black holes. It is worth reviewing x-ray astronomy to see how analysis of present observations, supplemented by new data, will improve our understanding of the way that stars die, and the possible presence of black holes in some x-ray sources.

What types of animals are in the cages of this zoo of x-ray sources? Some x-ray sources pulsate, and are identified with neutron stars in binary systems. (I discussed these in Chapter 3.) Others flicker; one of the flickering ones is our best black-hole candidate, Cygnus X-1. Eruptive x- and gamma-ray sources flare up to extremely high luminosities and then subside in periods ranging from seconds to weeks. Several x-ray sources, including many of the bursters, are identified with globular star clusters. What general patterns can be found in these phenomena?

Many x-ray sources involve dead stars—neutron stars or black holes—in double-star systems. Mass dumped onto the dead star swirls around it, forming an accretion disk, and emitting x-rays. A working hypothesis is that many if not all x-ray sources found in the Milky Way galaxy involve dead or dying stars, and so an extremely active area of research involves the violent and nonviolent ends to stellar life cycles. X rays tend to be produced in double-star systems, where mass is dumped onto a stellar corpse. Thus a related unifying thread of x-ray astronomy is the accretion disk.

So far, this chapter has concentrated on stellar black holes, the subject of Chapters 2, 3, and 5. Most research is directed toward these objects, since the chances for conclusively proving that black holes exist seem highest in this aspect of the search. But black holes come in all sizes, not just the 10-solar-mass regular size. Globular clusters may contain maxi-black holes, with masses of several hundred solar masses. Mini-black holes—black holes far smaller than stars—may also exist. A rather curious property of mini-black holes is that the classical theory of gravitation due to Einstein is insufficient to describe them completely. Consider the properties that quantum mechanics gives to these mini-black holes, and a surprising thing happens: They explode!

Frontiers: exploding mini-black holes

Prior to the mid-1970s, all exploration of black holes involved the use of Einstein's theory of gravitation in its pure, unmodified form. The singularity was treated like a point mass, but even if it were a tiny glob of matter, like an enormously dense billiard ball, the classical properties of black holes would not be different. The large-scale properties of black holes were the focus of attention, and the changes to the picture that the small-scale nature of matter produces were not considered (largely because no one knew how to calculate what they were).

Yet matter does not consist of tiny, billiard-ball-like particles. Electrons, protons, and neutrons, the constituents of matter at the atomic level, are not little tiny marbles. Electrons in atoms are more correctly visualized as clouds of charge. Recall the photograph of a molecule in Figure P-2; here the electrons are seen to be a few angstrom units across, not tiny points.

It was in the 1920s that physicists, seeking to understand atoms, developed quantum theory in order to explain the behavior of matter on a microscopic scale. It is not possible to locate a subatomic particle—an electron, for example—precisely. You cannot say that *now* it's on one side of a nucleus, and that at some subsequent moment it's on the other side of a nucleus. All you can say is that over a period of time, there is a certain probability of that electron's being in a particular place. To understand exploding mini-black holes, you don't need to understand all the details of quantum theory. The key concept is the concept of *uncertainty*—that you cannot precisely locate a very small particle, but can only specify probabilities that it is in one particular place. Readers interested in exploring quantum theory somewhat further will find Banesh Hoffman's book, *The Strange Story of the Quantum*,[5] a book written at the same level as this one, quite fascinating.

What makes black holes evaporate?

But this book is about black holes, not quantum theory. So let's see what happens when you apply quantum theory to black holes, exploring the discovery of the evaporation of tiny black holes by Stephen Hawking, a British physicist. There are two ways to visualize the way that this evaporation takes place, illustrated in Figures 6-2 and 6-3. Take Figure 6-2 first. Quantum theory states that on a small scale, particles are smeared out, looking like tiny clouds. From this viewpoint, the singularity at the middle of a black hole cannot be seen as an infinitely tiny marble, but must be seen as a cloud of matter. You can make statements only about the probability of where the singularity is at some instant in time.

For a 10-solar-mass black hole, the difference between the quantum and classical picture is relatively unimportant. The smeared-out singularity is deep within the event horizon, and the difference between this smeared-out singularity and the pointlike singularity of classical black-hole theory seems not to matter much. But for a mini-black hole, the smeared-out singularity is larger relative to the size of the black hole.

The shaded regions in Figure 6-2 do not show the full extent of the smeared-out singularity. The probability of finding it near the center of the hole, in the dark shading, is quite high. However, there is still a small—but nonzero—probability of finding some of the stuff in the singularity far away from the center of the hole. Someone taking a series of snapshots of the black hole might, once in the age of the universe, find some of the black-hole stuff far away from the center, even outside the event horizon. At such a time, this material might be able to head away from the black hole, tunneling out of the event horizon.

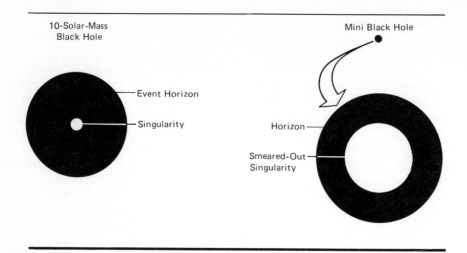

FIGURE 6-2   Quantum effects cause mini-black holes to evaporate. In a large, 10-solar-mass black hole (left), the smeared-out singularity is well within the event horizon. But in a mini-black hole (enlarged view at right), the singularity is reasonably large compared to the size of the event horizon, and there is a finite probability that some of the mass in the singularity will be found outside the horizon.

This process of evaporation would accelerate, for it is the small black holes that have event horizons closest to the singularity. Once a mini-black hole lost a little matter, it would become smaller and more susceptible to losing more. This process would accelerate until nothing was left.

It is possible to view the evaporation or explosion of mini-black holes in another way, a way that is a little harder to visualize but a little closer to the way that contemporary physicists conceive of the microscopic world. This view, shown in Figure 6-3, translates the somewhat vague concept of a "smeared-out singularity" into a cloud of *virtual particles* that surround any object that produces a strong force. These virtual particles are identical to real particles, save for the fact that they appear and disappear in very short periods of time. Objects interact with each other by the interaction of their attendant clouds of virtual particles.

What then happens to an evaporating black hole is not too hard to visualize, once you have made the conceptual leap (a leap that I find is not easy!) of viewing every particle in the universe as carrying along its escort of virtual particles. Again, it is the smallness of the event horizon in a mini-black hole that makes them evaporate in reasonable periods of time. There is a small probability that one of the virtual particles surrounding a mini-singularity will find itself outside the event horizon, on a path that will carry it far, far away from the black hole. Again, in this process the mini-hole loses mass, and the evaporation process accelerates. Eventually the mini-hole disappears in a cloud of high-energy particles, virtual particles that managed to find themselves outside the protective cloak of the event horizon.

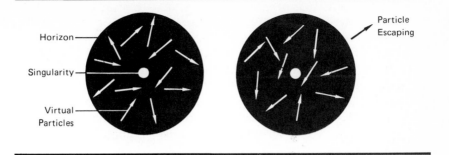

Horizon

Singularity

Virtual Particles

Particle Escaping

FIGURE 6-3   The particle viewpoint of the evaporation of black holes. Any mass is accompanied by a cloud of virtual particles (arrows). These particles are called virtual because they appear and then disappear so fast that they don't live long enough to be observed. It is possible that a virtual particle near the singularity of a mini-black hole will end up outside the event horizon of the hole. The hole will lose a bit of mass, and start to evaporate (right).

The concept of evaporating, or exploding, black holes would be nothing more than good fun for theoretical physicists were it not for the impact that Hawking's discovery had on our interpretations of the real world. An intriguing area of somewhat speculative research deals with the possibility of confirming the picture of exploding mini-holes by some kind of observations. In addition, this analysis has produced some radical revisions in our thinking about the conceptual impact of the idea of black holes on an overall, almost metaphysical view of the universe.

A search for mini-black holes

So what? So black holes evaporate, and the small ones evaporate more readily. Is this not yet another version of the Pygmalion syndrome, theory for the sake of theory? A skeptic might ask these questions. But it turns out that it is possible to observe mini-black holes evaporating, and that even a failure to observe such events could produce some insight into the early stages of the Big Bang, the explosion that marked the beginning of cosmic evolution.

What kinds of black holes might evaporate? The larger a black hole is, the slower the evaporation process is. The universe has been around for 10–20 billion years, and any black holes larger than 10^{15} grams would not have evaporated yet. Thus the search for evaporating holes must be confined to these tiny ones. The event horizon of a 10^{15}-gram black hole is 10^{-28} cm in radius—a very, very small distance. Now that we have specified the kinds of black holes that might evaporate now, let us ask the two questions involved in any search: How do you make them? and How do you find them?

Mini-black holes are made—if they are made at all—in the early stages of the exploding universe. 10^{15}-gram objects currently found in the universe have made their peace with gravity, since they are rigid enough to support themselves against gravitational forces. In the Big Bang, the matter that at present makes up the universe expanded and cooled. Did it expand smoothly, uniformly, like a high-speed version of the slow, languid flow of the Mississippi River? Or was the early expansion like a turbulent, white-water mountain stream, with small condensations of hot gas colliding, annihilating, and colliding again? Bernard Carr[6] has shown that unless the expansion was reasonably smooth, primordial black holes might well form. In fact, a very turbulent early universe might produce lots of them.

So now we go and look for them. An evaporating black hole produces a horde of particles tunneling out through its event horizon as it shrinks to nothingness. Calculations[7] show that many of these particles are gamma rays, so observers seek to find bursts of gamma rays in the universe. You are probably thinking that the gamma-ray bursts discussed earlier in this chapter might be just the gamma rays that are sought. However, the observed gamma-ray bursts are not rapid enough, nor are the gamma rays of high enough energy. Searches for the higher-energy, rapid-fire gamma-ray bursts from exploding black holes have been unsuccessful, and limits show that there must be fewer than a few thousand mini-black holes per cubic parsec. Under some circumstances, the limit might be reduced even further.

Another, far more speculative attempt to find mini-black holes was made by two theoretical astrophysicists who postulated that one of these objects collided with the earth in 1908. In June of that year, the Yenisei Valley in Siberia was devastated by a collision with some object from space—something that killed hundreds of reindeer (no humans) and scorched the forest for miles around. Because Russia was preoccupied with events that had a different kind of violent impact (the beginning rumblings of the Russian Revolution), the site, named Tunguska, was not scientifically explored until 1927.

Various people have proposed a number of possible objects as the source of this explosion. The theories range from the commonplace to the bizarre, and the commonplace theory is probably correct. It was probably a small comet, a chunk of ice a few hundred yards across, that collided with the earth in 1908. This model explains what happened. A small comet would evaporate in the atmosphere, leaving no remnant (as was observed); would have produced most of the damage from above (as was observed); and would have left some dust in the atmosphere (as was observed). Yet the success of the comet model has not deterred other speculations.[8]

In 1972, two theorists from the Center for Relativity Theory at the University of Texas speculated that a mini-black hole collided with the earth, sliced through it like a hot knife slices through soft butter, and came out the other side of the planet (conveniently, in the Atlantic Ocean where there were no trees, reindeer, or other objects to record the devastation). But the exit of this tiny object would have produced atmospheric shock waves, which were not observed. Further, its passage through the earth would have produced earthquakes, which again were not observed. But the strongest argument against the black-hole interpretation of the Tunguska event is that it is extraneous. There is no need to invoke a black hole to explain what happened. (Further, there is no need to invoke antimatter or alien spaceships either.)

To summarize: There is no observational evidence that mini-black holes exist. Searches for gamma-ray blasts from exploding mini-black holes have been unsuccessful. Even stranger proposals involving collisions between such objects and the earth have not withstood scientific scrutiny. The lack of abundance of these objects has allowed us to rule out some of the Big Bang models in which the early stages of the expansion were exceedingly turbulent. But you should keep in mind that these mini-black holes don't have to exist at all, for there are Big Bang models in which they never form.

Frontiers: testing general relativity

So far I have been assuming that the general theory of relativity, proposed by Einstein in 1915, is the correct description of the way that gravity works. This theory has not gone unchallenged; close to a hundred competing theories have appeared in the scientific literature since then. These theories have all been tested, and most (some believe all) have failed to withstand the scientific test of agreement with experiment. There are various types of tests that a theory must pass: It must be complete and self-consistent. It must produce effects on orbiting planets that agree with recent observations. It must agree with a number of other ways in which the nature of gravity has been probed by experiments.

The first requirement that any theory of gravitation must meet is that it be complete and self-consistent. A theorist must be able to predict the results of a well-defined experiment or observation from a clearly stated set of first principles. As new experiments are done, you cannot just keep making up new physical laws to account for the experimental results. The set of first principles, basic physical laws, must be so complete that any experimental situation involving the forces that the theory explains (in this case gravity) can be modeled. The results of the model calculation must agree with experiment, within the limitations of each. Self-consistency requires that various mathematical approaches to the model calculation produce the same result. An example of an incomplete theory is Milne's kinematical relativity, proposed in 1937. This theory is unable to make any statements about gravitational redshifts. Although you could postulate gravitational redshifts as an additional physical law, the need to make such an extension of the theory shows that it is incomplete.

A surprising number of theories turn out to be incomplete or inconsistent. Many laymen have ideas that appear to be original, at least in the sense that these ideas have not migrated to the nontechnical literature. These ideas sometimes are written up as "theories" and are then sent to scientific journals or scientific societies. But an idea by itself, whether original or not, is not a theory. These "theories" generally start and end in thin air, since they are often just isolated ideas. A theorist must be able to show how an idea can be interwoven with all the other physical laws that apply to that particular field of science. A theory must be able to produce models that can then be experimentally tested.

Yet completeness and self-consistency, the ability to model experimental situations, are not the only tests that a theory must pass. General relativity and other modern theories of gravitation agree with the laws of Newton as long as the forces of gravity are extremely weak. Thanks to centuries of observation of planetary positions, we know that where gravity is weak, Newton's laws describe the motion of objects to a high degree of precision. Any new theory must agree with Newtonian theory in situations in which gravitational fields are extremely weak. And if a new theory of gravitation appeals to a connection with some other forces, like electromagnetism, it must agree with the classical descriptions of electromagnetism in situations in which those classical descriptions have been tested.

Yet even if a theory is complete, self-consistent, and in agreement with Newtonian theory in situations in which gravity is weak, it has to agree with a number of additional experimental tests. Einstein's theory of gravitation does differ from Newton's in several important ways. Many of the critical differences have been tested by experiment. For example, Einstein's theory states that any small object, no matter what its nature, falls under the influence of gravity. Such an assertion can be tested inexactly by dropping hammers and feathers in the lunar vacuum and seeing them hit the ground at the same time. Or it can be tested more precisely by carefully designed experiments that confirm this assertion to one part in a trillion.

In addition, there are a variety of fairly subtle effects of general relativity on the motions of planets. For example, the long axis of Mercury's egg-shaped orbit moves about 43 seconds of arc more per century (two parts in ten thousand) in Einstein's theory than it does in Newton's. Experiments in the 1970s showed that these subtle effects agree with models based on Einstein's theory within a percent or so. The high precision of these experiments has ruled out a number of competing theories of gravitation.[9] Some readers may have heard of the Brans-Dicke theory of gravitation, for example. The Brans-Dicke theory (also proposed by Jordan) enjoyed a fair degree of popularity in the late 1960s. Now, however, it is out of favor, since it does not agree with experiment.

Experimental tests based on the motion of planets in the solar system and on various terrestrial experiments share the weakness that in all cases the gravitational fields are quite feeble. There are a few theories of gravitation that produce the same results as Einstein's in such situations. These theories cannot be tested by solar-system experiments because in such experiments they agree with general relativity. Many of these theories have been proposed as mere foils to Einstein, as simple demonstrations of the existence of such theories. Such theories are conceptually equivalent to Einstein's.

Another theory that agrees with Einstein's in solar-system experiments, one proposed by Nathan Rosen in the 1930s, has the virtue that it was proposed before solar-system tests became quite accurate. Rosen's theory does not permit the existence of black holes, but it does permit supermassive neutron stars. Were Rosen's theory correct, the companion to Cygnus X-1 would be a large neutron star rather than a black hole.

A test of a theory like Rosen's requires a situation in the real world in which gravitational fields are quite strong, not weak, as they are in the solar system. It would be tempting to appeal to black holes as a test of such a theory, but it is very difficult to observe them, and very hard to imagine a situation in which an observation of a black hole would test a theory of gravitation rather than somebody's model of the structure of an accretion disk.

A cleaner test of gravitation theories comes from observations of very close double-star systems, where the strong gravitational interaction between two closely orbiting, massive objects produces some unusual effects. The changing gravitational forces in such a system produce *gravitational radiation*, fluctuating gravity fields that travel through space and carry energy away from the system. Consideration of this radiation can rule out Rosen's theory of gravitation on the basis of observations of a double star, consisting of a pulsar and a non-pulsating neutron star, the binary pulsar PSR 1913+16.

Frontiers: gravitational radiation

Many of the astronomical phenomena described in this book were discovered because astronomers developed new ways of looking at the universe. You have already seen the key role that x-ray astronomy has played in the search for black holes. The centerpiece of Part 2 of this book, the quasar, was discovered largely because radio astronomers picked up and analyzed radio waves emitted by celestial objects. Thanks to this history, astronomers have a keen interest in detecting new forms of radiation from the cosmos. By the beginning of the 1980s, we have exhausted the electromagnetic spectrum; all wavelengths of electromagnetic radiation have been detected from objects outside the solar system. Gravitational radiation may be one of the next frontiers, though at the present time the direct detection of this form of radiation appears to be quite difficult.

The generation of gravity waves

How might gravitational radiation, *gravity waves*, be produced? Consider a collection of massive objects, objects the size of stars or larger. These objects could be individual stars or parts of one larger star about to evolve violently. Put these objects very close together. Now let these objects fly about madly, rapidly orbiting each other or chaotically collapsing toward each other. Such a violently evolving object or collection of objects would be a source of gravitational radiation.

What produces the gravitational radiation? Imagine standing very close to this event, and measuring the gravitational fields from this collection of madly collapsing or circling matter. The gravitational field you measured would change rapidly in both intensity and direction as the massive objects, each a source of gravity, moved around. Someone standing close to a radio transmitter would see a similar changing electrical field because electrons in the transmitting antenna are accelerated by electrical forces.

The scenario of the last two paragraphs does not involve any particular theory of gravitation. Such an event could occur in Newton's theory. However, in Einstein's theory of gravitation, the changing gravitational fields seen by someone close to the rapidly accelerating masses propagate through space at the speed of light in the same way that electrical disturbances propagate in the form of electromagnetic radiation. Someone anywhere in the universe could detect this gravitational radiation by measuring small changes in the gravitational field at a particular location.

Thus, in principle, gravitational radiation can exist. But as scientists we are interested in real, observable phenomena, not phenomena that exist only "in principle." Black holes and neutron stars remained in limbo for decades as objects that might, in principle, exist, but could not be observed with the techniques available in the first half of the twentieth century. Gravitational radiation has had a similar history, with a long dormant period that separated the theoretical discovery of its existence and the first searches for it. The development of the first pioneering gravity-wave telescopes and the realization that violent astrophysical events could produce detectable amounts of this radiation produced a resurgence of interest in this field.

The detection of gravitational radiation is a difficult job. Consider two masses separated by some distance, as shown in Figure 6-4. When a gravity wave passes by, the gravitational field in the vicinity of these two masses fluctuates slightly, and the masses start moving back and forth. You can detect a gravity wave by observing the tiny motions of these masses. It sounds simple, but in practice it is extremely difficult. Were a 10-solar-mass star to collapse to form a black hole at a distance of ten parsecs, and were gravitational radiation to be generated by this event at the maximum conceivable rate, the separation between two masses would change by one part in 10^{14}, ten parts per quadrillion. Detecting gravity waves is a difficult job.

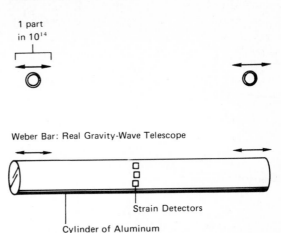

Ideal Gravity-Wave Telescope

1 part
in 10^{14}

Weber Bar: Real Gravity-Wave Telescope

Strain Detectors

Cylinder of Aluminum

FIGURE 6-4   Detectors of gravitational radiation. An idealized telescope consists of two masses some distance apart. A gravity wave causes these masses to oscillate. Because it is very difficult to measure the separations of masses with the required accuracy (one part in 10^{14} for the strongest conceivable gravity wave), most gravity-wave telescopes are long bars with strain detectors to measure the strain on the bar produced when the ends oscillate.

Large amounts of gravitational radiation might be produced in situations in which large masses move around each other or otherwise accelerate quite rapidly. Many of the events discussed in Part 1 of this book are good candidates as sources of gravity waves. Were a massive black hole at the center of a star cluster to eat a star, the final stages of this process would produce a rapid acceleration of the star, considerable distortion of the gravitational fields from it, and a detectable burst of gravitational radiation. A collision between two black holes—a very improbable event—would produce enormous gravity waves. The most plausible violent event that would produce gravitational radiation would be the final collapse of a massive star to the neutron-star or black-hole state. It is possible, but not certain, that this final collapse would occur chaotically, producing the violent motions needed to make gravitational radiation. Rapidly orbiting double stars are less violent forms of sources of this radiation, and radiation from one such system may have been detected indirectly.

Possible observation of gravity waves: the binary pulsar

In the late 1950s and early 1960s, several astronomers added some numerical estimates as flesh to the skeletal picture of gravity waves outlined above. These calculations showed that it would be extremely difficult to detect gravity waves from any reasonable source of them. A stroke of luck, such as a supernova explosion within a parsec or so of the sun, would be needed to produce enough radiation to be observable. Joseph Weber, pioneer of gravity-wave research, proceeded on undaunted and built the first gravity wave telescopes. Astronomers and physicists involved in the field give Weber enormous credit for his pioneering work, in spite of the fact that his claim to have actually detected gravity waves has been discredited because no one else has been able to duplicate his observations.

Weber's detector is the prototype of the advanced gravity-wave telescopes that are currently under construction. An idealized gravity-wave detector (Figure 6-4) would consist of two masses, which would oscillate with respect to each other when a ripple in the gravitational field passed by. Currently, real detectors are long bars of aluminum. Each end of the bar moves when a gravity wave passes by, and since the two ends are separated, they oscillate differently. The different motions of the two ends of the bar produce strains that are detected by detectors mounted around the belly of the bar. The most powerful astronomical source imaginable, a very chaotically forming 10-solar-mass black hole collapsing 10 parsecs away, produces a strain on the bar of one part in 10^{14}, or 10 parts in a quadrillion.

In the early 1970s, pioneer Joseph Weber startled astronomers with his claim that his detectors had actually seen gravity waves. Were his observations correct, an unbelievably powerful source of gravitational radiation would have had to be located at the galactic center, since that was where the radiation seemed to be coming from. As soon as his results were published many other investigators set up their own radiation detectors, but no one was able to confirm Weber's results, in spite of doing experiments that were far more sensitive. At this time, the consensus is that Weber's data-analysis procedures were at fault; few astronomers believe that he actually found gravity waves.

Despite the false alarm, gravity-wave astronomers have not given up. Current experimental efforts are all based on Weber's original idea. More sensitive telescopes use better materials and cool the bars so that thermal vibrations do not set off the strain detectors, mimicking gravity waves. Yet the prospects for immediate positive results are not good. Even the detectors of the early 1980s cannot see the gravitational radiation from the collapse of a massive star unless the star is within a dozen parsecs or so, and such an event is rather unlikely. Gravity-wave astronomy is not even in its infancy; it is in a stage of gestation.

Although direct detection of gravity waves is still in the future, gravity waves may have been indirectly detected by radio astronomers. Double stars can be good gravity-wave sources; the fluctuating gravitational fields produced by the rapid motion of two stars around each other can carry away energy. But for most stars, observations of the orbit are so imprecise and the motions are so slow that the tiny amounts of energy carried away by gravity waves could not be detected for centuries. But as soon as the binary pulsar, PSR 1913+16, was discovered in 1975, a spate of theoretical papers appeared that showed that gravitational radiation would carry away a detectable amount of energy from the system. And, in late 1978, a team of radio astronomers announced that they had seen an energy change that was equal to the change that would be expected as a result of gravity waves.

PSR 1913+16 consists of a pair of neutron stars, each with a mass of roughly 1.4 solar masses, orbiting each other in an elliptical orbit. On the average these stars are 2.8 solar radii apart, and it takes them 7¾ hours to complete their orbit. Their average orbital speed is about 0.001 of the speed of light. This rapid motion produces gravity waves that decrease their orbital period by 0.0001 second per year. Joseph Taylor, Peter McCulloch, and Lee Fowler of the University of Massachusetts announced their discovery of such an orbital decrease in late 1978, at a "Texas" symposium on relativistic astrophysics that was held in Munich, Germany.[10] (These "Texas" symposiums were originally held in Texas, but now they are held in various places; their proceedings are published by the New York Academy of Sciences.) Although this detection of gravitational radiation is only indirect, if it stands up to critical scientific scrutiny, it does represent an important confirmation of Einstein's theory of gravitation. Further, if gravity waves exist, Rosen's theory of gravitation would be shown to be wrong, since the binary pulsar would produce far too much gravitational radiation to be compatible with observations of its orbital change.[11]

Gravity waves represent a speculative research frontier. Einstein's general theory of relativity predicts that they do exist. However, they are very weak and difficult to detect. Direct searches for gravity waves have not been successful. The only real detection of this radiation has been an indirect one, based on careful observations of the orbit of a binary pulsar. But even if you believe that gravity waves have not yet been seen, they are still within the mainstream of scientific research. Their existence is firmly founded in theory and they can, in principle and quite possibly in practice, be observed. Fringe areas of black-hole research involve work that is less well founded in theory and is impossible to verify by observation, since these fringe areas involve the unknown and so far unknowable lower depths of the black hole, the region of space inside the event horizon.

Fringes

Papers dealing with speculative areas of research occasionally appear in the scientific literature. Any paper published in a scientific journal is generally critically examined by other scientists, who determine whether it has scientific value. Scientific referees and journal editors hesitate to reject all speculative papers, since something that appears to be far out on the fringes of science today may well turn out to be important tomorrow. Anyone trying to make an accurate assessment of the correctness of these speculative ideas has to be aware of the ideas themselves and criticisms of the ideas that appear in the literature.

These speculative topics, called fringes in this book, have great appeal to people who seek strange phenomena. Several books at the non-technical level have dealt with these fringe areas.[12] Unfortunately, authors of these books, when writing for their lay audience, present a rather one-sided view of the scientific research literature. This lay audience naturally has not read the papers in the scientific literature, including the more skeptical ones. Readers interested in more details on the topics in this section will find William Kaufmann's book, *The Cosmic Frontiers of General Relativity*,[13] interesting. Kaufmann does mention the skeptical viewpoint, but not as prominently as I do here.

An uncritical scan of the scientific literature can turn up a variety of strange beasts that are vaguely connected with the black-hole phenomenon. Two that are dear to the hearts of science-fiction enthusiasts are white holes and space warps, or wormholes. A *wormhole* is a tunnel through space: Fall into it, and you emerge somewhere else in the universe, or even in "another" universe (whatever that means). A *white hole* is a time-reversed black hole. A black hole forms when stuff falls inward through an event horizon; a white hole is the eruption of stuff outward through an event horizon. Wormholes and white holes are mathematical creatures, not real ones.

A proper appreciation of the distinction between mathematical models and the real world provides the foundation for understanding just how worm holes and white holes fit into the world of black holes and associated phenomena. Recall our discussion in Chapter 1 of how science works. People make observations of the real world and seek to interpret these observations by creating mathematical models, idealized mental pictures of what is going on. These models are essential tools in the development of an understanding of a natural phenomenon, since we can only observe or experiment with the results of natural forces, not observe the forces directly. A mathematical model consists of calculations of the structure and behavior of matter. One can do calculations to describe the behavior of a wide variety of objects. The calculations that are useful for scientific research are those that depict objects that do exist or could exist in the real world.

Survey the different ways in which solutions to the Einstein equations have appeared in the book so far. When the mathematical idea of a black hole was first developed by Karl Schwarzschild in 1916, black holes had not been discovered. Schwarzschild's calculations were interesting because black holes could exist, and we now believe that they do exist. Calculations of the structure of accretion disks are models of a real phenomenon, the phenomenon that produces x rays from sources in the Milky Way galaxy. People test various theories of gravitation by calculating the effects of these different laws of gravitation on orbiting objects like planets and the binary pulsar; here again, the calculations represent models of real objects. Hawking's calculations of the behavior of tiny, evaporating black holes are calculations of phenomena that could exist and could be observed.

Calculations or mathematical concepts can be irrelevant to the real world. Consider negative numbers. There are some circumstances in which negative numbers are useful and do correspond to something real. Suppose you are overdrawn at the bank by, say, $100. A useful way of stating this is to say that you have a negative balance, a balance of minus $100. But consider another situation. Say you are counting the number of apples in an apple orchard. You can, in principle, come up with a negative number. "There are −400 apples in the apple orchard." It's easy to write such a statement, but it doesn't mean anything. It doesn't have anything to do with real apples. (On the other hand, it *could* be relevant to reality, if you were an apple grower and sold somebody 400 apples that you didn't own.)

Now focus on the world of black holes. All the theoretical statements here and in Chapter 4 are based on solutions to the Einstein equations. These solutions must be idealized ones; a collapsing star is such a complex system that an exact solution that is valid for all times is impossible. The idealizing simplification is the assumption that the object at the center of the black hole is a point mass. Such a solution will be a valid representation of space and time after a star has formed to create the black hole.

To describe a real black hole, then, you take the Schwarzschild calculations of the behavior of space and time around a point mass and examine the part of these calculations that corresponds to times after the point mass—the singularity—has formed.

Examine other parts of your mathematics and you will find strange objects—wormholes. These wormholes, or space warps, are passages that connect our universe with a region of space and time that looks very much like a region of space outside a black hole, possibly another part of our universe. They are found by examining parts of the Schwarzschild solution to the Einstein equations that correspond to times before the black hole had formed. Imagine, say, an observer inside an event horizon, and visualize what this observer would see at earlier and earlier times. Go back far enough, and this observer would see the star that made the black hole collapsing under the influence of its own gravity. Go back further, ignoring that in the real world this observer would be inside a star, not inside a black hole. Eventually this observer would reach the point when the singularity would open up, no longer being a point mass. In this mathematical solution, the singularity would be connected to not one but *two* outside regions of space, each of which looks like our universe. In the Schwarzschild solution for a nonrotating black hole, the connection would remain open for only a millisecond before it pinched off. In a rotating black hole, the situation becomes a bit more complex, but the essence of the solution is the same. Wormholes are found only in parts of space and time in which a mathematical solution to the Einstein equations does not correspond to the behavior of real collapsing stars. Because wormholes are mathematical creatures, you can't go out and say they exist in the real world.

White holes are another creature lurking on the fringes of the black-holier-than-thou school of astrophysics. A black hole forms when a star collapses through its event horizon and forms a singularity. The Einstein equations say nothing regarding the direction in which time goes. A solution to these equations can be transformed into another solution simply by letting time run backward. As an analogy, consider the sinking of an eight ball in the final stages of a pool game. The cue ball travels toward the eight ball, strikes it, and the eight ball falls into the pocket, striking molecules inside the pocket and setting them vibrating in random directions. It is also remotely possible that the molecules inside the pocket would all move in the same direction, striking the eight ball, and erupting it out of the pocket so that it travels toward the cue ball, strikes it, and moves the cue ball back to its initial position.

Similarly, a white hole corresponds to the event that would occur when matter erupted from a black hole, outward through the event horizon, and gushed into the universe. The probability of a white hole really existing in the universe has not, to my knowledge, been calculated, but it is probably considerably less than the probability of an eight ball erupting out of the pocket of a pool table.

So white holes, too, are only mathematical objects, unrelated to the black holes that form from collapsing stars. But even if they did exist, what would happen? Douglas Eardley, then at Caltech, published some calculations of the behavior of a white hole in 1974.[14] Eardley found that matter around a white hole would be compressed. Photons would be compressed too, becoming bluer. The compressed, high-energy state of matter in the vicinity of a white hole is termed a "blue sheet." This blue sheet would rapidly form its own event horizon, rapidly converting the white hole into a black hole. For a 10-solar-mass white hole, this conversion would take place within a few hundredths of a second. It would be only under very, very special conditions that a white hole could avoid being swallowed by its own blue sheet and immediately converted into a black hole.

To summarize, then, white holes and wormholes (or space warps) are theoretical creatures that emerge from one particular solution of Einstein's equations. They are found by looking at the region of space and time in these equations which does not correspond to anything that we know of in the real world. In addition, white holes, even if they were to arise, would be immediately transformed into black holes. At the present time, these objects are theoretical speculation on the fringes of black-hole research. The status of these objects 10 or 20 years from now is anybody's guess. I find it extremely unlikely that they will be discovered in the real world. But I could be wrong: Some of today's speculation becomes tomorrow's fact. Much of it remains speculation forever.

Black holes and the universe

Another area of speculative current research deals with the significance of black holes in relation to the evolution of the entire universe. Here, too, theorists are dealing with speculations that are certainly difficult and probably impossible to test experimentally. Yet these speculations are more interesting than white holes and wormholes, from a theoretical viewpoint, because they help develop a deeper understanding of the meaning of Einstein's theory of gravitation. The peculiar nature of the event horizon and of the singularity within it focuses some questions regarding the one-way nature of cosmic evolution, the direction of the arrow of time.

The peculiar nature of the event horizon stems from its one-way character. Things can go into a black hole; they don't come out again. For a long time, this one-way property seemed to contradict a general tendency in the direction of energy dissipation that is noticed in all other cosmic processes. Sunlight, once a concentrated form of energy in hydrogen nuclei at the solar center, flows out and is dissipated into interstellar space. Drop this book onto the floor: Energy initially concentrated in your muscles ends up in the random, disordered motion of molecules on the floor and in the sound waves that die away. (These phenomena are both examples of the application of the second law of thermodynamics.) But when something forms a black hole, energy is concentrated in the singularity. The random motions of an enormously large number of molecules in a massive star are all hidden beneath the event horizon and end up in the singularity.

The singularity itself presents some potential problems. All the mass that falls into a black hole ends up there, in an object that has zero volume and infinite density, according to classical theory. Something traveling too close to a singularity would be subject to uncalculable forces. If a theory can't calculate something, it usually means that the theory needs help. So it is with classical general relativity; it breaks down near the singularity.

Hawking's work, showing that black holes evaporate eventually, provides some escape from the theoretical dilemmas involving the event horizon and the singularity. Energy falling down a black hole is eventually dissipated into the universe. It may take a while; a 10-solar-mass black hole like the one in Cygnus X-1 wouldn't evaporate for 10^{66} years. The singularity is no longer a point with zero volume. It has a finite size, or at least a cloud of virtual particles that make it look like it has finite size. The probabilistic nature of quantum theory is substituted for the absurdity of infinite, uncalculable forces that affect something passing too close to the singularity. The plight of a phantom astronaut — falling into a black hole, forever disappearing from sight, and subjected to unknown and unknowable forces near the singularity — is eased a bit. This resourceful rocket pilot will emerge to haunt us in the future—as a cloud of gamma rays! What happens near the singularity? Einstein never completely accepted the quantum theory because it never produced definitive calculations; as he put it, "God does not play dice." But here, as Hawking put it, "God not only plays dice but sometimes throws them where they cannot be seen."[15]

Can singularities be avoided? Roger Penrose, a British mathematician who has sometimes collaborated with Hawking, has asked this question. A large part of the answer is provided by a theorem the two proved in a jointly written paper, since known as the Penrose-Hawking theorem. The theorem states that—given some technical and apparently reasonable assumptions about the global properties of the universe—anything that ends up inside an event horizon must become part of the singularity that is inside the horizon. As long as Einstein's theory of gravitation holds, the trajectories of particles must end somewhere if they are ever inside the event horizon. Singularities must exist.

Singularities are rather uncomfortable places for a theorist to consider. Classical general relativity makes no predictions about what happens near them and we lack a quantum theory of gravitation which we need to produce an accurate description. But, fortunately, all these uncertainties are swept beneath the concealing carpet of the event horizon. The effects do not appear in the observable world outside.

Is this always true, even in black holes besides the standard ones discussed here? The "hypothesis of cosmic censorship," not yet proved but probably correct, says that a singularity is always hidden from the outside world, or clothed, by an event horizon. A singularity not inside an event horizon is called naked. A naked singularity is quite obscene, for a venturesome astronaut could go near it unaware and subject himself to unknown forces. If all singularities are clothed in event horizons, then the breakdown of Einstein's theory near one is unimportant to the theory of black holes. We encounter such a singularity only when we consider the evolution of the entire universe (Part 3).

A few caveats or warnings must be given in connection with the material of this last section. The relation between black holes and the universe is based on calculations involving Einstein's general theory of relativity. Einstein's theory has been tested only in the solar system, to some extent on the earth, and in the binary pulsar. It is possible (though unlikely) that it will be drastically different on a universal scale. Some of the technical, seemingly innocuous assumptions in theorems like the Penrose-Hawking theorem may invalidate the whole framework. Further, future progress on a quantum theory of gravitation may indicate that singularities behave in surprising and unexpected ways. Particle physicists—probers of the smallest constituents of subnuclear particles—have uncovered a set of fundamental particles that exhibit some striking symmetries. Consideration of these symmetries has produced a completely new approach toward unifying a theory of quantum gravity with theoretical descriptions of other forces of Nature. This new theory, called supergravity, is still at a very rudimentary stage, and it is unclear whether it will have any impact on the results discussed in this section.

The frontiers and fringes of research related to black holes cover a wide range. The launching of satellite x-ray observatories has stimulated the discovery and analysis of an enormous variety of x-ray sources. Most of these sources involve dying stars, and some may involve black holes. Theoretical work on the explosion or evaporation of black holes has prompted searches for these mini-black holes. General relativity has successfully stood up to a variety of experimental tests. Gravitational radiation is energetically sought, and may have been indirectly detected. On the fringe are speculations regarding the possible existence or nonexistence of white holes and wormholes. Another speculative area of thought involves contemplation and calculation of the relation between event horizons, singularities, and the entire universe.

Diverse as these frontier areas are, there are some common themes that unify them. One is the importance of observation, particularly the exploration of new types of radiation from the cosmos. In the 1950s, general relativity was an inactive field of research. The simple calculations had all been done, and there was no need for complex ones, since there were no observations to compare them with. X-ray astronomy has been a great stimulus to further work.The prospect of gravity-wave astronomy may play a similar role in the decades ahead. A second unifying theme is the role of cosmic violence in generating many of the phenomena that we observe. In particular, the violent end to the lives of massive stars, the catastrophic collapse to neutron-star or black-hole dimensions, produces sources of both x rays and gravitational radiation. There are places in the universe at which this type of violence occurs on a far larger scale. Huge galaxies, objects of 10^{11} solar masses rather than 10 solar masses, have cores that may also collapse catastrophically, producing a great deal of high-energy radiation. Those phenomena are the subject of Part 2.

Note added in proof. A bizarre object, discovered in early 1979, may be related to the topics discussed in this chapter. U.C.L.A. astronomer Bruce Margon and many of his California colleagues showed that this object, SS 433, contains two clouds of gas orbiting something at about 20 percent of the speed of light. These gas clouds may be high-velocity jets emerging from the neighborhood of a black hole.

Three types of stellar corpses may exist: white dwarfs, neutron stars, and black holes. A star dies when its nuclear fires stop providing heat to maintain the star's internal pressure, which has kept the star from collapsing under the weight of its outer layers. White dwarfs and neutron stars substitute another kind of pressure for heat pressure: degeneracy pressure. Degeneracy pressure is independent of temperature, so that white dwarfs and neutron stars can cool without collapsing. White dwarfs and neutron stars are known to exist in the real world, the former as stars and the latter as pulsars.

Black holes, if they exist, would form from the collapse of very massive stars. Neutron stars and white dwarfs are certainly no more massive than three solar masses and probably a good deal smaller than that. The central feature of black holes is the event horizon, a spherical boundary that separates the inside of the hole from the outside world. The black-hole phenomena that we can see from out here occur just outside of the event horizon, where something falling toward the black hole is compressed by tidal gravitational forces and appears to freeze just short of the event horizon. Some double stars exist whose invisible companions may be black holes emitting x rays from gas compressed as it is pulled down toward the event horizon, forming an accretion disk. Cygnus X-1 is the strongest black-hole candidate. In a few other double-star systems, the x-ray source may be a black hole.

The genuine frontiers of black-hole research were described in Chapter 6. X-ray astronomy has provided many observations illustrating the violent deaths of some stars, some of which may involve the formation of black holes. In any frontier area of science, lively controversy makes it difficult to regard all the exciting results as definitive. The following display summarizes the results of Part 1 and classifies them according to what is fact, what is working model (accepted by many people but not completely proved), and what is controversy and speculation. Working models represent viewpoints that most people believe in, but would not be surprised to see proved wrong. This field moves so fast that "facts" may be shown to be wrong later, but I certainly believe that what is called fact here will stand the test of time.

FACT	White-dwarf stars exist Neutron stars are pulsars and exist Evolution of stars through the red-giant stage Theoretical model of a classical black hole (possibly including rotation) Black holes have no hair
PROBABLE FACT	Cygnus X-1 is a black hole Low-mass stars → planetary nebulae → white dwarfs
WORKING MODEL	Medium-mass stars → supernovae → neutron stars Massive stars may become black holes Black holes evaporate (very slowly) Gravitational radiation exists Einstein's theory of gravitation (general relativity)
CONTROVERSY	Is Epsilon Aurigae a black hole? Are globular-cluster x-ray sources giant black holes or neutron stars? Are most x-ray sources in the Milky Way galaxy related to dying stars? How do pulsars produce radio emission? How massive can a neutron star be?
SPECULATION	Wormholes, white holes, and space warps

2 GALAXIES AND QUASARS

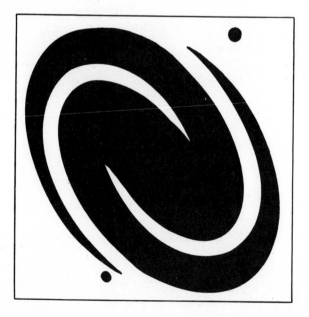

lack-hole studies are a theorist's paradise but an observer's hell. Models for black holes exist in profusion. But it is only recently that observations have appeared in reasonable quantities. The study of galaxies and quasars has a distinctly different flavor. Here observations abound, and theorists try to catch up, seeking to fit the observations into a logical pattern of galaxy and quasar evolution.

The focus of this part of the book is on violent events that occur in the cores of galaxies. Quasars were originally discovered as enigmatic radio sources that were identified with starlike objects. It has recently become clear that they are very distant galaxies, which have luminous cores that far outshine the stars surrounding them. Radiation from these cores is totally unlike starlight. High-speed electrons whizzing around magnetic fields produce radiation in all parts of the electromagnetic spectrum. Some additional visible light comes not from stars, but from tenuous clouds of glowing gas.

Given the more observational nature of this field of research, this part will be largely focused on the properties of these objects. The scene is set with a description of the explosive phenomena in the core of the Milky Way galaxy and of quasars in Chapter 7. Successive chapters treat radio emission (Chapter 8), emission in other parts of the electromagnetic spectrum (Chapter 9), and the location of emission in active galaxies (Chapter 10). Chapter 11 describes theoretical ideas regarding the powerhouses of quasars and active galaxies. A central assumption underlying the results presented here is that the distances of quasars can be determined from their redshifts and from the expanding universe. Chapter 12 summarizes past and present controversy regarding this point.

7 GALAXIES

The universe is filled with galaxies. These gigantic star swarms are the basic constituents of the universe. Each galaxy contains some billions of stars, and billions of galaxies are within reach of our contemporary telescopes. We live in a galaxy, the Milky Way galaxy.

But are galaxies just huge collections of stars, evolving only because the stars in them live, grow old, and die? The last two decades of astronomical research have shown that many galaxies do more than just sit there emitting starlight. The cores of these active galaxies contain violently exploding objects that spew out electrons traveling at speeds close to the speed of light. This chapter sets the stage for the more detailed discussions of galactic activity that follow. I start by describing the Milky Way, our home galaxy, and end with the quasars, the cosmic powerhouses that allow us to probe to the deepest reaches of the universe.

Our galaxy, the Milky Way

Stars

On a dark night, far from the interference of city lights, a faint band of light stretches across the sky. Every culture has had its own name for this band of light, but all names refer to it as a "way" or a "road." The Romans called it the *Via Lactea,* or as Ovid put it more poetically, the "high road paved with stars to the court of Jove."[1] The naked eye sees what looks like a band of light, but the naked-eye impression is misleading, as it is really the blurred image of countless numbers of stars.

When you direct a small telescope or a good pair of binoculars toward the Milky Way, this band of light resolves itself into individual stars. We live in a disk-shaped galaxy. When your line of sight is directed in the plane of the disk, you see vast numbers of stars, or the Milky Way. When you look out of the disk, fewer stars are in your line of sight. Thus the Milky Way is the galaxy itself, and that misty band of light marks the plane of the galactic disk. A composite illustration of the Milky Way is shown in Figure 7-1. The concentration of stars toward the plane of the galaxy is evident.

If we try to map our galaxy using optical telescopes, we are hindered by the dark dust lanes, plainly visible in the Milky Way illustration. As a result, our optical view of the Galaxy is a view of only a very small part of it. The eye can penetrate no further than 1000 parsecs in the plane of the

FIGURE 7-1   The Milky Way as drawn in a chart with the galactic equator at the center. (Chart by Martin and Tatiana Keskula, under the direction of Knut Lundmark, Lund Observatory, Sweden.)

Galaxy, and the telescope cannot penetrate too much farther because of the absorbing interstellar dust.

Dust does not absorb radio waves, however, so radio astronomers can perceive the entire Galaxy. They have been trying to map it. Although their work is not yet complete, they have sketched a tentative map, which is shown in Figure 7-2. Two views of the Milky Way are shown, a top view and a side view. The solid parts refer to areas that are fairly well mapped, while the shaded parts represent areas of the Galaxy about which our knowledge of the structure is much less certain.

The Milky Way is a disk galaxy, about 15 kiloparsecs in radius and 100 parsecs, 0.1 kiloparsec, thick. We are on the outer edge, in the thin disk, and about 10 kiloparsecs from the center. The disk is composed of spiral arms, which are a characteristic feature of other spiral galaxies (Figure 7-3). The only part of the spiral pattern that we are sure of comprises the spiral arms in the vicinity of the sun; we are not sure how the pattern is put together. Is the Orion arm an extension of the Cygnus arm, or does the Cygnus arm connect with the Carina arm, with the Orion "arm" just an offside spur? Our data on the more distant arms are much more

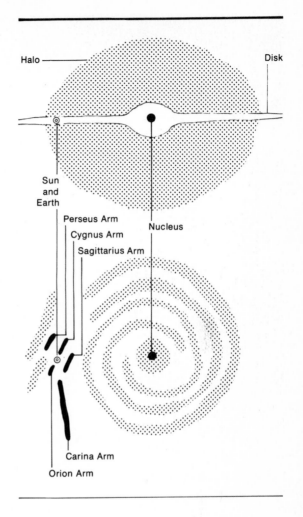

Halo ——————

Disk

Sun
and
Earth

Perseus Arm

Cygnus Arm

Sagittarius Arm

Nucleus

Carina Arm

Orion Arm

FIGURE 7-2
Two views of the Milky Way
galaxy. The shaded spiral
arms are drawn in schematically; their exact positions
are not known.

scanty. The shaded arms that I have drawn represent a schematic view
rather than a precise model. A great deal of insight regarding the structure
of our own galaxy has come from examination of photographs of other
galaxies.

The spiral structure of the galactic disk is embedded in a tenuous,
spherical cloud of stars and star clusters called the *galactic halo*. This halo
contains old stars, stars that have been around ever since the galaxy
formed. There are no more massive stars in the galactic halo; they have all
burned out. The only stars that remain are the tiny, low-mass, dim stars,
stars that consume their nuclear fuel very, very slowly. Such small stars can
survive for a very long time.

FIGURE 7-3   The spiral galaxy Messier 51. (Hale Observatories)

Evolution of stars in galaxies

The picture of the Milky Way galaxy that is presented by its stars is a very placid one. Stars in any galaxy evolve at the slow, stately pace of stellar evolution, punctuated by an occasional violent event like a supernova explosion. The story is not yet completely clear, but enough is known about the life cycles of stars so that a reasonably accurate picture of the evolution of a spiral galaxy like the Milky Way has been produced. Related to the evolution of stars is the maintenance of the spiral arms in a galaxy like ours and the production of heavy elements—the carbon atoms that the paper of this book is made of.

What makes the spiral arms of a galaxy different from the halo? Stars in the spiral arms are younger. They don't all date back to the beginning, to the days when the galaxy had just coalesced from primeval chaos. Look at the spiral arms; there are many massive, short-lived stars in them. A double-star system like Cygnus X-1, a system that formed less than ten million years ago, is found in the Cygnus arm of the Galaxy. The spiral arms contain the gas that is needed for formation of stars. Stars form from this gas when gas clouds collapse. The greatest uncertainty about star formation is understanding the way that the collapse of a gas cloud starts. Currently we believe that star formation is triggered by the passage of a shock wave through the interstellar medium. Such a shock wave could be produced by the spiral arm itself, or it could be produced by a supernova explosion—that occasional violent blast that punctuates the otherwise relatively quiescent cycle of star birth and star death.

When a star is born, fuses hydrogen on the main sequence, runs out of hydrogen and becomes a red giant, and finally dies, it produces starlight from the fusion of nuclei to make heavier nuclei. We have been able to make considerable progress in understanding the past evolution of stars in the Milky Way by observing the results of the chemical processing of elements in the Galaxy. The first generation of stars contained very little of the heavy elements. When these stars died, supernova explosions ejected material that had been through the nuclear furnaces of stellar interiors. This material was enriched in the elements that are the end products of fusion reactions. Subsequent generations of stars contain larger quantities of heavy elements, and future generations will contain still more. Ten billion years ago, many of the carbon atoms that this book contains had not been produced yet. They are the result of fusion reactions in stellar cores.

In the late 1970s, astronomers made considerable progress in filling in the details of the general picture of stellar evolution in the Galaxy outlined in the last two paragraphs. We gain confidence in this picture when detailed models of the evolution of stars and the production of heavy elements begin to agree with observations. People have analyzed the chemical compositions of many stars, and the relative abundance of heavy elements fits the values that are calculated from the presumption that they were produced in supernova explosions. Close analysis of the kinds of atoms in meteorites—interplanetary rocks that fall to the earth—show that a short-lived radioactive element, a particular isotope of aluminum, was present in them when the solar system was formed. This short-lived element had to come from the bowels of a supernova explosion, and that supernova explosion occurred only a few million years before the solar system formed.

The story of stellar evolution in the Milky Way is a relatively peaceful one. Stars are born, grow old, die, and make heavy elements. If that was all that happened in galaxies, the second part of this book would be far shorter and involve no really new phenomena. But galaxies do more than just produce starlight. Many galaxies contain some sort of object at the galactic core that produces other types of radiation, types of radiation that indicate that cosmic violence is occurring on a galactic scale. The core of the Milky Way galaxy is a nearby, weak example of this type of activity, of something in a galaxy that goes beyond the production of starlight, beyond the phenomena that are associated with the birth, lives, and deaths of stars.

The galactic core

Figure 7-1, the panorama of the Milky Way galaxy, shows no signs of a core in the Galaxy. Dust particles in interstellar space limit our vision; visible light cannot penetrate that far through interstellar space. A first approach to the appearance of the core of our own galaxy, then, can best come from an examination of a picture of another galaxy. Messier 51 (Figure 7-3) is a galaxy not unlike the Milky Way in appearance. It doesn't have as many spiral arms, and one of the arms is connected to a companion galaxy, but otherwise it is fairly similar to our own galaxy. There is, apparently, a concentration of stars near the core. Were you to live on a planet in the neighborhood of one of these stars, the night sky would be spectacular. At midnight, the sky would be as bright as it is at twilight, from the light of all of those stars. The nearest star would be as bright as the full moon.

It might seem as though the galactic core would be an optical astronomer's paradise. You could learn a great deal about stars, for they would be very close. But a radio astronomer would have a hard time. On a planet around the galactic center, the only thing that a radio telescope could detect would be some tremendously powerful radio emission from the core itself. We have mapped this radio emission from our distant vantage point on the galactic rim. A tremendous cloud of radio-emitting material, producing a total of 10^{37} ergs/sec of energy, lies at the heart of the galactic core. Our sun is a far weaker source of energy, producing 4×10^{33} ergs/sec, or only 1/2500th of the power of the galactic core. The center of this radio-emitting cloud is about as large as the solar system. This central region emits 10^{33} to 10^{34} ergs/sec of radio power, roughly comparable to the sun, but producing energy in the form of radio waves, not light.

An infrared astronomer would also have problems working at the center of the galaxy. Most of the radiation from the galactic center is emitted in the infrared part of the electromagnetic spectrum. An enormous cloud, about 300 parsecs across, emits 10^{42} ergs/sec, close to one billion times as much energy as the sun produces. At shorter wavelengths, several smaller clouds produce a total of 10^{39} ergs/sec of radiation.

Farther from the core are some other forms of galactic excitement. The two spiral arms nearest to the center are expanding outward at high speeds—53 and 135 km/sec. More complex motions of spiral arms, not yet unraveled, fill out the picture of violent motions occurring near the heart of the galaxy.

This wide array of phenomena—radio emission, infrared emission, and rapidly moving clouds of gas—is all part of the grand story of galactic activity. Galaxies do more than just produce starlight. The story of Part 2 of this book is the exploration of galactic activity: understanding the observations, explaining them, modeling these various clouds of emission, and eventually trying to see why some galaxies become that way. At the galactic core, we have seen three phenomena: infrared emission, radio emission, and rapidly moving spiral arms. Current models have it that the infrared radiation comes from a huge cloud of dust grains that are heated by light from the enormous collection of stars in the galactic core. Radio emission comes from clouds of electrons, madly spiraling around magnetic field lines at speeds close to the speed of light. The rapid motion of the inner spiral arms arises either from the complex distribution of matter near the center of the galaxy or from some tremendous cosmic explosion. Details of the ways in which matter produces these types of radiation will be explored in future chapters. This exploration begins with an examination of other types of galaxies, an examination of different possible sites for galactic activity.

Galaxies

Shapes

Our galaxy is one of billions of galaxies in the universe. Only three of these billions of galaxies are visible to the naked eye—the two clouds of Magellan, visible only to residents of the Southern Hemisphere, and the Andromeda galaxy, a faint fuzzy patch in the autumn sky. All the other galaxies are telescopic objects. Some were discovered by Charles Messier, an eighteenth-century astronomer who compiled a list of 103 nonstellar patches of light to help him search for comets. The Andromeda galaxy is thus known as Messier 31, or M 31. Fainter galaxies were discovered somewhat later.

Anyone who seeks to understand galaxies must first classify them, and the most obvious way to do so is according to their shape. It is assumed that galaxies with similar shapes are similar in other respects also: size, total brightness, stellar content, and evolutionary history. The three principal shape classifications are spiral, elliptical, and irregular.

Spiral galaxies have much the same shape as our Milky Way. Messier 51, shown in Figure 7-3, is a somewhat more loosely wound spiral than ours, but the overall appearance is similar. The spiral arms of Messier 51 contain a lot of gas, dust, and young stars, as the arms of our galaxy do. Color photographs show that the spiral arms are somewhat blue, as you would expect if they contain young stars. The blob at the end of one of the arms of M 51 is a companion galaxy. Current theory indicates that the gravitational interaction between this companion and the main galaxy is at least partly responsible for the existence of the spiral pattern. Some spiral galaxies have bars in the middle, with the spiral arms attached to the ends of the bars, while others have rings at the centers of the arms. We are a long way from understanding how these structural features of spiral galaxies originate.

FIGURE 7-4   The Magellanic Clouds, with the Large Cloud at the left. This is a wide-field view of the sky, as the Large Cloud is twelve degrees across. The bright star at the upper right is Achernar, a star in our own galaxy. (Harvard College Observatory)

If you consider a spiral galaxy without its halo and nucleus, only the spiral arms remaining, you have a good idea of the stellar content of an irregular galaxy. The Magellanic Clouds, landmarks of the Southern sky, are both irregular galaxies. Figure 7-4 shows both of them. The Large Magellanic Cloud has a trace of a one-armed spiral structure which you can detect in the photograph if you look at it for a while. These clouds are composed primarily of blue stars, and contain a great deal of gas and dust. They are both small, much smaller than our own galaxy; this is true of most irregular galaxies.

Now consider a spiral galaxy without the disk and spiral arms. The halo and nucleus remain, and you have an approximate idea of what the third main type of galaxy, the elliptical, looks like. An elliptical galaxy resembles the halo and nucleus of a spiral galaxy in that it is roughly spherical and has quite old stars in it. Elliptical galaxies present a variety of

FIGURE 7-5   NGC 205, a dwarf elliptical galaxy. Note that individual stars are visible.
(Hale Observatories)

shapes, from spherical galaxies that look circular from our vantage point to flattened ones. A photograph of a dwarf elliptical galaxy, one of the flattened ones, is shown in Figure 7-5. It is evident that there is no dust, and radio observations show very little gas.

Galaxies come in all sizes. The smallest galaxies contain only a few million stars or so; they are impossible to discover unless they are extremely close to us, within a megaparsec or two. The largest galaxies are a few hundred times larger than our own galaxy, containing 4×10^{12} (4 trillion) solar masses of material. Elliptical galaxies can be found in all sizes, small or large. Spiral galaxies tend to be large, with more than, say, 10^{10} solar masses of material at a minimum, and irregulars tend to be small.

Activity

For the most part, most of the power output from the Milky Way galaxy is in the form of starlight. The spectacular events at the core produce a total of about 10^{42} ergs/sec of power, mostly in the infrared part of the spectrum. The radio power is about 10^{37} ergs/sec. Both these numbers pale before the total stellar output, estimated to be 3×10^{44} ergs/sec, 300 times greater than the infrared emission and 30,000,000 times greater than the radio emission. Thus the activity of our galaxy, defined as nonstellar radiation, is at a comparatively low level when the amount of starlight that comes from our galaxy is considered. Were we not right next to the core of our galaxy, this activity would be so feeble as to be undetectable.

A small fraction of galaxies, approximately one percent of the total, show signs of activity on a considerably larger scale. The range of activity and the various forms it takes will be explored later, in Chapter 10. The most active galaxies are those in which light from the explosive events at the nucleus outshines the starlight from the surrounding galaxy. Put one of these objects far, far away, and the starlight would be invisible. Such an object would be a point source of light, since only the tiny nucleus would be visible. It would be a "quasi-stellar object," something that at first glance looks like a star, but is, in fact, an extremely distant, hyperactive galaxy. These objects were originally discovered because they emit radio waves, and thus were named *quasi-stellar radio sources,* or QSRS. The name is easier to pronounce if a couple of vowels are added to it, and so the astronomical community came up with the name *quasar* to describe these galaxies that are so active that only the explosive events in the core are visible.

But if a quasar is a point source of light in the telescope, how can we determine how far away it is? We cannot see any stars in a quasar and determine the distance by comparing their brightness to the brightness of other, similar stars near to us. The determination of the distance to the quasars depends on the fact that the universe expands. I shall defer a detailed discussion of the expansion of the universe until Part 3, in which I'll describe the origin and consequences of this phenomenon. Here it is primarily of interest as the way to find how far away a very distant galaxy or quasar is.

The expanding universe

As we probe further and further into the depths of intergalactic space, the spectra of more and more distant galaxies are increasingly red-shifted, indicating that these galaxies are moving away from us. The speed of recession increases with the distance of the galaxy from us. Figure 7-6 shows a series of photographs of galaxies and their spectra. As your eye travels down the page, you note increasingly distant galaxies. Their apparent angular size and their apparent brightness both decrease as their distance from our galaxy increases. In itself, this is not surprising; it is quite natural. But look at the spectra. The red wavelengths are on the right-hand side of the page, the blue wavelengths on the left. The two dark lines visible in each spectrum are characteristic of galactic spectra, the so-called H and K lines of ionized calcium. They are found at the extreme blue end of the spectrum unless the redshift is large. This is the case with the galaxy at the top of the page, in the Virgo cluster, where the redshift is quite small. Farther and farther out in the universe, the redshift becomes larger and larger.

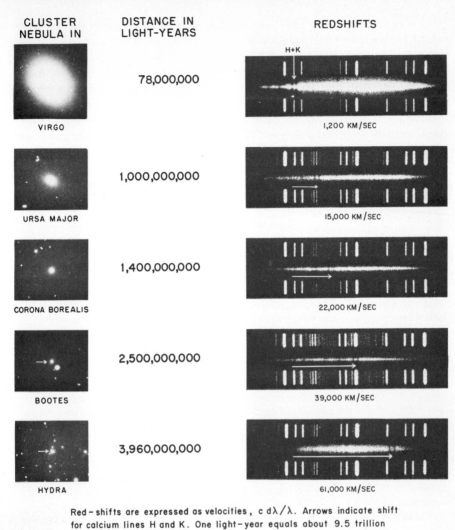

CLUSTER NEBULA IN	DISTANCE IN LIGHT-YEARS	REDSHIFTS
VIRGO	78,000,000	1,200 KM/SEC
URSA MAJOR	1,000,000,000	15,000 KM/SEC
CORONA BOREALIS	1,400,000,000	22,000 KM/SEC
BOOTES	2,500,000,000	39,000 KM/SEC
HYDRA	3,960,000,000	61,000 KM/SEC

Red-shifts are expressed as velocities, $c \, d\lambda/\lambda$. Arrows indicate shift for calcium lines H and K. One light-year equals about 9.5 trillion kilometers, or 9.5×10^{12} kilometers.

Distances are based on an expansion rate of 50 km/sec per million parsecs.

FIGURE 7-6   An illustration of Hubble's Law: The redshift increases with increasing distance. (Hale Observatories)

The galaxies seem to be moving away from only our particular Galaxy; in fact, the galaxies are all moving away from each other as a result of the expansion of the universe. The velocity of recession and the distance of a galaxy are correlated: The greater the velocity, the larger the distance. This relation was discovered by Hubble and Humason in the 1930s, and is known as *Hubble's Law*. This relation can be written: Velocity of recession equals Hubble's constant times distance (or V = *HD* if you like to keep the shorthand expression in mind).*

Thus, to measure the distance to an extragalactic object, you measure the redshift, and Hubble's Law gives you the distance. Such a procedure works as long as the redshift is due to the expansion of the universe. Hubble's constant has been measured many times with differing results; the currently accepted value is 50 kilometers per second per megaparsec. The measurements on which this number is based will be discussed in more detail in Chapter 14; the exact value of the Hubble constant is not a prerequisite to understanding quasars. The increase of redshift with distance has been well established for ordinary galaxies, but the objects with the largest redshift, and therefore presumably the greatest distance, are the enigmatic quasars.

---

*The mail I have received since the first edition of this book was published indicates that there is a great deal of confusion regarding the applicability of Hubble's Law to objects with extremely large redshifts. The simple Doppler formula discussed in Chapter 5 and Hubble's Law, shown here, cannot be naively used for high-velocity objects, for several reasons. (1) The simple Doppler formula, $\Delta\lambda/\lambda = v/c$ in the notation of Chapter 5, needs to be replaced with its special-relativistic version which, if you're interested, is $\Delta\lambda/\lambda = z = [(1 - v^2/c^2)^{-\frac{1}{2}} \times (1 + v/c)] - 1$. For small values of v/c, this can be reduced to the simple Doppler formula. As a result, values of z larger than 1 do not mean that an object is traveling faster than light, as the simple Doppler formula might indicate. (2) At large distances, Hubble's Law is no longer valid; measurement of distance requires a presumption of what the expansion rate was in the past. For the purposes of this book, I assume that the expansion rate does not change; observations (Chapter 16) show that that is a reasonable if uncertain assumption. (3) At large distances, the very notion of "distance" becomes ambiguous. Is "the distance" the distance between us and a faraway galaxy now, or the distance at the time the galaxy emitted the photons we see now, or some intermediate number? You use different measures of distance to calculate different properties. This book uses the distance that light had to travel as the measurement of the distance to an object. Thus if a distant quasar is referred to as being 10 billion light-years away, light left the quasar 10 billion years ago. Readers interested in details should refer to Steven Weinberg, *Gravitation and Cosmology* (New York, Wiley, 1972), pages 415–451. I thank readers of the first edition for their comments on this matter.

Quasars

Now go back to a hyperactive galaxy, one in which radiation from violent events at the core outshines the starlight. Such an object would be visible at large distances, distances at which the less powerful radiation from stars would not show up. At large distances, this object has a large redshift. The strict definition of quasar is that it is an object that looks like a star on photographs taken with the 48-inch telescope at Palomar, and that

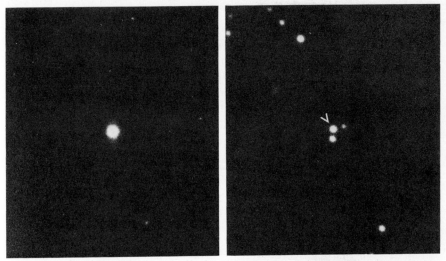

3C 48   **3C 147**

3C 273   **3C 196**

FIGURE 7-7   Four quasars. (Hale Observatories)

it has a spectrum with emission lines with large redshifts. (The spectra of ordinary stars show very small redshifts.) The term quasar has incompletely replaced a confusing variety of terms found in the literature of the 1960s: QSO (or quasistellar object), QSS (quasistellar source), BSO (blue stellar object), interlopers, and QSRS. People sometimes draw finer distinctions between these classes of objects, but as far as I can tell these details are not relevant to the nonspecialist (and they may not even be relevant to the phenomenon itself).

A photograph of four quasars is shown in Figure 7-7. The object 3C 273 is the brightest quasar known. Although 3C 273 is a twelfth-magnitude object, visible in smallish telescopes, its peculiar nature was undiscovered for a long time, since it looks like a star. There are about ten million stars of comparable brightness in the sky, and it is not surprising that this remarkable object remained unnoticed for a long time. One of the other quasars in Figure 7-7, numbered 3C 48, also was important in the discovery of these objects.

Identification of quasars

In the 1950's, radio astronomers discovered a multitude of objects in the sky that emitted large quantities of radio-frequency radiation. Because radio telescopes cannot easily pinpoint the location of a radio source (for reasons described in Chapter 8), it was not known what sort of visible object, if any, corresponded to the radio sources. A few radio sources allow their positions to be determined more accurately as the moon passes in front of them and cuts off the radio waves. If you know the moon's orbit, you can measure exactly when the moon cuts off the radio noise and get a much better idea of where the radio source is in the sky. Identification of radio sources is also easier if the source lies far from the Milky Way, for then there are fewer stars near the position of the radio source.

The first quasar to be recognized as a strange object was 3C 48. Thomas Matthews and Allan Sandage of the Hale Observatories noticed an unusual stellar object at the location of this radio source, number 48 in the 3C catalogue prepared by the Cambridge University radio astronomers. Several astronomers obtained spectra of this object and saw something very strange: broad emission lines at wavelengths that did not correspond with any features normally seen in spectra of other objects. Such a spectrum is not at all like the spectrum of a star. These emission lines meant that this object was concentrating its photon emission at certain wavelengths, but these were not the wavelengths normally seen in spectra of other objects in the galaxy.

The solution to the 3C 48 puzzle came a few years later. In 1962, the moon passed in front of another radio source, 3C 273. Cyril Hazard and his Australian colleagues were able to time this occultation and to determine accurately the position of the radio source. It was then known which one of the many stars near the radio-source position was producing the radio noise, and it was a bright one, the brightest quasar ever discovered. California astronomer Maarten Schmidt then obtained a spectrum of this quasar, and noticed the same peculiarities that were noticed with 3C 48. Once again, 3C 273 had a bright-line spectrum, quite unlike the usual stellar spectrum. It was emitting light at only a few wavelengths, and these wavelengths fitted no pattern. Hydrogen, the most common element in the universe, emits light at just a few wavelengths when it is observed in a glowing gas cloud, but such were not the wavelengths seen in 3C 273.

Schmidt puzzled over his spectrum for a while. As he was writing a report on the results, on February 5, 1963, the answer appeared in one of those flashes of intuition that are the supreme rewards of a successful scientific career. He noticed that the emission spectrum of 3C 273 did correspond to the hydrogen spectrum if he assumed that this starlike object had an enormous redshift—a redshift corresponding to a recession velocity of 47,000 kilometers per second, more than one-tenth the speed of light. No star in the galaxy could possibly move that fast; it would have escaped from the galaxy long ago. Furthermore, a star would not have an emission spectrum like the one of this object. By 1974, the spectra of more than 200 quasars had been analyzed, all of them having very large redshifts. Such redshifts characterize the quasars and give them their mysterious nature.

The redshift of a quasar is usually denoted by the letter z, which equals the shift in wavelength of a spectrum line ($\Delta\lambda$) divided by the wavelength that that line had when it left the quasar (λ_0; thus $z = \Delta\lambda/\lambda_0$). It can also be expressed as a velocity by the Doppler shift formula. Quasar redshifts range from relatively small numbers, like 0.158 for 3C 273, to large, like 3.53 for OQ 172, the most distant quasar known at this time (OQ = Ohio State University radio survey, List Q). The simplest way to explain these redshifts is to assume that the quasars are extremely distant objects that follow Hubble's Law, and are thus the most distant objects known. Most astronomers accept this explanation, and the bulk of Part Two will be based on it. Alternative explanations for the redshift will be considered in Chapter 12.

Properties of quasars

If the redshifts of quasars are caused by the expansion of the universe, they are very luminous objects indeed. The most powerful one, 3C 273, is six trillion times as luminous as the sun. Other quasars are comparable (Table 7-1). You can see 3C 273 in a four-inch telescope even though it is 820 megaparsecs away. The light you see from 3C 273 left the quasar almost three billion years ago, when no land life existed on the earth. It would take a 60-inch telescope to spot a giant elliptical galaxy, were the elliptical galaxy as far away as 3C 273 is.

As the name indicates, many quasars are radio sources. The total amount of energy emitted in the radio range is somewhat less than the

TABLE 7-1   PROPERTIES OF QUASARS

QUASAR	REDSHIFT, Z	VISUAL MAGNI-TUDE	RADIO BRIGHT-NESS*	DISTANCE (MEGAPARSECS)	VARIABILITY	LUMINOSITY (SUNS)
3C 48	0.367	16.2	47	1610	Occasionally	3×10^{12}
3C 147	.545	16.9	58	2120	None detected	3×10^{12}
3C 273	.158	12.8	67	820	Yes, on time scales of years	6×10^{12}
3C 196	.871	17.6	59	2790	Yes	2×10^{12}

*In flux units; 1 flux unit equals a brightness of 10^{-26} watts per square meter per hertz at the surface of the earth.

optical luminosity, and in some cases the radio emission is much less, being undetectable. Recent radio measurements indicate that the radio emission comes from very small, compact clouds of high-speed electrons producing synchrotron radiation as they gyrate around a magnetic field.

The radiation from quasars in the optically visible part of the electromagnetic spectrum is not all in the form of emission lines from a cloud of hot gas. Some of this radiation is not concentrated at any particular wavelengths. Such continuous radiation is also seen in the infrared—it may be that most quasars put out most of their radiation in the infrared part of the spectrum.

The continuous radiation from quasars is variable with time. Most of the observations of variation have been made in the optical part of the spectrum. Most quasars vary relatively slowly, brightening or dimming over periods of a year or so, but a few are considerably more violent in their variations, doubling their brightness in periods of a day or so.

Most of what we see in quasars is present in a much less powerful form at the center of our own galaxy. An almost bewildering array of objects fills in the middle of the scale of galactic activity, so that any one property—say radio emission—ranges from the weak, feeble hiss of spiraling electrons in our own galaxy to the tremendous blast from a powerful quasar like 3C 273. Infrared emission varies on a similar scale. Some quasars and active galaxies vary violently, and others are more quiescent.

In the next two chapters, we shall move down the electromagnetic spectrum, considering in some detail the way in which quasars and active galaxies produce various kinds of radiation. At this time, enough links are available so that quasars are no longer regarded as a unique class of object, but as the upper end of the scale of galactic activity. These chapters will focus on explanations of what has been observed. Quasars are very distant.

The more numerous but less powerful active galaxies are somewhat closer. Study of active galaxies can tell us just where these violent events take place, providing us with some clues to the nature of the powerhouse, the engine that provides the energy for the explosive phenomena seen in active galaxies and quasars. The general approach is to take the observations and work backward, using the tools of inference to obtain information on the causes of these violent phenomena, the most powerful sources of radiation in the universe.

Most of the objects in the extragalactic universe are galaxies like our own. Every galaxy is a collection of myriads of stars, held together by gravity. They come in various sizes and shapes, from the small dwarf elliptical galaxies to the giant elliptical galaxies, ten times larger than our own galaxy, the Milky Way.

Yet galaxies are not the most luminous objects in the universe. That honor belongs to the quasars, which are hundreds of times as luminous as galaxies. This chapter just touched on the properties of these strange objects; the chapters that follow will describe them in more detail.

8 RADIO WAVES FROM HIGH-SPEED ELECTRONS

Quasars were originally discovered because they were intense sources of radio radiation. Any complete model of a quasar or active galaxy must be able to explain where this radiation, which exceeds the optical luminosity of our galaxy in its power, comes from. Recent observations with pairs of radio telescopes on opposite sides of the world indicate that the source of this radio noise is some object a few light-months across. How can such a small volume produce so much energy?

Before considering detailed models for the radio-emitting region, we must first ask how we know that the region is so small. Here we encounter the vast difference between the techniques of optical astronomy and radio astronomy. One difference is that a single radio telescope cannot focus radio waves as well as an optical telescope can focus light. As a result, a single radio telescope gives a very fuzzy view of the radio sky, but pairs of radio telescopes can be combined to sharpen the perception.

In addition, the source of radio waves is often very different from the source of light waves. Light from stars is generated by the heat of the gas that makes up a star. Such thermal radiation can, in some cases, produce radio noise, but the radio waves from quasars come from a quite different source—electrons spiraling around magnetic lines of force at speeds close to the speed of light. When astronomers analyze and describe the radiation from quasars, they begin with the radio waves because these high-speed electrons are directly or indirectly responsible for quasar radiation in all other parts of the electromagnetic spectrum.

Fuzzy views of the radio sky

Radio astronomers face the disadvantage that their telescopes give a very fuzzy picture of the radio sky—somewhat like the view of the world that a nearsighted person gets when without glasses. You can measure the sharpness of any picture of the sky by noting its resolution, which is the size of the fuzzy, blurred image of a pointlike object, or equivalently, the distance by which two objects need to be separated to be perceived as separate objects. It is the limited resolution that prevents a person with eyeglasses removed from distinguishing the letters on the bottom lines of the eye chart. They look like blurs.

The resolving power of a telescope, or the human eye, is measured in angles, as size for all images is expressed in angles. The size of the moon in the sky is half a degree, or 30 arc-minutes. A single radio telescope generally has a resolving power of about 30 minutes of arc, so that a radio view of the sky, from a single telescope, would show many blobs the size of the full moon. The human eye can do much better than this, as its resolution is about one minute of arc. If you want to see what a fuzzy view of the sky looks like, you can get it if you are nearsighted. Just take off your glasses and go outside and look up. You will be amazed at what you cannot see. Such is the fate of the radio astronomer who has not discovered *interferometry,* or the art of combining pictures from two telescopes to obtain a sharp representation.

Radio telescopes have poor resolving power because of the long wavelength of radio waves. Light waves are very short, with wavelengths of less than one micrometer (1 micrometer = 10^{-6} meter or 1/25,000 of an inch). Radio waves have wavelengths ranging from a few centimeters to some tens of meters. A telescope can focus well only if its dimensions are considerably greater than the length of the waves that it is focusing. This condition is satisfied with optical telescopes, which are several meters across, large compared with the small length of light waves, but it is difficult to satisfy this condition with a radio telescope unless it is thousands of meters —kilometers—in diameter. Building a single radio telescope that was a few kilometers or miles in diameter would be a monumental (and expensive) undertaking.

To understand why telescopes have to be so much larger than the wavelengths that they are focusing, consider a somewhat simpler problem. Instead of worrying about focusing, just worry about how sharp a shadow an object can cast. If something can cast a sharp shadow, it can also focus well. Light shadows are quite sharp, because walls and trees and buildings and people—objects that generally cast shadows—are much bigger than light waves. Such light waves can also be well focused by people-sized objects, like telescope mirrors. But think of the shadows that breakwaters cast, as they try to cast "shadows" in their preventing ocean waves from reaching into a harbor. These shadows are not sharp, but fuzzy. Immediately behind the breakwater, the water is quite calm (Figure 8-1). However, soon the water waves manage to turn the corner, around the breakwater, and the area of calm water has a fuzzy boundary. Sound waves do the same thing, for they can turn corners around buildings. The reason that sound waves and water waves can turn corners around obstructions and cast fuzzy shadows is that their wavelengths are long—a few feet in both cases. But light waves are short, and cast very sharp shadows. The upshot is that light waves, with their short wavelength, are easy to focus, while long-wavelength radio waves are not.

For a first look at the radio sky, fuzzy pictures were not bad at all. Radio astronomy pioneers wanted to find out whether there was anything to be seen at all before trying to get sharp pictures. The first maps of the

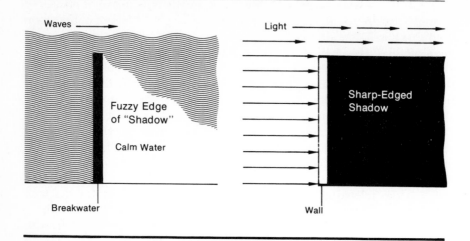

Waves ——→   Light ——→ —→ —→ —→

Fuzzy Edge
of "Shadow"

Sharp-Edged
Shadow

Calm Water

Breakwater   Wall

FIGURE 8-1   As water waves approach a breakwater from the left, the breakwater is unable to cast a sharp shadow since the waves eventually turn the corner and penetrate the space behind the breakwater.

radio sky, obtained by Karl Jansky of Bell Laboratories, the man who discovered that radio emission from the sky existed, and by the American amateur Grote Reber, were quite fuzzy, only good enough to indicate the presence of a few blurred concentrations of radio emission. Some of these concentrations came from obvious places, like the galactic center. But other sources of radio noise had no obvious cause. You cannot go to an optical astronomer and say, "There's a radio source in the constellation Cygnus— what do you think it is?" You need better positions for radio sources, and to get them you need a sharp picture.

Interferometry

You could, of course, use the brute-force method and build a radio telescope a few kilometers in diameter, if you could persuade someone to foot the billion-dollar bill. Yet there are ways to be a little more clever than that. You can use a pair of radio telescopes a few miles apart to obtain some of the information that you could get from a huge antenna, or dish. (When you see a picture of a radio telescope, you can see why it is called a dish.) Because you combine the signals from the two dishes and watch the signals interfere with each other, such a combination is called an *interferometer*. Eventually, a pair of radio telescopes can obtain a sky view equivalent to the view obtained by a single telescope of the size of the distance between the two telescopes. If you wish to speed up the process, so you do not have to spend three months on one object, you can add more radio telescopes and create a number of pairs. The principle that governs the working of all interferometers or multitelescope arrays is the same, and is easiest to understand in the context of the basic building block of an array—a two-element interferometer.

How interferometry works

Figure 8-2 shows a two-element interferometer, observing radio waves with a wavelength of ten centimeters and with a separating distance of three kilometers between the two dishes. Both antennas are looking at the radio source simultaneously, as the source passes overhead. The signals from the source come in waves, which you can imagine as consisting of crests and troughs, alternating. (For electromagnetic fields, these are not really crests and troughs like water waves, but the analogy is still reasonable.) The signal picked up by each antenna is displayed in the figure and then fed into a mixer, which combines the two signals and displays the output. The mixer watches the two waves interfere with each other; hence the name interferometer.

This antenna seeks to determine the exact location of a point source of radio waves by determining exactly when the source is overhead. At that time, in Figure 8-2(a), crests in radio waves from the source reach both antennas at the same time. These signals are in phase, and when the two separate signals are added, the strength of the signal is increased. An ordinary radio telescope would continue to see a strong signal for a long time, since it would see the source as a large blob. But our pair of telescopes can pinpoint the location of the source much more accurately. As the source moves across the sky, it soon reaches a point [Figure 8-2(b)] at which a trough from the left-hand telescope reaches the mixer at the same time as a crest from the right-hand telescope. The two signals mix and cancel, and zero signal comes out of the mixer. We now know that the source is no longer overhead. It seems now that we can pinpoint the source, for a pair of telescopes three kilometers apart can locate the source within a few seconds of arc.

One measurement will not quite do the job, however. Look at Figure 8-2(c). The source has moved a little farther along, and once again two crests reach the mixer simultaneously. The signals add again. It is not the same crest that reaches the mixers. However, someone who can see only the output from the mixer cannot tell whether it is the same crest or two different crests that add. How is the observer to tell that the source is not overhead now? If he were to draw a map of this source, he could say only that the source either is overhead now or was overhead a short while before, when the same crest arrived at both scopes. Such a map would look

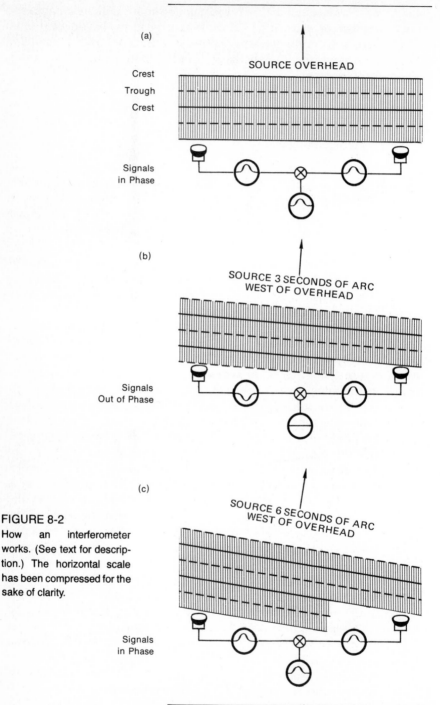

(a)

SOURCE OVERHEAD

Crest
Trough
Crest

Signals
in Phase

(b)

SOURCE 3 SECONDS OF ARC
WEST OF OVERHEAD

Signals
Out of Phase

(c)

SOURCE 6 SECONDS OF ARC
WEST OF OVERHEAD

FIGURE 8-2
How an interferometer works. (See text for description.) The horizontal scale has been compressed for the sake of clarity.

Signals
in Phase

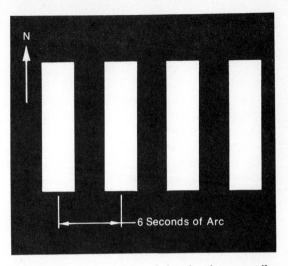

FIGURE 8-3
A map of the radio source
detected by the interferometer of Figure 8-2.

like Figure 8-3. The white stripes indicate the parts of the sky that contribute some signal to the interferometer when the antenna looks at them. If
the source is a single point, all that can be said is that it lies in one of the
light stripes. Which stripe is still doubtful. The reason that the stripes are
elongated in the North-South direction is that our simple interferometer is
no better than an ordinary radio telescope in a direction perpendicular to
that line between the two telescopes, and it still sees a blur.

What the interferometer has done is to take the blurred picture of a
radio source provided by a single telescope and slice it into a series of
stripes, as in Figure 8-4. It is somewhat difficult at this point to imagine

Single Radio Telescope

Interferometer

FIGURE 8-4   A comparison of the beam patterns of an interferometer and a single
radio telescope.

what you might do with such a striped picture of the radio sky. It doesn't seem to be much better than a blur. This striped pattern is known as the interferometer's *beam pattern*. The interferometer has told us that the radio source we are seeking is located somewhere in one (or more) of those stripes.

The usefulness of these striped pictures of the radio sky can be appreciated a little better if you go back to our basic challenge. We know that there is a radio source up there somewhere and we are trying to find out where it is. With a single telescope all we can do is take a sky chart, draw a big circle on it, and tell our optical astronomer friends that the source is somewhere in that circle. They will then tell us to go home and do better, as there are thousands of stars in that circle and they certainly aren't going to look at them all to see if they can find the optical counterpart to the radio source. With an interferometer, we can draw some long skinny boxes on the map and say that the source is in one of those boxes. If we can now overlay our first set of boxes with a second set of different boxes, maybe we can get somewhere. Consider a small part of the beam pattern to see how this procedure works.

Figure 8-5 demonstrates this procedure. The left panel shows a sky map of the part of the sky in which the radio source lies. It shows four stars (A, B, C, and D) and four galaxies (1, 2, 3, and 4). We want to know which one, if any, is the radio source. A single radio telescope would tell us just that the radio source lies somewhere on that map, which isn't much help. One interferometer would slice the map into stripes, as shown in the second panel of the figure. It provides some help, as we can rule out galaxies 2 and 4 and star B as a possible source. But this is not good enough. We then use a second interferometer, whose telescopes are at right angles to the

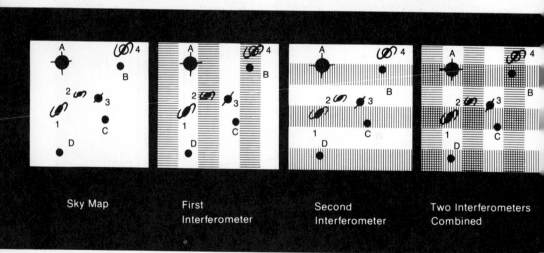

Sky Map

First
Interferometer

Second
Interferometer

Two Interferometers
Combined

FIGURE 8-5   How two interferometers combine to provide a precise location for a radio source. (See text for details of the procedure.)

first. We could also adjust the spacing between the telescopes to obtain wider stripes than the first interferometer had. This second interferometer rules out galaxy 1 and stars A, B, C, and D. Putting the two pictures together, we see that if the radio noise is coming from any of the objects on the star map, it must be coming from galaxy 3 (as the right-hand panel shows). More measurements, with still different spacings or orientations, could serve to pinpoint the source location even better. Whenever you change the distance between the two telescopes or the direction of the line connecting them, you change the width or orientation of the stripes in the beam pattern.

Types of interferometers

In the early days of radio astronomy, the measurement of precise positions of radio sources was extremely laborious. A typical radio observatory had a few radio antennas, and one or two of them were mounted on railroad tracks. You had to put the telescopes in one position, and accumulate data. Then put them in another position, and accumulate some more data. Drag them along the railroad tracks again, and accumulate still more data. The whole procedure could take months.

You may well be asking whether there is an easier way to do radio astronomy. If you have an array of many radio telescopes, you can use each telescope in tandem with each of the others in the array to make observations with many pairs all at the same time. In the early 1980s, the Very Large Array (or VLA), a collection of 27 identical radio telescopes, will be complete (Figure 8-6). The instrument, built in an isolated area on the Plains of Saint Augustin in New Mexico, will be able to map radio sources in a day or so, a vast improvement over the months required to extract similar information from collections of two or three antennas.

Thus radio astronomers have used the ingenious techniques of interferometry to overcome nature's limitations on our ability to locate the sources of celestial radio emission. Brute force, the construction of bigger and bigger pieces of equipment, was not the route to the solution. Rather, it was the discovery that two radio telescopes can be a great deal better than one if you use them in tandem. In the early days of radio astronomy, two-element interferometers showed that the technique would work and that it would provide useful information. The development of the VLA will now make it possible to assemble a much larger body of data for theorists to ponder over.

Most of the observations described in this book are rather straight-forward to understand. They involve simple measurements of how much radiation comes from some object in the sky. I spent quite a few pages on interferometry because it is a little more subtle. Further, the small amount of space in this book devoted to observational techniques might imply that

FIGURE 8-6   Aerial view of the Very Large Array Radio Telescope taken in January of 1978, showing 13 antennas. The telescope is being built by the National Radio Astronomy Observatory, under contract with the National Science Foundation. When it is completed in January 1981, it will have 27 antennas arranged in a Y-shaped configuration. Figure 10-2 shows some results obtained with seven antennas in late 1977. (Photograph courtesy John H. Lancaster, VLA project)

observational astronomy was easy. You can see from the discussion of interferometry that wresting information from the universe is sometimes a difficult job.

Where does the radio emission come from?

Interferometers have been used to observe a wide variety of quasars and active galaxies. There does not seem to be any obvious difference between the radio structure of the two types of objects. Consequently I shall describe a couple of typical radio sources. One of these is Cygnus A, a radio galaxy that is one of the strongest radio sources in the sky. Another is the quasar 4C 39.25.

Cygnus A, a large radio galaxy

Shortly after World War II, electrical engineers turned their talents from designing radar sets to surveying the radio sky. Karl Jansky, the Bell Laboratories pioneer, had discovered the first two radio sources—the sun and the center of the Milky Way galaxy—in 1933. During World War II, radio engineer Grote Reber had constructed a radio telescope in his back yard outside Chicago, and discovered a concentration of radio emission in the Cygnus region. A number of Australian radio astronomers used interferometers to locate the radio emission more precisely. As a result, by the early 1950s, Rudolf Minkowski was able to obtain a photograph and spectrum of the object. It was surprising that one of the brightest radio sources in the sky was identified with a tiny little galaxy (Figure 8-7). Minkowski measured the redshift of this galaxy. Hubble's Law places it at a distance of 340 megaparsecs.

In the 1960s, interferometers were used to map the location of the radio emission more precisely. The contour lines in Figure 8-7 show that the radio waves come from two giant clouds on either side of the visible galaxy. Each cloud is about 17,000 parsecs across, almost as large as our own Milky Way galaxy. Each is about 50 kiloparsecs on either side of the main galaxy. The total radio luminosity is 1.5×10^{45} ergs, nearly ten times what our galaxy puts out in starlight. About three-quarters of all radio galaxies and quasars show this type of source structure.

With Cygnus A as a prototype, you might think that all the radio emission from quasars comes from huge, extended objects. This is not the case, however. A small amount of the emission from Cygnus A comes from a source centered on the galaxy itself. Even the VLA would show nothing more than a blur of radio emission when this central source was examined. However, radio astronomers have been able to use interferometry to resolve some tiny, tiny radio sources, objects with 1/1000th the angular size of objects that optical astronomers routinely resolve.

Transcontinental radio astronomy

The width of the stripes in an interferometer's beam pattern depends on the spacing between the radio telescopes. The further apart the telescopes are, the smaller the width of the stripes. There is no technical reason why you cannot use a pair of radio telescopes thousands of miles apart in order to try to resolve very small radio-emitting clouds. However, it is usually not possible to use a sufficient number of spacings to understand the distribution of radio emission in full detail. The observations must be interpreted in terms of simplified models for the radio emission, as we shall show in a moment.

Use of interferometers where the two telescopes are separated by thousands of miles—at opposite ends of a continent or even on different continents—is called *Very Long Baseline* (VLB) *interferometry*. A VLB interferometer works in much the same way as the two-element interferometer described earlier, but the two telescopes cannot be connected by cables. Instead, the signals are recorded on magnetic tape at each telescope, along

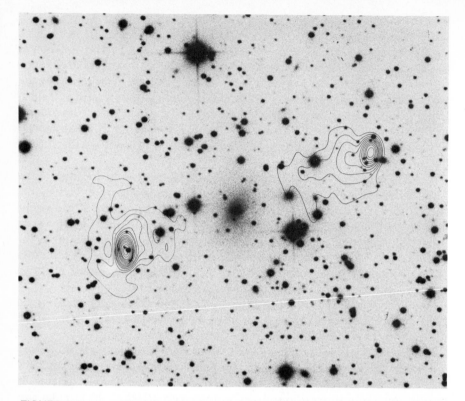

FIGURE 8-7   The radio galaxy Cygnus A. The photograph was taken by Walter Baade at the Hale reflector. The radio contours were observed in 1969 with a 6 × 10 arc-second beam. (From A. Moffet, "Strong Nonthermal Radio Emission from Galaxies," in A. Sandage, M. Sandage, and J. Kristian, eds., *Galaxies and the Universe,* published by the University of Chicago Press. © Copyright 1975 by the University of Chicago. All rights reserved.)

with time ticks from an atomic clock. They are then mixed later when the tapes are brought together and processed by a computer.

The international aspects of VLB interferometry provide an interesting sidelight to this business. Most VLB work is done with very large telescopes, and few countries have more than one or two. In addition, the geographical size of a country limits the possible length of the baselines you can use. In the United States, for example, the longest baselines, or telescope separations, are 3000 miles. Thus VLB astronomers can improve their results if they can develop a working relation with scientists from other countries. Perhaps the most noteworthy intercontinental experiment involved collaboration between American and Russian radio astronomers. These experiments presented a unique opportunity for achieving the highest revolution possible on the earth, as only the American and Russian telescopes were constructed well enough to observe at the shortest wavelengths, at which the highest resolution can be obtained. Aside from the scientific value of the experiment, it was very satisfying to see international cooperation working in the form of a research paper in the *Astrophysical Journal* with both American and Russian authors.

To give you an idea of how VLB interferometry works, I shall examine a particular quasar and show how a VLB astronomer extracts information on the quasar from the data he or she gathers.

A close look at a quasar: 4C 39.25

The quasar 4C 39.25, so named because it is source number 39.25 in the Fourth Cambridge catalogue of radio sources, is a seventeenth-magnitude quasar with a redshift of 0.698. David Shaffer, in his thesis work at Caltech, found that this quasar had a well-determined structure at a wavelength of 2.8 centimeters, and so I have selected it as an example of the use of VLB interferometry.[1]

For this particular picture of this quasar 4C 39.25, three different antennas were used for the observations. With three antennas, there were three pairs, and as the earth rotated, the line connecting each pair rotated as projected on the sky. With each change of baseline, more information regarding the position of the radio-emitting clouds was obtained. But even the overlap of various striped-beam patterns would not produce one single picture of where the radio emission was coming from.

As a result, Shaffer had to interpret his measurements by making some assumptions about the location of the radio emission from this quasar. Scientists make the assumptions, see how well they work, and then judge how valid they are. Based on experience with other radio sources, stretching back to the early analysis of Cygnus A (Figure 8-7), Shaffer assumed that the radio radiation from this quasar 4C 39.25 came from a few circular clouds. He then adjusted the size and position of the clouds of his model so that the emissions from the model clouds would fit with the interferometer measurements—would correspond most closely with the striped patterns that the interferometers produced.

The final model is shown in Figure 8-8. The radio emission comes from two clouds, 3 and 5 parsecs across, and 16 parsecs apart. It should be stressed that the information from the VLB observations is fragmentary, and that Figure 8-8 is a model, not a direct observation. Yet it is an enormous achievement to resolve such tiny clouds at such distances. To do the same thing optically, an astronomer using a telescope focusing visible light would have to photograph an object seven inches across placed on the surface of the moon.

VLB astronomers have detected an extremely small source of synchrotron radio radiation at the core of the Milky Way galaxy. A group of VLB astronomers at the National Radio Astronomy Observatory, analyzing data obtained in 1974, apparently found that a small fraction of the radio

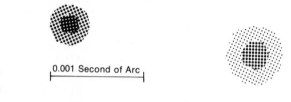

FIGURE 8-8
A sketch of the model of 4C
39.25 derived from the VLB
observations of Figure 8-7.

emission from the galactic core came from a tiny region only 10 astronomical units (AU) across. The luminosity of this core, 10^{33} ergs/sec, is quite modest by quasar standards, but its size is exceptionally small. Subsequent observations made in 1976 by a VLB group at Caltech showed no signs of the presence of the very tiny core, but did show that about 10^{34} ergs/sec of radio emission came from a slightly larger region, about 140 AU across.[2] For comparison, the orbit of Pluto, the outermost planet, is 80 AU in diameter. It is not clear whether there is a discrepancy between the results of the two groups, since the Caltech group would have only barely been able to resolve the 10 AU core if it was there. It is also possible that the tiny core may vary in intensity or in luminosity. In any event, these observations show that a very small source of radio emission, not much larger than the solar system, exists at the galactic core.

Thus the radio emission from quasars and radio galaxies comes from clouds that are more or less round (Figures 8-7 and 8-8). In many cases, the clouds are extremely large, as large as galaxies. In other cases, the clouds are small—only a few parsecs across, 0.001 of the size of a galaxy. In the galactic core, the radio source is as small as the solar system, only one ten-millionth as large as the large clouds. Some objects, such as Cygnus A, contain both sizes of clouds. But what makes these clouds emit radio waves?

Radio emission from high-speed electrons

When radio astronomers used interferometry to locate radio sources, people immediately abandoned the term "radio stars" because it became clear that radio emission came from completely different sources.

The radio emission was so powerful that it could not come from heated objects—thermal sources like stars. A thermal source of radio radiation would have to have a temperature of a few kelvins—far cooler than any star. But if the radiaiton was not thermal, what was it?

In the 1950s, the Russian astronomer Iosif S. Shklovsky showed that the radio-frequency radiation from the Crab Nebula was a form of radiation called *synchrotron radiation,* coming from high-energy particles spiraling in a magnetic field. It was soon realized that most strong radio sources were emitting synchrotron radiation. We now know that all galaxies, including our own, are weak radio sources, and in our galaxy we have observed the raw materials needed for synchrotron emission. Fast particles are seen as cosmic rays, and we have observed the weak magnetic field of the galaxy in several ways. Thus we have directly verified that the radio emission from our galaxy is synchrotron radiation. Before we examine a detailed model for the production of synchrotron radiation in quasars, it will be useful to know about the synchrotron process in a little more detail.

A synchrotron is a particle accelerator. Inside it, electrons or protons are accelerated so that they move at speeds close to the speed of light. Magnets in the synchrotron confine the motions of the fast particles so that they move in a circle. As these particles spiral around the circular path, they emit a ghostly light known as synchrotron radiation. This radiation is given off because the particles are forced to travel in circles, and the nature of the radiation depends on the energy of the spiraling particles. The higher the energy of a particle, the harder the magnetic field must work to hold it in its spiral and the higher the energy of the emitted radiation (Figure 8-9). High-energy radiation is also short-wavelength, high-frequency radiation, so that electrons of extremely high energy give off x rays, lower-energy electrons give off optical radiation, and still-lower-energy electrons give off radio radiation. All these electrons must move at speeds close to the speed of light in order to give off any synchrotron radiation, so even the low-energy electrons are moving quite fast.

Synchrotron radiation can be distinguished from other types of radiation because it is polarized. The electrons are being accelerated only around the magnetic field lines, not parallel to the field lines. Electromagnetic waves, whether they are light waves, radio waves, or something else, consist of electric and magnetic vibrations in space. In polarized radiation these vibrations operate only in one direction. Acceleration of the electron causes the electric field near the electron to vibrate in sympathy with the electron that is moving and in the same direction. Thus, in Figure 8-10, the electron is seen to move horizontally and the electric field to vibrate horizontally. When you investigate the synchrotron radiation with two antennas, one oriented horizontally and the other vertically, only the horizontally oriented antenna can pick up the polarized radiation. The antennas can be considered as pieces of wire with electrons free to move inside them, and only the electrons in the horizontal antenna can move in the same direction as the vibrations of the electric field. The vertical

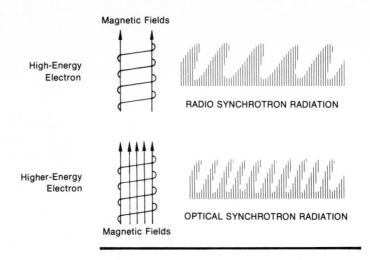

Magnetic Fields

High-Energy
Electron

RADIO SYNCHROTRON RADIATION

Higher-Energy
Electron

OPTICAL SYNCHROTRON RADIATION

Magnetic Fields

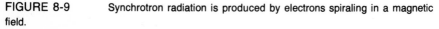

FIGURE 8-9   Synchrotron radiation is produced by electrons spiraling in a magnetic field.

antenna does not pick up any radiation at all, because the electrons cannot move horizontally without moving out of the antenna. Another way of visualizing the situation is to think of the electrons in the two antennas as moving in the same directions as the electrons spiraling around the magnetic field. The synchrotron radiation is only a medium of communication between the two sets of electrons. Because the electrons in the vertically oriented wire antenna have no corresponding vertically moving electrons in the synchrotron radiation source, they do not move. Radiation that results from the acceleration of electrons in one direction only is polarized.

The polarization of the synchrotron radiation is the key to its origin. There are very few ways to produce polarized radiation in Nature. The radiation from quasars is polarized, so one can conclude that the radiation from quasars is synchrotron radiation. In addition, the variation of the radio intensity with wavelength is consistent with what you expect from a synchrotron source.

You can now visualize how these small clouds in quasars emit radio waves. They contain large numbers of electrons moving close to the speed of light. (High-speed protons are not energetic enough to radiate much energy.) These electrons are spiraling around a fairly weak magnetic field, which causes them to emit radio waves by the synchrotron process. The amount of energy contained in these fast electrons is tremendous, equal to the amount of energy radiated by an entrie galaxy in millions of years.

This model sounds nice, but it is very qualitative. Before a reasonable comparison of model and reality can be made, we must try to be more definite. With no information about where these clouds come from, all we

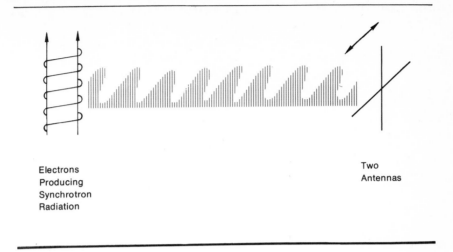

Electrons
Producing
Synchrotron
Radiation

Two
Antennas

FIGURE 8-10   Synchrotron radiation is polarized.

can do is to try to fit the experimental data to a synchrotron model of the kind described above. Such a fitting procedure begs the superimportant question of where these energetic electrons come from. That question underlies the deep mystery of quasars and active galaxies, so it is deferred until we know a little more about them.

The active galaxy 3C 120

The clouds of high-speed electrons are not static, unevolving objects. In fact, it would be surprising if the small clouds such as those found in the quasar 4C 39.25 did not expand, move around, or do something besides just sitting there radiating. It turns out that these clouds expand, and that when they expand the intensity of their radiation varies. For definiteness, let's consider what happens to one extremely well-observed object: the active galaxy 3C 120. We shall discuss this galaxy's emission in other parts of the spectrum a little later in connection with its status as one of the most powerful of all the active galaxies. If you want to see what it looks like optically, look ahead to Figure 10-9. This object has been carefully monitored by radio astronomers for several years.

Figure 8-11 shows the results of a recent analysis of the observations of 3C 120 by George Seielstad, an astronomer at the Owens Valley Radio Observatory in eastern California. The observations (dots) illustrate the changes in the radio intensity, with 3C 120 increasing and declining as the years pass. This behavior is typical; most quasars and active galaxies fluctuate from one year to the next.

Now imagine what should happen to the radio intensity according to the synchrotron model. Pairs of radio-emitting clouds are produced by a central source. At first, these clouds contain many electrons of very high energy, and these electrons radiate at high energies or short wavelengths. Gradually, the electrons lose energy and begin to radiate at longer wavelengths. Further, when the cloud expands, longer-wavelength radiation can escape more easily. Detailed numerical calculations produce specific predictions of the way that radio emission at different wavelengths varies with time. Figure 8-11 shows some of these curves.

But we do not stop with a generalized, qualitative model. Astronomers fit the detailed calculations to the observations by adjusting the time that the clouds were produced, their expansion rate, and the total intensity of radiation produced by particular clouds. The lesson of Figure 8-11 is that the theoretical model fits the data quite well, explaining not just the general shape of the data but also the bumps and wiggles. The timing of variations at different wavelengths fits the theoretical model. The quality of the fit between model world and real world indicates that our understanding is probably pretty complete. In this case, we astronomers can do more than just wave our arms, talk loudly, and gesticulate wildly in an attempt to persuade our colleagues that the model is right. (There are cases, described elsewhere in this book, in which the role of verbal persuasion is larger and the quality of agreement between theory and data is considerably less.)

3C 120 has also been observed by VLB astronomers. They have shown that it has a structure somewhat similar to that of 4C 39.25. At any one time, two clouds of radio emission, each a few parsecs across, are visible. It is not yet clear that these clouds are exactly the same clouds that Seielstad's analysis calls for, but the fit is good, for the most part. There is one considerable surprise, however, VLB observations indicate that these clouds are moving apart at speeds that appear to *exceed* the speed of light. This apparent faster-than-light expansion has been observed in several other quasars.

Faster than the speed of light: an illusion

"Faster than a speeding bullet . . . Able to leap tall buildings at a single bound." Superman—a long-lived comic book, television program, movie, and media-hype hero—can travel quite fast. But not even he can travel faster than light. According to Einstein's special theory of relativity, the speed of light is the ultimate limiting velocity. This theory has been verified countless times in particle accelerators, in which the application of force to particles traveling at near-light speeds gives them more energy, but fails to break the light barrier. But observations of objects like 3C 120 show

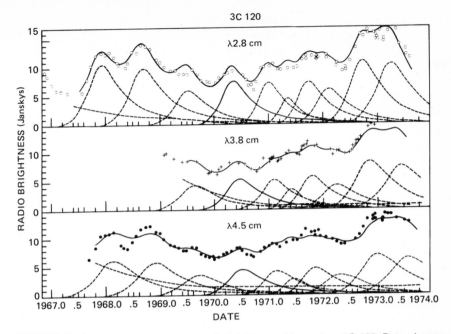

FIGURE 8-11   Variation in the radio brightness of the galaxy 3C 120. Dots: observations; solid line; theory; dashed line: emission from an individual cloud. (From G. A. Seielstad, "The Rapidly Variable Radio Source 3C 120," *Astrophysical Journal*, 193, p. 56, Fig. 1, 1974. Published by the University of Chicago Press. © by the American Astronomical Societies. All rights reserved.)

that the compact clouds of high-speed electrons are moving apart at speeds that apparently run to five or ten times the speed of light. How can this be?

Before these observations were very extensive, a number of astronomers (including myself) suspected that this apparent expansion was caused by a simple illusion. This illusion, called the "Christmas-tree" illusion, was produced by the assumption that two radio sources observed at different times were in fact one and the same cloud. Imagine a Christmas tree with some flashing lights on it. A light flashes on on one side of the tree, and then turns off. A light of similar color then flashes on on the other side of the tree. If you didn't know what was going on, and had only looked at the tree a few times, you might think that the light was traveling across the tree. Were the dimensions correct, as they could be in distant radio sources, the apparent motion could exceed the speed of light even if the clouds weren't moving at all.

But now that the VLB observations are more complete, the Christmas-tree illusion no longer explains the observations. Were this type of illusion to be the cause of the apparent velocities, the radio clouds would move in all sorts of directions—diverge, converge, and move sideways— not just move apart. The velocities observed in any one source would not be consistent. By the end of the 1970s, it was clear that the clouds were always separating at high speeds. However, another kind of illusion turns out to be the explanation for these rapid motions.

This more complicated illusion is produced by the rapid motion of expanding clouds of high-speed electrons toward us here on earth. These radio-emitting clouds are chasing their own photons. Two successive radio maps produced a year apart, in which a year passes from the viewpoint of radio telescopes on earth, may in fact represent the positions of radio clouds at time intervals that considerably exceed a year from the viewpoint of the clouds. A specific numerical example illustrating this illusion is shown in Figure 8-12.

The upper panel of Figure 8-12 shows what the radio observers see when they look at the quasar in pictures taken two years apart. Apparently the clouds have separated by 10 light-years in these two years, seeming to separate at a speed of 10 light years/2 years or 5 times the speed of light. But suppose they were traveling toward us at 13/14ths, or 0.93 of the speed of light, as shown in the bottom panel. They are first seen when they were just produced by the central powerhouse. Radio waves travel at light speed toward us and take some time, say 1000 years, to reach us. We'll see these clouds ejected 1000 years after the central energy source produced them.

Now look again, after 14 years have gone by in the quasar's time frame. These clouds have traveled 13 light-years in that time interval. (Thirteen light years = 13/14 light-years per year × 14 years. The speed of light is 1 light-year per year.) These clouds race toward earth, chasing but not quite keeping pace with the photons they emit. Further, they are not traveling exactly toward us; they move 12 light-years toward us and 5 light-years across our line of sight in that time interval. The photons emitted at this time will only take 988 (= 1000 − 12) years, not 1000 years, to reach us. They will thus reach us 14 + 988 = 1002 years after the central energy source produced the clouds, or 2 years after the first picture was taken. In those two years, each cloud has moved 5 light-years to one side. The two clouds have moved 10 light-years apart in what seems to us to be 2 years, but in 14 years in the time frame of the quasar.

Don't feel that you need to follow the arithmetic in detail in order to understand the essential idea.[3] The principle of the illusion is that an apparent faster-than-light expansion can be produced by objects that chase their own photons, speeding toward earth. Nature can produce the illusion of faster-than-light travel if the radio clouds travel at speeds that approach, but do not exceed, the speed of light.

But is Nature playing the role of magician or master illusionist? The secret of this particular trick is the motion of the clouds of high-speed electrons toward the observer. Must the clouds always move that way? Or is it that the clouds that move toward us are the easiest ones to see? Further analysis produces more complex models to explain this illusion. It is, of course, possible that there is some fundamental problem with our understanding of active galaxies and quasars, that these faster-than-light motions

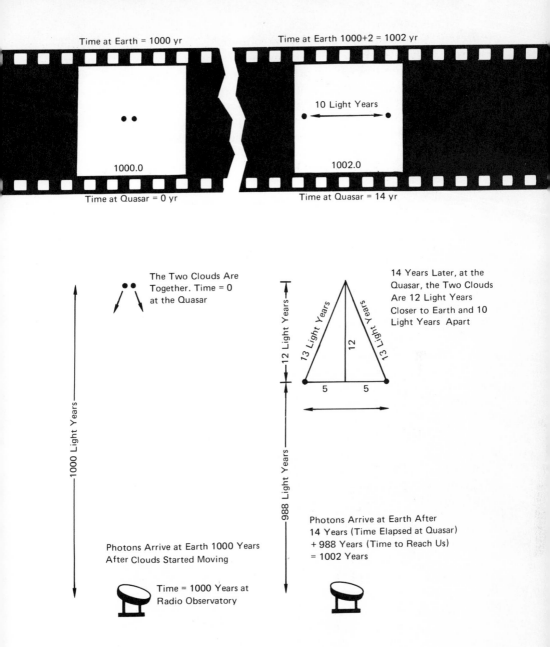

FIGURE 8-12   This illusion can produce radio sources with two radio-emitting clouds that expand at speeds apparently exceeding the speed of light. The movie at the top shows what an earth-based observer would see with a radio telescope. Two years elapsed between these two pictures, and the radio-emitting clouds separated by 10 light-years, apparently traveling at 5 times the speed of light (10 light years/2 years). However, at the quasar, 14 years passed between the two pictures. See text for details.

represent some real problem for astronomers. However, the model in Figure 8-12 has the virtue of explaining the essential features of the observations within the context of presently known laws of physics. We work within the current paradigm until it must fail. Complex as the model of Figure 8-12 is, it does explain the observations. This particular puzzle can be solved without appealing to new laws of physics, even though the idea of motions faster than light might seem to indicate that we need them.

Observations of quasars by radio astronomers that indicate that the following model explains the radio properties of quasars:

1. The radiation from quasars in the radio part of the electromagnetic spectrum comes from high-speed electrons, traveling at speeds close to the speed of light, spiraling in magnetic fields.

2. Interferometer observations indicate that these electrons are localized in clouds. ranging in size from a few parsecs to a few thousand parsecs.

3. The variations in the total intensity of active galaxies and quasars can be explained by the expansion of clouds of high-speed electrons. The apparent motions that seem to exceed the speed of light are probably illusory.

Yet you should keep in mind that these conclusions are based on fitting models to observations. The observers are ahead of the theoreticians in the quasar business, and these models are explanations. The whole synchrotron idea leaves the important consideration of the ultimate source of all this energy undetermined. Yet there are other ways that a quasar puts out energy. Let us next turn our attention to these.

9 RADIATION FROM QUASARS: INFRARED THROUGH GAMMA-RAY

What do high-speed electrons do besides whip around magnetic field lines and emit radio waves? They can also radiate, by the synchrotron process, in other parts of the electromagnetic spectrum, from the infrared through the optical, untraviolet, and x-ray ranges, all the way out to the gamma-ray part of the spectrum. These electrons can collide with photons and give the photons more energy. Synchrotron radiation can also be intercepted by atoms or by dust grains, and the atoms or dust grains can radiate in other parts of the electromagnetic spectrum.

It was radio astronomers who first discovered quasars. The nature of quasars as extremely active galaxies was established by optical work. But by the end of the 1970s, astronomers were able to examine virtually all of the electromagnetic spectrum at some level of sensitivity. They have detected many quasars in all wavelength ranges. Analysis of the data indicates that many of the phenomena described above could in fact be occurring in many of these objects. This chapter will describe the nature of radiation in these diverse parts of the spectrum and the theoretical interpretations of the data.

The mysterious continuum

Radiation coming from an astronomical object can be characterized by the way that the intensity at different wavelengths is distributed. *Continuum radiation* is radiation that is not concentrated at any particular wavelength. Look at a particular wavelength, and look at a wavelength that is just slightly longer or shorter, and you will see almost the same amount of radiation. Synchrotron radiation is continuum radiation. But when atoms emit or absorb radiation, they tend to do so at particular photon energies, or particular wavelengths. Concentration of photons at particular wavelengths are referred to as *emission lines*. All quasars have emission lines in their spectra. Where there is a shortage of radiation at some particular wavelength, an *absorption line* is said to be present in the spectrum of the object. In this chapter, we shall describe these three types of radiation in order, since the different types of radiation come from quite different objects.

Figure 9-1 shows the spectrum of continuous radiation from the quasar 3C 273, which is one of about four or five active galaxies and quasars that have been detected at a wide variety of energies, from the radio to the gamma-ray region. (The gamma-ray measurement shown in Figure 9-1 is tentative.) In general, most quasars and active galaxies have continuous spectra that resemble that of 3C 273, once you allow for the optical emission from stars.

At first, there seems to be a more or less uniform distribution of radiation all the way from the radio to the gamma-ray ranges of the spectrum. Such a smooth distribution might indicate that the synchrotron process described in Chapter 8 could produce all of this radiation. The synchrotron interpretation is one possible interpretation of the continuous radiation. To verify or disprove this model, particular spectral regions must be considered in detail. I begin with the best-observed part of the spectrum, the optical region.

Variations in brightness

The optical radiation from quasars varies, just as the radio emission does. About four-fifths of all quasars vary in much the same way that 3C 273 does, brightening and fading over periods of years. Such behavior is very similar to the radio behavior of quasars, and can be interpreted in much the same way. A quasar brightens when a new cloud of relativistic high-energy electrons is produced, and then fades as these electrons lose energy. The optical bursts are not the same bursts as the radio ones, since the optical and radio variations do not take place at the same time. However, the processes are similar.

About a fifth of all discovered quasars vary more dramatically. These quasars, called *optically violent variables,* change brightness rapidly. The best-studied one, 3C 446, has been known to double in brightness in a matter of a day or so. Such quasars may vary in the radio range as well, but the necessary observations have not been made.

This group of quasars comprising the optically violent variables presents a problem. The rapid variations of these quasars indicate that these objects are just a few light-days across. The volume of space responsible for the variation must receive the signal to vary in less than one day, and since no signal can travel faster than light, the object must be less than two light-days across. (Recall the discussions of Cygnus X-1 in Chapter 5; in that case, the variations were much faster, showing that the object producing them was very small.) If 3C 446 were any larger than two light-days, it would be like the brontosaurus—it just could not coordinate itself well enough to manage to cause a large gas cloud to start emitting in a short span of time.

What is wrong with having an object about two light-days across putting out so much radiation? It is difficult to construct a model that allows light to escape from the object without hitting an electron first.

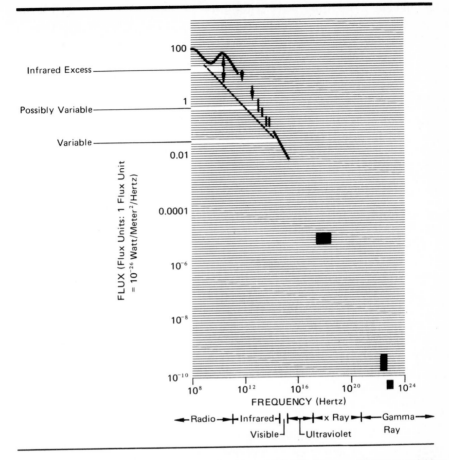

FIGURE 9-1   The spectrum of the continuum of 3C 273. Because 3C 273 is variable and the different measurements were taken at different times, these data represent an average spectrum, illustrating the gross features.[1]

Synchrotron radiation is produced by electrons and magnetic fields, and radiation trying to escape from a small, dense cloud of relativistic electrons will hit the electrons before it can manage to escape. This difficulty is not faced by the synchrotron model for the radio emission, because the radio clouds are known through VLB measurements to be larger, and they vary much less rapidly than the optically violent variables. You encounter the size problem only when you try to explain the optical radiation from the rapidly varying sources.

When the size difficulty was first pointed out in 1966, some investigators thought that the problem was serious enough that the cosmological distances of quasars should be questioned. If you brought the quasars closer, they would not need to be as luminous and the size difficulties could be alleviated. Yet other ways around the problem have since been discovered that leave the quasars at great distances. If, for example, the cloud of electrons is expanding relativistically, at a speed close to the speed of light, you can produce a model that explains the rapid variations. Although the situation is still unclear at present, most theorists feel that the small sizes suggested by the rapid variations in light do not present a serious threat to the practicality of the synchrotron model. Problems arise only for the most rapidly varying sources, according to a recent treatment of the problem.[2] The size difficulty is really a quasar puzzle, not a fundamental problem.

The infrared puzzle

Another puzzling aspect of quasar continuum radiation comes from observations of infrared radiation. Look at Figure 9-1 again. Between the short-wavelength, high-frequency end of the radio spectrum and the visible there seems to be an excess of radiation. A line drawn through the radio observations extends down to the optical observations, but short-wavelength radio waves and infrared radiation lie above this line. Such an irregularity is rather strange. Does this mean that this excess infrared radiation comes from some source that differs from the source of radio emission? First we must ask whether this excess of infrared is really there.

Making infrared observations is notoriously difficult because the earth's atmosphere is uncooperative. Infrared radiation is absorbed strongly by water vapor in the atmosphere, but if you can put your telescopes high enough you can obtain some useful data (see Figure P-4). The near-infrared observations, which can be obtained from the ground, indicate some excess of radiation, but the real power seems to be coming in at a wavelength of some tens or hundreds of micrometers. Initially, the confidence that could be put in these infrared observations was fairly low, but when it was discovered that many other objects, such as the nucleus of our own galaxy, were emitting large amounts of energy too, it became more reasonable to accept the reality of the infrared radiation from quasars. An excess of infrared like that in Figure 9-1 has now been observed from many quasars and active galaxies, not just from 3C 273. There is no question that such infrared excesses exist. Measuring just how *much* of an excess there is is not an easy task.

There are several possible sources of infrared radiation from quasars and active galaxies. In any given object, these different sources each contribute to some degree or other. One possible source is synchrotron radiation. Electrons with energies just a little higher than those energies producing radio emission could produce infrared radiation. However, we have to invoke a different cloud of high-speed electrons to explain why the infrared continuum is not just the extension of the radio continuum. For the less powerful objects, cool stars can produce radiations in the near-infrared part of the spectrum, close to the optical range. But it is hard to explain the far-infrared radiation as stellar radiation.

The strongest competitor to the synchrotron model for explaining the infrared excess is the dust-grain model. Infrared radiation at wavelengths like those of the infrared excess, in the ten- to hundred-micron region, can be produced by objects heated to perfectly ordinary temperatures. This book in your hands is an infrared source, for it has a temperature of 300 K, room temperature, 300 degrees above absolute zero. In a galactic core, the energy required to heat dust grains is there—it comes from synchrotron radiation generated by high-speed electrons. The dust grains are there too, for they are present in many other galaxies, like our Milky Way. We cannot see our own galactic core in visible light, because dust gets in the way. These dust grains could intercept synchrotron radiation, heat up, and reradiate this energy in the infrared part of the spectrum.

Is the infrared radiation from dust or is it from synchrotron radiation? The way that radiation intensity varies with wavelength differs for the two sources. Thus, for any particular quasar or active galaxy, sufficiently precise observations in the infrared might enable us to determine which mechanism produces the infrared excess in any particular object. In two objects where this type of detailed analysis has been done, two answers have appeared. The infrared radiation from the active galaxy NGC 1068 is thermal emission from heated dust grains. In another active galaxy, Centaurus A, the infrared radiation comes from a second cloud of high-speed electrons, and is synchrotron radiation.

Moving down the electromagnetic spectrum in Figure 9-1, we see nothing unusual. The optical and ultraviolet radiation is, apparently, a smooth extension of the radio radiation. It's possible that it comes from the same cloud of high-speed electrons. 3C 273 also emits in the x-ray part of the spectrum, and it may well emit some gamma rays as well. A naive interpretation might be that all of this radiation comes from the same source—synchrotron radiation. Satellite observatories operating in the 1980s will provide enormous amounts of information about these x rays. Thus it behooves us to consider possible sources of x rays and gamma rays in more detail.

X rays and gamma rays from quasars and active galaxies

When it comes to the x rays, we ask the same question that we asked about infrared radiation. If they don't come from synchrotron radiation, where do they come from? X rays are high-energy photons, and high-energy phenomena are required in order to produce them. One possibility is that there is a huge cloud of million-degree gas in the cores of active galaxies, producing the x rays. Such a cloud could be heated by synchrotron radiation—or by other means.

Another possibility: What happens to a very small cloud of high-speed electrons emitting synchrotron radiation? In this cloud are hordes of fast electrons and low-energy radio photons, emitted by the electrons spiraling around magnetic fields. If one of the low-energy photons collides with a high-speed electron, it can gain a considerable amount of energy and be transformed from a low-energy radio photon into a high-energy x ray or gamma ray. These photons could be responsible for the x-ray and gamma-ray emission. Recent work indicates that this process (called *inverse Compton scattering*) is at work in at least some objects.

Some recent x-ray observations show that some of the x radiation from a few objects is concentrated at particular wavelengths in emission lines. X rays and gamma rays have energies that are millions of times larger than the energy-level spacings in atoms, which are responsible for emission lines in the optical part of the spectrum. It is not yet clear just what physical process is responsible for the production of these emission lines. Atomic nuclei are put into a high-energy state somewhere, somehow. Then, as they drop down from a high-energy to a low-energy state, they emit x-ray or gamma-ray photons of a precisely defined energy. We understand the production of emission lines in the optical part of the spectrum far better, and so let us next focus our attention on those features of the spectrum.

Hot gas clouds and emission lines

If quasars emitted only a continuous spectrum, it would be very difficult to verify their extragalactic nature. You would just see a smooth rainbow, and there would be no way to establish a redshift or a distance. Several such objects exist. Some have been shown by distance measurements to be nearby stars—white dwarfs. But some of these objects with continuous spectra vary in the same way that quasars do, and they have in many cases been identified with radio sources. They are called *Lacertids,* in honor of the first one discovered, BL Lacertae, located in the constellation Lacerta (the Lizard).

Any bona fide quasar, however, must have emission lines in its spectrum, for a quasar is a starlike object with a large redshift, and you need emission lines to verify the existence of a redshift. Emission lines appear as vertical streaks in a spectrum (see Figure 9-2), indicating that light emission is being concentrated at certain wavelengths. Recognition of a familiar pattern, such as the hydrogen spectrum in 3C 273, allows you to measure a redshift, as you then know what the wavelength of the emission lines should be if the quasar were not moving away from us. (If you like this kind of activity, you might want to measure the wavelengths of the lines of the quasar spectrum in Figure 9-2, compare these with their rest wavelength,

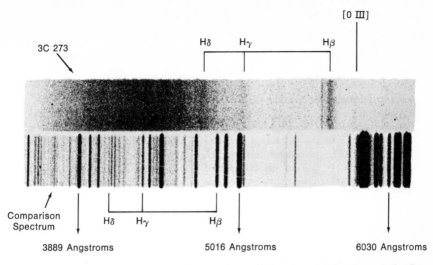

FIGURE 9-2   The spectrum of 3C 273 in the optical region, showing the emission lines of hydrogen (Hδ, Hγ, Hβ) and [O III] at the top. The bottom spectrum is a comparison spectrum of a helium-neon-argon lamp, exposed when the 3C 273 spectrum was obtained, to give a wavelength scale. The figures on the bottom give the wavelengths, along with the unredshifted positions of the hydrogen lines. (Spectra from a Lick Observatory photograph.)

and see how close you come to the published value of $z = 0.158$. The rest wavelengths are 4101, 4340, 4861, and 5006 for Hδ, Hγ, Hβ, and [O III] respectively.)

Sources of the emission lines

What else can be determined about a quasar from these emission lines besides the redshift? Emission lines are an indication of the presence of a cloud of low-density gas that is heated by some source of high-energy radiation. A cloud of low-density gas is called a *nebula* (for example, see Figure 2-9). Emission lines detected anywhere in the universe betray the presence of low-density gas, whether in a nebula in our own galaxy, in an active galaxy, or in a quasar. Because astronomers had encountered emission-line spectra before quasars were discovered, in nebulae in our galaxy, techniques for analyzing these types of spectra existed. They soon applied these techniques to quasars in an effort to deduce properties of the regions in the quasar that were responsible for the emission lines.

To determine where these emission lines come from, consider an atom sitting around in space, some distance from a source of high-energy radiation (see Figure 9-3, frame 1). All atoms have energy levels. Their internal structure is such that an electron in the atom can exist only in certain energy states. It is like a staircase, where you can stand on the steps at only certain heights above the ground. When an electron moves between energy states, it gains or loses energy, just as you lose energy when you climb stairs and gain some energy of motion when you fall down stairs. In the case of the electron, the energy gained or lost is usually in the form of a photon. Our simplified atom has four energy levels, labeled by numbers.

The emission-line story starts when the atom is zapped by a high-energy photon (frame 2). The atom absorbs the photon, and the energy in the photon is used to kick the electron out of the atom, up to the shaded energy level marked "ion." Once the electron has escaped from the atom completely, it can have any amount of energy it wants to, so the "ion" energy level is not narrowly defined. What is left is an atom with one electron missing—an *ion*—and a free electron wandering around in space.

The clouds of gas that produce the emission lines from quasars contain a lot of atoms sitting around in space. These atoms are illuminated by a central source of high-energy radiation—the part of the quasar that is responsible for the optical and x-ray continuum. Some of this continuum radiation is energetic enough that it can be absorbed by one of the atoms in the cloud, as in frame 2 of Figure 9-3. The electron escapes from the atom and flies off into space.

Eventually this electron collides with an atom, as this gas cloud is filled with electrons and ions. The atom and electron recombine (frame 3), as they get back together again in what should be their natural, neutral state. However, the electron is not necessarily captured directly into the lowest energy level (No. 1 in this case); often it is captured into a higher energy state, No. 4 for instance. When the electron and atom recombine, the electron loses some of its energy. Because energy is neither created nor destroyed, a photon is emitted and carries away this extra energy that the electron gave up.

But will the electron sit around in level 4 forever? No, because electrons are lazy and seek the lowest possible energy state. The electron now cascades downward through the different energy levels of the atom (frame 4). It can drop through the levels sequentially, as shown here, or it can skip over a few levels, just as a person going down a flight of stairs can take the steps one, two, or three at a time. Eventually the electron ends up in level 1 again, ready to be ionized by another high-energy photon and start the sequence over again.

The emission lines (shown in frame 4) are produced as the electron cascades downward through the atom's energy levels. Since energy is neither created nor destroyed, the electron must get rid of its energy as it drops down these levels. As it goes from, say, level 3 to level 2, it loses an amount of energy equal to the difference in energy between levels 3 and 2, or 1.89 electron volts if this is a hydrogen atom. This energy is given off in the form of an emitted photon, which eventually may find its way to planet Earth and our telescopes. When this photon leaves the quasar, it has an

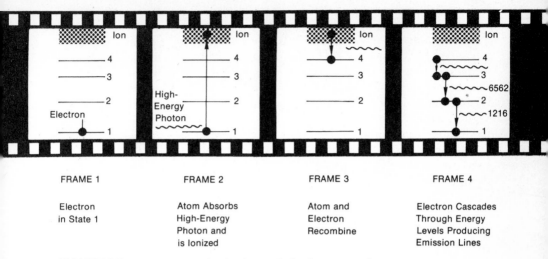

FRAME 1	FRAME 2	FRAME 3	FRAME 4
Electron in State 1	Atom Absorbs High-Energy Photon and is Ionized	Atom and Electron Recombine	Electron Cascades Through Energy Levels Producing Emission Lines

FIGURE 9-3   A movie showing how emission lines are made.

energy exactly equal to the difference in energy between levels 3 and 2. Since the energy of a photon is related to its wavelength, this photon has a wavelength of 6562.8 angstroms when it leaves the quasar.

Any atom of a given element has a distinct pattern of energy levels, different from the pattern of any other atom or any other ion. This pattern of energy levels shows up in the pattern of energies of photons emitted as the electrons cascade downward through the atom. Hydrogen, for example, produces photons with wavelengths of 6562.8 angstroms (level 3 to level 2, as shown), 4861.3, 4340.5, and 4101.7 angstroms (levels 4-2, 5-2, and 6-2 respectively), and 1216 angstroms for a jump between levels 2 and 1. (This list does not exhaust the hydrogen spectrum.) Because hydrogen is the most common element in the universe, this pattern of emission lines produced by descents to the second level is familiar to most astronomers: 6562.8, 4861.3, 4340.5, and 4101.7 angstroms. The last three of these are visible in Figure 9-2 as the lines Hβ, Hγ, and Hδ.

Whenever you see an emission-line spectrum from an astronomical object, you know that the object contains a cloud of low-density gas that is subject to ionizing radiation. There is no other way to produce an emission-line spectrum. A quasar contains a source of ionizing radiation—the clouds of high-energy electrons that are producing synchrotron radiation. The continuum is being produced at all wavelengths from radio waves to x rays. In order for hydrogen atoms to be ionized, ultraviolet photons are needed. There are probably plenty of such photons around as a result of synchrotron radiation.

Spectra of quasars

Completing our outline, we can now add an emission-line region to our picture of a quasar. Gas clouds surround a central source of continuum radiation. This gas acts as a converter, absorbing ultraviolet radiation from the central source, ionizing its atoms, and then producing emission lines. It is generally believed that this gas is concentrated in clumps or filaments (see Figure 9-4). The reason we believe that the gas is concentrated in clumps is that some of the ionizing radiation escapes to the outside and can be seen through our telescopes when it is redshifted into the visible spectrum in the quasars with the highest redshifts. Furthermore, it is often thought that quasars are a kind of super-Crab Nebula (Chapter 3), and that the Crab contains filaments too.

We should not stop here with the process of constructing a generalized model for the emission-line region of a quasar. The schematic model depicted in Figure 9-4 must be made more quantitative and applied to particular quasars. We must enter the normal science process of making detailed, not just generalized, models of particular objects and seeing whether they fit the real world of observations. You must consider such questions as, Can you produce the right kind of emission spectrum? Are the relative intensities of different emission lines in reasonable agreement with what the model says they should be? Are the atoms whose spectra are found in quasars the kinds of atoms that you would expect to find there? Figure 9-4 is just a skeleton; you need to add some flesh to it. This process is going on now. There are enough data in the emission-line spectra of quasars that the data are amenable to this kind of detailed model building. Studies of the intensities of emission lines in quasars are particularly satisfying in that we can get away from generalized arguments. Theoreticians need not be content with explaining general trends; they can calculate emission-line strengths from models of hot gas clouds subject to ionizing radiation. These calculations are now going on, and the model agrees with the observations. Scientists have observed hydrogen, some helium, carbon, oxygen, neon, magnesium, and argon in quasar spectra, and analyses of the models of these spectra indicate that the relative abundance of these different elements is, as far as we can tell, entirely normal. The model agrees.

One striking characteristic of the emission lines in quasar spectra is their breadth. If the filaments of gas in the quasar were at rest relative to each other, someone looking at the quasar from a distance would see every emission-line photon produced by a hydrogen atom cascading from level 3 to level 2 appear at the same wavelength, 6562.8 angstroms. The atoms within the filaments are moving, however, because the filament is hot. This motion would cause these emission-line photons to be Doppler-shifted, some to the red, and some to the blue. The shifts produced by these atomic motions are due to the heat of the gas in the clouds causing the recombina-

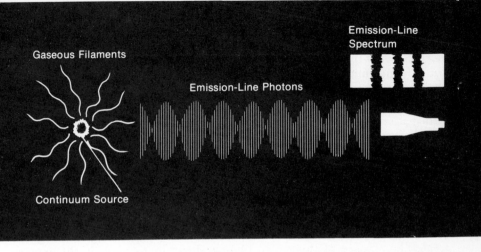

FIGURE 9-4   A generalized model for the regions of a quasar that are producing the
emission lines.

tion radiation. Yet these thermal Doppler shifts are quite small. For the
temperatures derived for these gas clouds from analysis of the emission-
line spectra, the hydrogen lines would be about 0.1 angstrom wide on the
spectrum as a result of thermal broadening. In fact, the hydrogen lines are
smeared out, some 50–100 angstroms wide in the spectrum.

The tremendous breadth of these emission lines indicates that the
emitting atoms are moving relative to each other at great speeds, some
thousands of kilometers per second. Some of the emitting atoms are moving away from the observer, and some are moving toward the observer,
relative to the center of the quasar. As a result some emission-line photons
are redshifted and some are blueshifted, relative to the average photon.
The cause and consequences of this motion have not been fully understood. Is it due to rotation of the quasar? Are the filaments exploding away
from the central source? Are these motions random or systematic? Whatever they are, they are indications that the filaments producing the emission lines are in rapid, even violent motion.

Thus the emission lines—concentrations of photons at particular
wavelengths—have provided us with a good deal of information about
quasars and active galaxies. In the quasars, they are the source of the
redshift of the object. Their presence indicates that something in the heart
of the galactic nucleus is producing ultraviolet radiation, photons that have
an energy sufficient to ionize atoms in gaseous filaments. These spectral
features are understood fairly well. There are some other spectral features
in quasars that are far less well understood: the absorption lines.

Quasar absorption lines

In virtually all quasars that have been examined in sufficient detail, there are gaps in the spectrum, particular wavelengths at which less radiation is being emitted. In all cases, the redshifts of the absorption lines in the quasar spectrum are less than the redshifts of the emission lines. Spectral features, like emission lines and absorption lines, contain a great deal of information if only we can interpret them. Analysis of emission lines provides a great deal of information about quasars, but the absorption lines are a disappointment because we don't even know where they come from.

Absorption lines are familiar features of the spectra of stars; they are formed in the same way in the spectra of quasars. As the quasar spectrum passes through a cool gas cloud between us and the quasar, a hydrogen atom, for instance, steals a photon of 1216-angstrom wavelength from the radiation from the quasar, and this atom moves its electron to a higher level. The absence of this stolen photon shows up as a gap in the quasar spectrum. (You might wish to refer to the discussion of stellar spectra in the Preliminary section.) The wavelength of the gap is not shifted as much as the wavelengths in the emission-line spectrum of the quasar itself, indicating that the absorption-line cloud is not moving away from us as fast as the quasar is.

But where are the cool clouds that produce the absorption lines? There are two possibilities, each favored by different investigators. One model has it that these gas clouds were once part of the quasar itself and were driven toward us by the tremendous pressure of radiation from the quasar. A gas cloud's motion relative to us is the sum total of the quasar's motion away from us and the cloud's motion away from the quasar, toward us, so that the absorption-line redshifts are less than the emission-line redshift, as observed. But what makes these gas clouds move so fast?

Another possibility is that light from the quasar passes through the halos of galaxies that lie between us and the quasar. These galaxies are nearer to us than the quasar is, and therefore the redshifts of their absorptions are lower than the redshift of the quasar, in accordance with Hubble's Law. A growing feeling that galaxies have large, massive halos that could absorb light lends some support to this viewpoint. The interpretation of quasar absorption lines is ambiguous. Either model more or less explains what data there are, but both models are somewhat generalized. Until we know where the absorption lines are produced, it is premature to try to analyze them.

A comprehensive model

The last two chapters have discussed different types of radiation produced by quasars and active galaxies. There is continuum radiation in the radio, infrared, optical, ultraviolet, x-ray, and gamma-ray ranges of the

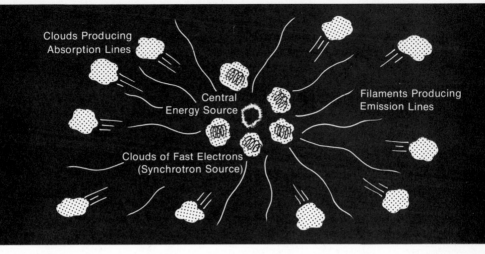

Clouds Producing
Absorption Lines

Central
Energy Source

Filaments Producing
Emission Lines

Clouds of Fast Electrons
(Synchrotron Source)

FIGURE 9-5   Generalized model of a quasar (see text).

spectrum. Emission lines are seen in the optical and ultraviolet parts of the
spectrum. Many quasars contain absorption lines in the spectrum as well.
One can produce a generalized, comprehensive model to account for the
phenomena. Figure 9-5 shows such a model. (If you think Figure 9-5 looks
a little like the Crab Nebula—Figure 2-1—you're right. The Crab Nebula
is the high-energy astrophysicist's model for many high-energy phenomena.)

Start from the inside out. At the center is the unknown powerhouse—a source of high-speed electrons. There are a number of small
clouds of high-speed electrons, clouds that produce radio radiation, optical
continuum radiation, and possibly continuum radiation in other parts of
the spectrum. Filaments of gas, ionized by high-energy continuum radiation, convert this radiation to emission-line photons, photons that are concentrated at particular wavelengths. Outside may be fast-moving clouds,
which may produce absorption lines as they capture photons from the
beam of radiation that travels toward us. Overlaying all of the gas clouds
shown in Figure 9-5 are stars and dust, objects not associated with high-
energy events. Dust produces infrared radiation, and, in some of the less
powerful objects, stars produce continuum radiation.

Another way to review what we know about quasars is to go down
the electromagnetic spectrum and see where the radiation could come
from. The *radio continuum* comes from synchrotron radiation. The *infrared
continuum* comes from synchrotron radiation, heated dust, or starlight. The
optical continuum comes from synchrotron radiation or starlight. The *x rays
and gamma rays* come from synchrotron radiation, hot gas, or collisions
between radio photons and high-energy electrons. The *emission lines* come
from gas filaments.

The last two paragraphs and an uncritical glance at Figure 9-5 might convince you that we know all about quasars. It all sounds so simple when you sweep the unknowns under the rug. Remember that we don't know where the energy comes from. Further, we don't really know where these various clouds are located in the quasar. We don't know where these gas clouds are, how they are related to each other, how they evolve, or what all this has to do with the much larger clouds of radio emission that are found in many objects. Most of those questions are unanswered. They are at the cutting edge of contemporary research. To answer them, astrophysicists turn away from a generalized analysis of high-energy phenomena in active galaxies and quasars and turn to specific objects. We know most about the active galaxies, and so in Chapter 10 we examine a few active galaxies in detail.

Figure 9-5 summarizes the last two chapters. In a generalized way, we understand how the radiation in quasars is generated. Infrared and optical continuous radiation comes from synchrotron radiation, dust, or starlight in the less powerful objects. Emission lines come from clouds of gas, where the atoms are ionized by high-energy photons and then recombine, bouncing down the energy stairway and emitting photons as they go. Absorption lines come from cooler gas clouds at the outskirts of the quasar or active galaxy.

10   ACTIVE GALAXIES

The generalized picture of activity in quasars and active galaxies developed in the last two chapters seems pretty comprehensive, but in many respects it is tantalizingly incomplete. Analysis of different types of radiation—continuum, emission lines, and absorption lines in different parts of the electromagnetic spectrum—provides the skeletal picture illustrated in Figure 9-5 and described in the closing section of Chapter 9. Yet questions remain. The clouds responsible for synchrotron radiation are different from the clouds responsible for emission lines. In many objects, more than one cloud of high-speed electrons emits synchrotron radiation. Are these clouds related to each other? If so, how? Are these clouds ejected in random direction or does the ejection mechanism tend to spew them out along a particular line? There is a wide variety of power levels involved in active galaxies, ranging from the comparatively feeble activity at the center of our own Milky Way to the tremendous power of the quasars. Are all these forms of galactic activity related to the same phenomenon?

Perhaps the most important question deals with the ultimate source of all this energy, the "quasar engine," as some theorists call it. But the quasars themselves do not provide much insight into this question or into the more detailed questions regarding the location and relationship of the various clouds that emit gas. Quasars—quasistellar objects—look like points of light because the active core emits so much more light than the surrounding galaxy. Because the quasars are so far away, we cannot distinguish the individual clouds of electrons and their relationship to the surrounding galaxy. VLB astronomers can tell us that there are several distinct radio-emitting clouds, but cannot tell us whether they are ejected from the galactic core or not. The active galaxies, on the other hand, are sufficiently close to us that we can make out some details. We can see the various clouds and obtain a better picture of just what goes on in these objects. And further, there is no question regarding the distance of the active galaxies. We need make no assumptions regarding the nature of their redshifts in order to perform the analysis.

Signs of activity in galaxies

An active galaxy does something more than just produce starlight. This activity is exhibited in a bewildering variety of ways: synchrotron radiation, broad emission lines, peculiar optical appearance in photographs, and infrared radiation. Nearly all galaxies, including the Milky Way, are active. It is only the irregular galaxies and the loosely wound spirals that seem to contain nothing more than shining stars. This activity tends to originate in the nucleus of each galaxy, and since irregulars have no nucleus, it is not surprising that they are not active. How is the activity displayed?

Optical appearance

The most active galaxies often look peculiar in photographs. You might want to go back to Chapter 7 and look at the photographs of ordinary galaxies, comparing them with the peculiar galaxies of this section. A number of different peculiarities are evident.

Two active galaxies and one quasar are associated with jets. Compare the photographs of 3C 273 (Figure 7-7), Messier 87 (Figure 10-5), and the Seyfert galaxy NGC 1275 (Figure 10-8). Each one has a jetlike protuberance. It is curious that all three of these objects are x-ray sources. Jets are seen only in active galaxies.

The only analogue of galactic activity that we see in our own galaxy, the Crab Nebula, displays many of the same characteristics as active galaxies, only on a much smaller scale. The Crab Nebula was so named because through a telescope the filamentary structure looks like crab legs. This filamentary structure is also characteristic of active galaxies. Compare the photographs of Messier 82 (Figure 10-1) and NGC 1275 (Figure 10-8) with the Crab Nebula (Figure 3-1).

Some active galaxies show prominent dust lanes. The prototype here is NGC 5128, the radio source Centaurus A (Figure 10-4); Messier 82 contains much dust, portrayed to some extent in Figure 10-1.

Nearly all active galaxies have bright nuclei. It is difficult to pick out these nuclei on photographs, which must be overexposed so that the rest of the galaxy can be photographed. Figure 10-7 shows some N galaxies, which have extremely luminous nuclei that look like tiny stars surrounded by the fuzz of the surrounding galaxy. I shall return to these later.

The different peculiarities of active galaxies are of great use to astronomers seeking to discover new ones. You look at photographs of the sky and make a list of objects that look peculiar and seem worthy of closer attention. The rewards of search work are that your name is attached to this list of peculiar objects, and these objects are named after you. A disadvantage of this historical process is that a bewildering variety of terminology has crept into the literature, presenting a confusing array of terms describing different types of active galaxies.

Yet optical peculiarities, useful as they may be to people trying to discover active galaxies, do not tell us very much about what is going on in these galaxies. How can you investigate such activity once you have discovered it?

FIGURE 10-1   Messier 82 in the light of the Hα emission line. (Hale Observatories.)

Spectroscopic characteristics

Activity in galaxies is most clearly shown in a spectrum of the galaxy in question. A galaxy that is radiating starlight produces a spectrum that looks like the spectrum of a star: a continuous band of light with a few dark lines crossing it (recall Chapter 5 and the Preliminary section). Active galaxies generally show emission lines in their spectra. Where there are emission lines, there are clouds of hot gas, and this gas is ionized by some source of energetic ultraviolet radiation.

Synchrotron radiation

Where you find synchrotron radiation, there is some object that can accelerate electrons to the high speeds necessary to produce it. Synchrotron radiation can show up at all wavelengths, but it is most obvious at those wavelengths at which stars do not radiate very much: x ray, infrared, and especially radio. Many active galaxies have been discovered because they were counterparts to radio sources.

Classification of active galaxies

With such a wide variety of phenomena showing activity in galaxies, how can you bring some order to the observations so you can even begin to attack the problem? Many classification schemes exist, but most are based on the optical peculiarities of the galaxies rather than on any underlying properties. The scheme I use to order the discussion in this chapter is based on the relative scale of galactic activity.

Galaxies can do two things: They can contain stars that shine or they can contain more exciting objects that explode, produce synchrotron radiation, or do something else that is violent. Most galaxies show some signs of violent activity, which can be classified by examining the relative amount of effort that the galaxy puts into nonviolent (stellar) and violent radiative processes. This ratio—the ratio of nonstellar to stellar luminosity—is a ratio that I use as a guide to arranging different kinds of objects in order, ranging from the least powerful to the most powerful. But precise numerical values are hard to establish, for much of the radiation from the violence of galactic activity is produced in the infrared region of the spectrum, where measurements are hard to make. Furthermore, gamma-ray astronomers are beginning to discover that gamma rays often carry a great deal of energy away from these objects.

Yet the order in which different types of objects fall is reasonably well established. At the bottom end of the scale of galactic activity are normal spiral galaxies like our own Milky Way galaxy. We described the core of the Milky Way some time ago, in Chapter 7, as a way of introducing the active-galaxy phenomenon. Our galaxy is a radio source, like the quasars, but it is far feebler. Far-infrared radiation is emitted from a giant dust cloud, but this dust cloud is probably heated by stars rather than by some kind of high-energy process. There are some gas clouds producing emission lines, but they, too, are probably found around stars, not around high-speed electrons producing synchrotron radiation. There is also an x-ray source—probably synchrotron in origin, possibly from a hot gas cloud. The only unambiguous sign of activity in our galactic core is the synchrotron component of the radio radiation, which shows that something in the core produces high-speed electrons.

Most spiral galaxies show signs of activity at much the same level as our own galaxy. Weak radio sources are found in the central regions. In some cases emission lines from low-density gas clouds can be observed from the vicinity of the nucleus. The tiny galaxies, the smallest ellipticals and the irregulars, show no activity. But we seek to move up in the scale, not down. We encounter another galaxy that is a weak radio source, but somewhat stronger than our galaxy in terms of total power. This one has often been presented as an example of an exploding galaxy. Other interpretations, however, are possible.

Messier 82: a hyperactive "normal" galaxy

Messier 82 is the nearest galaxy to show activity on a scale comparable to the stellar luminosity of an average spiral galaxy. Figure 10-2 shows its irregular, Crab Nebula-like appearance in a photograph taken in the light of an emission line, so the only radiation that registers on the photograph is from clouds of ionized gas.

The peculiarities of Messier 82 first became evident when C. Roger Lynds noticed that it contained a weak radio source at its center. In the early 1960s, the large telescopes at Lick Observatory and Mount Palomar were turned on this object. They discovered that much of the luminous energy from the galaxy was in the form of emission lines, indicating that there are vast quantities of hot gas subject to high-energy, ionizing radiation. Doppler shifts in these emission lines seemed to indicate that the filaments, shown in Figure 10-1, were expanding away from the center at speeds of 1000 kilometers per second. It was thought that M 82 was rocked by a violent explosion that tore the galaxy apart. Recently, infrared telescopes turned on the galaxy revealed that somewhere around 10^{44} ergs/sec of energy were being emitted as infrared radiation. It is clear that something very unusual is occurring in M 82, but we don't know exactly what.

In the last few years, Alan Solinger of MIT suggested that maybe M 82 is not exploding at all. His idea is that the visible filaments are really made of dust, which scatters the emission-line photons that are produced somewhere else. This dust is moving, but it does not need to be moving very fast to explain the Doppler shifts of the emission lines. If the dust model is correct, then the connection between M 82 and the quasar phenomenon is not so direct. Long-exposure photographs of Messier 82 and its companion galaxy Messier 81 show that the two galaxies are surrounded by a cloud of gas and dust, and Solinger, Tom Markert, and Philip Morrison argue that it was the encounter between M 82 and this dust cloud that is responsible for the radio emission from this enigmatic galaxy. The radio emission, the only unambiguous sign that something more than the production of starlight is occurring, comes from a tiny cloud about 25 light-days across. This radio source is only ten times as powerful as the radio source in the center of the Milky Way.

Thus Messier 82 is an enigma. It could be an exploding galaxy, a nearby, weak form of the same type of activity that occurs in quasars. But it need not be this textbook example of galactic activity, for the Doppler shifts in the filaments could come from the much slower motion of dust. Something unusual is going on in Messier 82. But are these unusual events connected to the radio-galaxy phenomenon? We do not know.

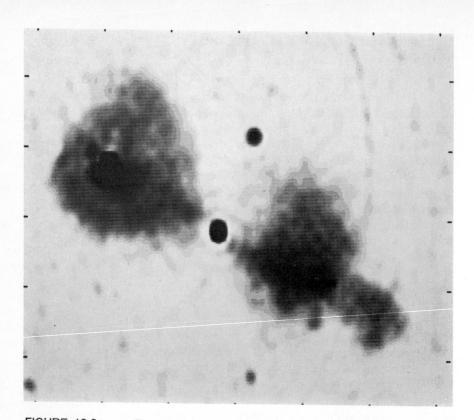

FIGURE 10-2   The distribution of radio emission in five radio galaxies. Most radio galaxies are double, as in two of the largest, 3C 236 (opposite) and DA 240 (above). The inner regions of the head-tail galaxy NGC 1265 are taken from observations with the VLA. The illustrations of 3C 236 and DA 240 are synthetic "photographs" produced by computer, based on data from the Westerbork radio telescope. [Illustrations of 3C 236 and DA 240 taken from Willis et al., *Nature* 250 (1974), p. 625, and used courtesy of the Leiden Observatory and G. K. Miley. The shapes of M 87 and 3C 465 are freely sketched from information in A. Moffet, "Strong Nonthermal Radio Emission from Galaxies," in A. Sandage, M. Sandage, and J. Kristian, eds., *Galaxies and the Universe,* Chicago, University of Chicago Press, pp. 211-283. The VLA map of NGC 1265 is from F. N. Owen, J. O. Burns, and L. Rudnick, "VLA Observations of NGC 1265 at 4886 MHZ," *Astrophysical Journal Letters* 226 (1978), L119-L123. The illustration of NGC 1265 is copyright © 1978 by the American Astronomical Society. All rights reserved.]

Radio galaxies

Normal spiral galaxies and objects like Messier 82 are not examples of especially energetic activity in galaxies. The spirals emit most of their energy in the form of starlight. The radio emission from clouds of high-speed electrons at the core is just a sideshow. The so-called "radio galaxies" are those in which a much larger portion of total power output is in the

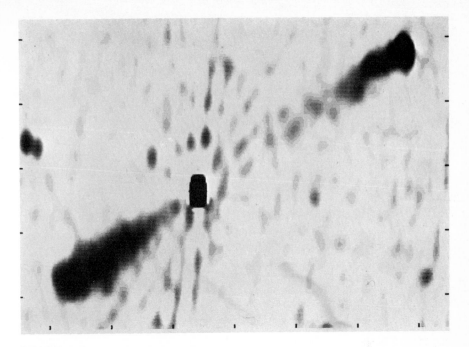

FIGURE 10-2   (Continued) 3C 236

M 87
(Core-Halo Galaxy)

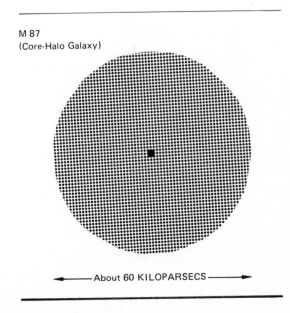

←——— About 60 KILOPARSECS ———→

FIGURE 10-2   (Continued) M 87

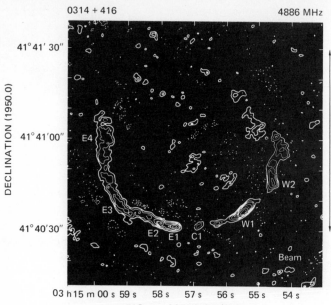

0314 + 416   4886 MHz

FIGURE 10-2   (Continued) NGC 1265

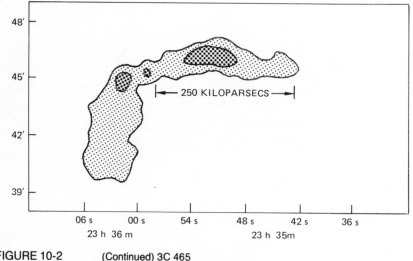

FIGURE 10-2   (Continued) 3C 465

form of radio radiation, radiation that unquestionably comes from high-speed electrons. For the sake of definiteness, radio astronomers have set 10^{42} ergs/sec of radio power as a cutoff. Any radio galaxy must have a radio luminosity that exceeds this figure. 10^{42} ergs/sec is about one-tenth as much energy as the most powerful galaxies emit in the form of starlight.

Twin clouds of radio emission

Most radio galaxies resemble Cygnus A, the prototypical radio galaxy shown in Figure 8-7. Two large clouds of radio emission are symmetrically placed on either side of a galaxy. The distance of 50,000 parsecs from the outer edge of one of these clouds to the central galaxy found in Cygnus A is also typical; observed sizes range from 15 kpc to 500 kpc. The distribution of radio emission within the clouds is not uniform; most of the emission is concentrated at the outer edges.

Some of the classical double radio galaxies are truly enormous. In 1974, Willis, Strom, and Wilson at the University of Leiden discovered an extremely large radio galaxy that now holds the record for being the largest object in the universe. In this object, 3C 236, the two clouds of radio emission are each two megaparsecs long. The total extent of this object, measured from the tip of one radio cloud to the tip of the other, is six megaparsecs. The total volume of these radio-emitting clouds exceeds the total volume of most medium-sized clusters of galaxies. This object remained undiscovered for so long because the radio emission from these huge clouds is very faint and easily overpowered by the intense emission from a stronger source located much closer to the optical galaxy.

The double structure, although common in radio galaxies, is modified in several cases, as the montage of Figure 10-2 shows. Some unusual galaxies are the "head-tail" galaxies, in which a series of clouds are produced by a galaxy. These clouds trail behind the galaxy because the galaxy has moved since it ejected the clouds. The V-shaped galaxy 3C 465 is particularly strange. Although it is possible that again the radio clouds could trail behind the moving, parent galaxy, the shape is unusual. Some radio galaxies contain a small, compact core surrounded by an extended halo. There are examples of yet more complex and peculiar shapes. Figure 10-2 illustrates all the shapes that are observed relatively often. When the VLA becomes complete, it is likely that we shall see an increase in the number of radio galaxies with complex distributions of radio emission.

Associated with many of the extended radio galaxies are small, compact clouds of radio emission. These clouds are observed by VLB radio astronomers. The quasar 4C 39.05, discussed in Chapter 8, is again typical, in that we observe two clouds, each a few parsecs across. In closer objects, still smaller clouds have been seen. The smallest known radio source in a galactic nucleus is the one in the core of the Milky Way, which is about 10 astronomical units across. The compact radio sources generally contain far less energy than the extended sources.

A remarkable feature of the positions of compact and extended radio sources, when both are detected in the same object, is their alignment. A line connecting the compact sources points in the same direction as a line connecting the extended sources, and in general this line passes through the core of a galaxy. The most dramatic example of alignment of radio sources involves the radio galaxy NGC 6251. Here there is a radio jet about 200 kiloparsecs long aligned with the outer lobes of radio emission (Figure 10-3). VLB observations published in 1978[1] showed that there is a compact jet within the larger one. The end of this jet coincides with the nucleus of an elliptical galaxy, and the jet is perpendicular to the long axis of the galaxy. The compact jet, the 200-kpc jet, and the double structure of the entire source are all oriented in the same direction within a few degrees, even though the outer lobes are five million times bigger than the compact jet. Because of its appearance, this galaxy has sometimes been called the "blowtorch" galaxy.

The optical galaxies that are associated with the radio galaxies are relatively unremarkable objects, considering the enormous power radiated by these huge clouds of high-speed electrons. At one time people thought that the central object in Cygnus A was a pair of galaxies colliding with each other. Closer examination shows that it is an elliptical galaxy with a dust lane across it. The dust lane makes the galaxy appear to be double. The closest extended radio galaxy, Centaurus A, also contains a dust lane (Figure 10-4). The optical counterparts of all these radio galaxies seem to be giant elliptical galaxies with extended envelopes, called D galaxies. These D galaxies are the most luminous ones known. About half of all radio galaxies show emission lines in their spectra, indicating some form of activity in the central radio galaxy in addition to the extended lobes of radio emission.

One radio galaxy in particular has attracted much attention from the astronomical community in recent years. Although Centaurus A is the nearest radio galaxy, the dust lane obscures the central core (Figure 10-4), defying any attempt to uncover what lies at the heart of this giant elliptical galaxy. So, to examine the nucleus of a radio galaxy, we go to the second nearest radio galaxy: Messier 87. This galaxy is about 20 megaparsecs away, and is the dominant galaxy of the Virgo cluster of galaxies.

Messier 87: a radio galaxy with a compact nucleus

The radio emission from M 87, all synchrotron radiation, comes from three distinct regions. Most of the emission comes from a large halo about 60 kiloparsecs across, roughly the same size as the visible galaxy. A remarkably compact core, only 2.5 light-months across, also produces radio noise. This core is the most compact radio source known to be connected with the radio galaxy phenomenon. The optical jet contains several radio sources.

The most obvious sign of optical activity is the jet extending from the nucleus of M 87, visible on the short exposure photograph of Figure 10-3. On heavily exposed photographs, we see a counterjet extending in the opposite direction. This jet is a string of highly condensed regions, each of which emits a lot of polarized, presumably synchrotron radiation. Since a high-energy electron could not travel from one end of the jet to the other

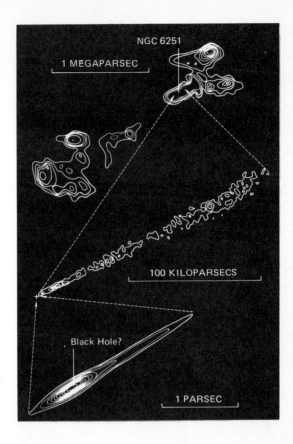

FIGURE 10-3
A contour map showing the distribution of radio emission in the radio galaxy NGC 6251, sometimes called the "blowtorch" galaxy. The upper two contour maps were produced by an interferometer at Cambridge University in England. The lower map is produced from Very Long Baseline observations by a Caltech group. [From A. C. S. Readhead, M. H. Cohen, and R. D. Blandford, "A Jet in the Nucleus of NGC 6251," *Nature* 272 (9 March 1978), 131–134].

without losing its energy, it is generally supposed that the knots in the jet surround individual sources of high-energy electrons.

Most of the nonstellar radiation produced by M 87 is in the form of x rays. Its x-ray luminosity is about 10^{43} ergs/sec. It is not yet known whether these x rays are produced in the jet, in the halo, in the compact central source, or in all three locations, since x-ray telescopes cannot produce such high-resolution photographs. These x rays are possibly synchrotron radiation as well, or they could be the result of collisions between high-speed electrons and low-energy photons.

Recently a large team of astronomers using the latest generation of photon detectors on some of the world's largest telescopes has uncovered some evidence that may show that a massive object—possibly a black hole—is located within M 87's core. Take another look at Figure 10-5, examining the center of the elliptical galaxy. The printed page cannot show it absolutely clearly, but there seems to be a point source of light at the center of this galaxy. These recent observations show that that light source is, in fact, there, and that the motions of stars near the core can be interpreted with a model in which there is a massive object that may be responsible for all the excitement involving jets, radio emission, and x rays.

FIGURE 10-4   NGC 5128, Centaurus A. (Hale Observatories.)

Consider, for a moment, what would happen in a galactic nucleus if there were a large mass at the center. Stars in the galactic core would be sucked toward this large mass. There would be more stars in the central regions than there would be if the central mass were not there. Further, these stars would fall rapidly toward the central mass, whiz around it, and be tossed out into the outskirts of the galaxy. Observations of the motions of stars near the core would indicate that they move very quickly.

FIGURE 10-5
Messier 87, showing the jet.
(Hale Observatories.)

In May 1978, a large group of California astronomers reported that they had made observations indicating that the phenomena described above were, in fact, seen in the core of M 87. The bright point of light in the center of the circular blob, the galaxy M 87 shown in Figure 10-5, seemed real. The Doppler shifts of objects in the galactic core were larger than they should be if there were nothing but stars in the core of this galaxy.

Yet these observations must be interpreted with some caution. They were widely touted as showing that there is a black hole in the core of M 87, while in fact all that they show is that there is a large mass there. Further, it is possible that these observations can be interpreted in other ways. I have heard colleagues mention other possible interpretations, though nothing has appeared in print yet.

Explaining the shapes of radio galaxies

What makes radio galaxies look the way that they do? The enormous variety of shapes seems to be—and is—bewildering to theorists who seek to interpret what is going on, rather than merely describe what is there. So far, no one has tried to go too much beyond the interpretation of the shapes of the most abundant type of radio galaxy, the double one like Cygnus A. Two key questions have emerged as ones that theoretical models must answer. Why are the radio sources so precisely aligned? How can you generate high-speed electrons at large distances from the parent galaxy? The second question can be illustrated most dramatically with the enormous object 3C 236. There the megaparsec-sized radio-emitting clouds were produced over a billion years ago. You could not generate high-speed electrons a billion years ago and expect to see the electrons still radiating.

Figure 10-6 shows three possible models to explain the shapes of double radio galaxies. Currently the most exciting one is the relativistic-beam model proposed by Roger Blandford and Martin Rees, in which something in the central galaxy directs a beam of high-speed particles or possibly electromagnetic radiation into intergalactic space. The effects of this beam are seen when the beam of particles encounters gas in the intergalactic medium, outside the galaxy. This beam then accelerates electrons in the intergalactic gas, and presto! We have a radio galaxy. The alignment of large and small radio sources is explained by the alignment of the beam itself. This model is not without its problems. The detailed calculations needed to precisely explain just how a beam of high-speed particles produces high-speed electrons when it collides with intergalactic gas have yet to be done. But, at the moment, it is the best model we have.

Two other models are currently less fashionable, not because they are unambiguously ruled out by observation, but because the difficulties they have in explaining the double radio sources seem more fundamental. One model is dubbed the *slingshot*. In this model, collisons between massive objects in the core of a galaxy shoot these massive objects out in opposite directions. These massive objects are then responsible for generating high-speed electrons, and thus lie at the centers of the two double clouds. In yet another model, a galaxy ejects enormous clouds of hot gas. These clouds of gas expand and move outward, and turbulent, swirling gas currents within them produce the high-speed electrons. However, the slingshot must be aimed rather precisely, since large and small clouds—which are presumably produced by different collisions within a galactic core—are aligned in the same direction. In the model in which hot gas clouds are ejected, there is no clear way of creating fast electrons from hot gas.

Thus we have no definite explanation for the appearance of even the most common form of radio galaxy, the double radio galaxy. However, we do have some ideas that may or may not be right. But, despite this theoretical uncertainty, radio galaxies do tell us that the ejection of something in particular directions is a key part of galactic activity. The alignment of the small radio jet, the large jet, and the double radio source in the blowtorch galaxy, NGC 6251, is the most dramatic piece of evidence supporting this view, although it emerges from the analysis of many other types of radio galaxy as well. However, these radio galaxies only rarely provide any direct information regarding the core of the galaxy that produced the high-speed electrons. The next group of galaxies I discuss is a group in which the visible activity comes entirely from the core. These galaxies, called *N galaxies* (N = nuclear), resemble the quasars in many ways.

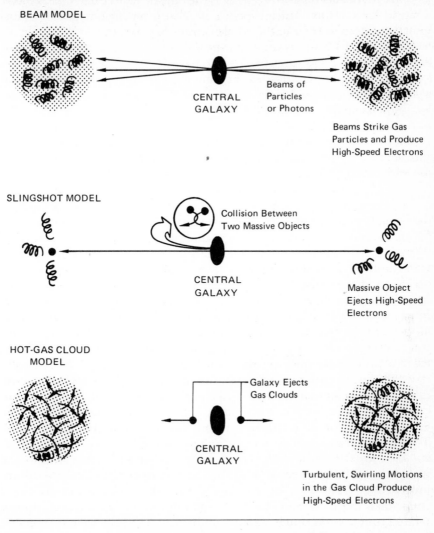

BEAM MODEL

CENTRAL
GALAXY

Beams of
Particles
or Photons

Beams Strike Gas
Particles and Produce
High-Speed Electrons

SLINGSHOT MODEL

Collision Between
Two Massive Objects

CENTRAL
GALAXY

Massive Object
Ejects High-Speed
Electrons

**HOT-GAS CLOUD
MODEL**

Galaxy Ejects
Gas Clouds

CENTRAL
GALAXY

Turbulent, Swirling Motions
in the Gas Cloud Produce
High-Speed Electrons

FIGURE 10-6   Three models for producing high-speed electrons in radio galaxies

Galaxies with bright nuclei

The N galaxies are the optical equivalents, in a sense, of the compact radio sources, having most of their luminosity contained in small, brilliant, almost stellar nuclei. A list of their properties reads very much like a list of quasar properties: spectra with broad emission lines, rapid variations, large infrared luminosity, and association with compact radio sources. There are some reasons to believe that quasars are N galaxies that are so far away that the surrounding galaxy is invisible.

A particular type of N galaxy is the Seyfert galaxy, an N galaxy with a particular type of nuclear spectrum. The term Seyfert galaxy is also sometimes used to refer to the whole class of N galaxies.

Figure 10-7 shows the brightness of the nuclei of N galaxies. This figure shows two N galaxies (which also have Seyfert spectra) in short- and long-exposure photographs. In the long exposures, they look like normal galaxies rather than N galaxies. Yet the short-exposure photographs at the left show something that looks very much like an N galaxy: a starlike nucleus with a little bit of fuzz around it. The nucleus of the Seyfert galaxy is much brighter than the rest of the galaxy. A one-minute exposure is sufficient to show the nucleus, while it takes a ten-minute exposure to show the underlying galaxy.

The spectrum of an N galaxy that belongs to the Seyfert subclass contains emission lines, at least some of them quite broad. The breadth of these lines indicates that the clouds emitting them are in rapid motion. When the emission-line photons leave these clouds, some are blueshifted as they come from clouds moving toward us (relative to the galaxy), and some are redshifted, as they are moving away from us. (Recall the discussion in Chapter 9 about the breadth of emission lines from quasars.) The result is a smeared-out emission line coming from clouds moving toward and away from the observer. The breadth of these lines indicates that the clouds are moving at velocities of several thousand kilometers per second. The entire emission line is redshifted by the expansion of the universe; it is the internal motions that are under examination here.

One of the puzzles, not yet solved, of Seyfert galaxies is that in some cases not all the lines in their spectra are smeared out. Some emission lines can be produced only in low-density regions, while others, such as the hydrogen lines, can be produced by both low- and high-density gas clouds. The lines that can be produced only in low-density regions are called *forbidden lines,* for the densities required are so low that it is impossible to see these lines in the spectrum of a gas in a terrestrial lab; the lab densities are just too high. The forbidden lines are often, but not always, much narrower than the hydrogen (*permitted*) lines. Presumably there are two different groups of gas clouds, some high-density clouds in rapid motion that produce the broad hydrogen lines and some low-density clouds, moving more slowly, that produce the low-density forbidden lines.

In general, the spectra of N galaxies and Seyfert galaxies can be explained in the same way that the spectra of quasars can be. They produce emission lines, and the relative intensities of the emission lines can be interpreted successfully, with models of gas filaments struck by ultraviolet ionizing radiation producing the correct line intensities. These objects produce continuous radiation similar to that emitted by quasars. Models like those discussed in Chapter 9 provide a successful explanation of this radiation. Seyfert and N galaxies also have an infrared excess. Again in some cases this excess radiation comes from dust, and in some cases it comes from synchrotron radiation.

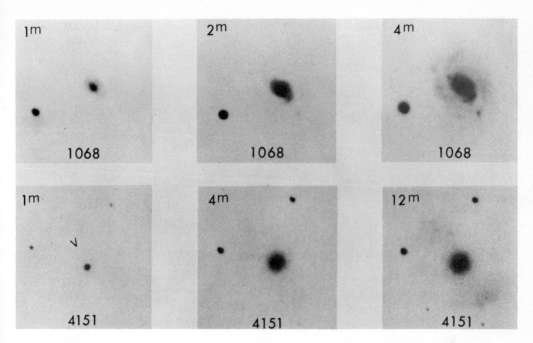

FIGURE 10-7   The Seyfert galaxies NGC 4151 and NGC 1068, with varying exposures showing that the nuclei are much brighter than the surrounding galaxy. The short-exposure photographs can pick up the bright nuclei but not the fainter surrounding galaxy. (Yerkes Observatory photograph.)

Another characteristic that distinguishes the N galaxies, in some cases, is their disturbed optical appearance. A photograph of NGC 1275 (Figure 10-8) shows an elliptical galaxy in the background, with some small clumps or knots of optical emission and a jet in the foreground. Such optical jets are also visible in 3C 273 (Figure 7-7) and Messier 87 (Figure 10-5). Deep photographs of Centaurus A—photographs with extremely long exposures taken with large telescopes—show some faint, possibly jetlike features located within one of the radio-emitting clouds, perpendicular to the dust lane seen in Figure 10-4. Not all N galaxies are optically peculiar. For example, 3C 120 (Figure 10-9) looks like a normal spiral galaxy. Compare Figure 10-9 with Figure 7-3.

The N galaxies are, in many respects, like quasars. Yet they are not as energetic as quasars are. Were N galaxies to complete the collection of active galaxies, the linkage between active galaxies and quasars would still be a bit debatable. All the same phenomena occur in both types of objects, but there is still a gap in luminosity. The recent discovery of a few hyperactive N galaxies begins to bridge this gap. 3C 120 is one of these hyperactive galaxies that nearly rivals the least powerful quasars. A number of others

FIGURE 10-8   The Seyfert galaxy NGC 1275. (Hale Observatories.)

have been discovered in recent years. But the strongest connecting link
between active galaxies and quasars comes from a group of objects in which
the underlying galaxies are only barely visible: the BL Lacerta or BL Lac
objects.

BL Lacerta (BL Lac) objects

BL Lacerta is a rather unusual name for a galaxy, for the name is
that of a variable star. Variable stars are named by the constellation that
they appear in, using single letters starting with R, then pairs of letters, and
then the designation "V" followed by a number. What is BL Lacerta doing
as the name of a particular class of active galaxies? The story of the discovery of this particular class of object provides most of the essential information about the properties of these beasts.

J. Schmitt, in 1968, noticed that a variable radio source (named VRO
42.22.01, if you like to collect the various aliases of strange galaxies) was in
the same location as the variable star BL Lacertae, also known as BL Lac for
short. BL Lac had been in the catalogue of variable stars for some time,
known as a "star" that changed with respect to the amount of light that it
produced. BL Lac does not behave like most variable stars. It is a genuinely

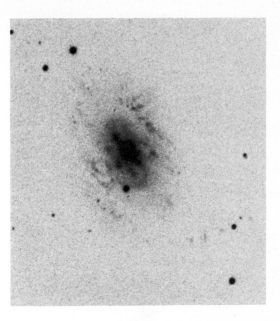

FIGURE 10-9
The N-galaxy 3C 120 photographed with the 4-meter Mayall telescope of Kitt Peak Observatory. (Hale Observatories.)

irregular variable, brightening and becoming fainter with no apparent pattern. But there are lots of irregular variables, and until it was identified as a radio source it was not regarded as being particularly interesting.

Spectroscopists then turned their telescopes on BL Lac to see what kind of a spectrum this "radio star" had. Surprise and disappointment were the result; the optical spectra were featureless. No emission lines, such as those you might find from a quasar. No absorption lines, such as those you might find from most stars. A few white-dwarf stars have featureless spectra, but these stars do not vary in brightness. In the early 1970s, people kept observing BL Lac, but no one knew what it was. By 1972, Peter Strittmatter at the University of Arizona and a number of colleagues had identified four other objects in the sky, all similar to BL Lac. This group of objects is now known, perhaps awkwardly, as the BL Lac (or BL Lacertae) objects.

The breakthrough in the identification of these objects came in 1974. A number of people had noticed that BL Lac did not look exactly like a star; there was some faint fuzz around the point source at the center. Caltech astronomers James E. Gunn and J. B. Oke used a rather ingenious method to examine the spectrum of this fuzz. They used a ring-shaped opening to prevent light from the core from reaching their instruments, and so were able to obtain a spectrum of the fuzz. The fuzz showed the spectrum of a normal elliptical galaxy, and the spectrum contained two absorption lines at a redshift of 0.07. Such a redshift, a Doppler shift, indicates that BL Lac is moving away from us at a speed of 21,000 km/sec. Interpreting this redshift as being due to the expanding universe, we can say that BL Lac is 420 megaparsecs away.

With BL Lac at that distance, at its brightest the core has a luminosity of 10^{46} ergs/sec. This object, an object that was once thought to be a faint, feeble variable star in our own Milky Way, is in fact a very distant galaxy with as much power as a typical quasar. It took more than just one measurement to confirm this interpretation. The Caltech astronomers repeated the measurement, and Joseph Miller and Stephen Hawley were able to confirm the result using different instruments at Lick Observatory. Spectral lines have now been discovered in other BL Lac objects. Although the discovery of a line in any particular object (other than BL Lac itself) may be questioned, the discovery of a whole class of objects with common properties indicates that our general interpretation of BL Lacs as galaxies with extremely bright nuclei is correct. Nature may be subtle, and the interpretation of observations in any one object may turn out to be mistaken. But it is unlikely that Nature would try to fool us in exactly the same way in a whole group of objects. Even astronomers who would like to be skeptical about the interpretation of BL Lacs as elliptical galaxies with superbright nuclei agree that this is exactly what they are.

The name of the BL Lac objects is unfortunate. The lack of an easily pronounced name clearly identified with either galaxies or quasars perhaps gives some people the impression that they are just an astronomical oddity rather than a crucial link between galaxies and the quasar phenomenon. At a recent conference on BL Lacs, Columbia University's Edward Spiegel quipped, "Perhaps they would have attracted more attention if they had a dramatic name like blazars." At the same conference, radio astronomer Kenneth Kellermann argued that the name BL Lac should "never be used again."[2] This name is not dramatic, and worse, it does not stress the essential fact that BL Lacs are a crucial part of the active galaxy phenomenon. However, once a name becomes entrenched in the astronomical literature, it is very hard to dislodge it. I suspect that we are stuck with the name BL Lac objects to characterize these hyperactive galaxies.

The importance of the BL Lac objects stems from their identification with objects that unquestionably contain stars. The tremendous power associated with the quasars has now been seen in something that is unquestionably a galaxy. You can't see the galaxy underlying a quasar and get a spectrum of it; the synchrotron radiation from the quasar is too bright. But you can see the galaxy that is associated with the BL Lac objects.

Apart from their lack of emission lines, BL Lac objects are very similar to active galaxies and quasars. They are radio sources, and the radio power varies. A few of them contain small clouds of high-speed electrons producing radio waves, as verified by VLB observations. No one has yet observed BL Lacs in the far infrared, but they become brighter and brighter in the near infrared in the same way that quasars do.

The one difference between BL Lacs and quasars is the variability of their light. BL Lacs are almost violently variable, changing their light output by tenfold in matters of months. Some BL Lacs are quite bright, and their variations can be followed by small telescopes. Variable objects like these need to be monitored from night to night, and the large telescopes at major observatories usually are being used for other tasks in which you need the enormous light-gathering power that such a telescope has. Small telescopes, such as those owned by small colleges, Eastern universities with unfavorable weather, and some amateurs can be used to monitor these objects. There are lists of likely objects that people interested in such a project can consult.[3]

Links between active galaxies and quasars

Our discussion of the BL Lac objects completes our quick look at various classes of objects in which there is activity similar to that seen in quasars. Table 10-1 summarizes the general characteristics of these various objects. Running down the electromagnetic spectrum from long wavelengths to short, we see that quasars produce radio emission, infrared radiation, optical continuum radiation that far exceeds the starlight produced by most average galaxies, emission lines, and x rays. Narrow absorption lines similar to those seen in quasars are not seen in any other type of object. However, all the other phenomena seen in quasars are present in active galaxies, to some degree or other.

The nonstellar luminosity in active galaxies generally increases as we go from the top of the table to the bottom. There is some overlap between the luminosities of the various categories. The most important place at

TABLE 10-1   PROPERTIES OF ACTIVE GALAXIES AND QUASARS

TYPE OF OBJECT	RADIO EMISSION	INFRARED RADIATION	NONSTELLAR OPTICAL CONTINUUM RADIATION	EMISSION LINES	X-RAY RADIATION
Spirals	Weak	Weak	None seen	None seen	Some
M 82	Weak	√	None seen	In filaments	None seen
M 87	√	√	√	Not broad	√
Classical radio galaxies (like Cygnus A)	√	Some objects	Some objects	Some objects	Cen A (which also emits gamma rays)
N galaxies	√	√	√	√	√
BL Lac objects	√	Short-wavelength	Overpowering	None seen	None detected
Quasars	√	√	√	√	√

which this overlap occurs is the bottom of the table, between the N galaxies and BL Lac objects and the quasars. Without such an overlap, it could well be argued that these quasars, these enigmatic stellar objects, were in fact something else. Yet there are a few extremely powerful N galaxies that would be called quasars if they were not fuzzy in appearance. The most luminous N galaxy known at this time is ESO 113–IG 45, with an optical luminosity of about 10^{45} ergs/sec, only one-third as much as the luminosity of 3C 196, one of the four quasars discussed in Chapter 7. (The name means "intervening galaxy No. 45 in European Southern Observatory field No. 113.") Many objects called quasars are fainter than this one. The BL Lac objects also overlap the quasars in luminosity.

Active galaxies are like the galactic x-ray sources discussed in Chapter 6: There is a wide diversity of phenomena that come under the same general subject heading. Yet here again there is a unifying thread. All phenomena that are seen in quasars are seen in galaxies, and there seems to be a continuous progression from the low level of activity in the Milky Way galaxy up to the tremendous power of the quasars.

The study of active galaxies contributes enormously to our understanding of those cosmic powerhouses called quasars. Because we can see stars in these galaxies, we have some confidence in our estimate of the distance to them. We can therefore estimate their total power output fairly accurately. Further, they are close enough so that we can figure out just where this power comes from. A particularly important discovery was that clouds of high-speed electrons are ejected in the same direction. Radio galaxies exist in which two clouds of radio emission, up to megaparsecs apart, are aligned with the small, parsec-sized clouds in the nucleus.

One question remains: What produces all this energy? An attack on this question involves abandoning the relative security provided by observational data and striding forth into theoretical territory. Because observations directly pertaining to the nature of the central energy source are relatively few, speculations abound. Chapter 11 examines our current ideas regarding the nature of the central energy source.

11   THE ENERGY SOURCE

At last, I confront you, the reader, with the central problem. What is the powerhouse of quasars? Enormous luminosities have been invoked in recent chapters, and although the mechanisms to produce electromagnetic radiation seem clear, the nature of the ultimate source of the energy has been swept under the rug, postponed until now. The most luminous quasar, 3C 379, was so bright in the late 1930s that its visual power was 10^{47} ergs/sec, hundreds to thousands of times more luminous than the stellar output of the Milky Way. What kind of object can produce such enormous amounts of energy at such rapid rates?

Another way to look at the energy problem is to consider the energy bursts in modestly powerful objects like 3C 120. In 3C 120, new clouds of relativistic electrons are created once a year. The energy in these clouds can be estimated reasonably accurately, since the clouds seem to fit the simple expanding-cloud model. The energy required is 10^{52} ergs per cloud in the form of high-speed electrons. Scientific notation is sometimes deceptively innocent. The number 10^{52} does not seem so very big. Why, 52 is just the number of cards in a deck, and you can hold that in one hand. Let us write it out. Then 10^{52} ergs becomes 10,000,000,000,000,000,000,000,000,000,000,000,000,000,000,000,000,000 ergs. That equals the amount of energy put out by our sun in 100 billion years, if the sun should live that long. Once every year, the central energy source in the middle of 3C 120 shoots off that much energy in the form of high-energy electrons. It must take a remarkable object to do that. Before we consider specific models, let us briefly review what we really know from observations about this energy source.

General properties

We have mentioned this energy source, the powerhouse of quasars and active galaxies, repeatedly in recent chapters. We have made a great many observations, but very few of them provide information that bears directly on the source of the tremendous luminosity that is associated with the most powerful objects, the quasars, or even on the source of energy of weaker objects. We can, however, draw several conclusions about the nature of the energy source. I summarize these here, giving the observational evidence supporting each generalization and the possible exceptions to it.

1. *Energy produced at rates up to 10^{47} ergs/sec in the form of high-speed electrons directly or indirectly produces the radiation we observe.* High-speed electrons produce synchrotron radiation, which is unquestionably observed in the form of radio waves (Chapter 8) and may be present in other spectral regions (Chapter 9). Infrared emission comes from dust in some cases and emission lines come from low-density gas clouds, but in each case dust or gas converts energy from synchrotron radiation to some other form. A possible exception to this generalization is the x-ray radiation that in some models comes from hot gas.

2. *High-speed electrons are produced in bursts that are separated by time intervals varying from days to years.* Most objects that have been observed vary in continuum intensity from one year to the next (Chapters 8 and 9). Specifically, the radio variations in 3C 120 can be successfully modeled by presuming that clouds of high-speed electrons are generated once a year or so (Figure 8-11). Some objects vary far more rapidly, showing that the outbursts of activity recur more often than once a year.

3. *The powerhouse tends to beam its energy in particular directions.* The most dramatic example is the blowtorch galaxy, NGC 6251 (Figure 10-3), with its radio jet that preserves its alignment from the compact, one-parsec jet in the core to the alignment of the radio-emitting lobes over a distance of 1,000,000 parsecs (1 megaparsec). The alignment of large and small components of radio emission and the existence of optical jets in 3C 273, M 87, and NGC 1275 also support this argument.

4. *The powerhouse is a small object.* The rapid variations of some objects indicate that the object is less than a light-day across, since one can tell the size of an object by the time it takes to double in brightness. In some lower-powered objects such as a galactic nucleus, we can directly verify a small size, a size that is comparable to the size of the solar system. In most cases it is hard to precisely define the word "small" in this generalization, but it is clear that we are dealing with something far smaller than a galaxy.

5. *Rapid motion is associated with the production of energy.* The breadth of the emission lines observed in some Seyfert galaxies and in quasars indicates that the filaments producing these lines move quite quickly. The rapidly expanding spiral arms near the center of the Milky Way galaxy, the possibly explosive nature of M 82, and the rapid motions of radio-emitting clouds are evidence for rapid motions. However, the evidence here is quite circumstantial. In some objects, rapid motion may not occur.

6. *Whatever is responsible for the activity is found in the nuclei of galaxies.* We can show a direct link between galactic activity and a galactic core in only a few of the weaker cases: the Milky Way, other spiral galaxies, M 87, and Centaurus A. The bright nuclei of the N galaxies also indicate a role for their cores. For this generalization to be true, we must assume that all types of galactic activity are based on the same fundamental phenomenon.

There is one additional piece of information that can also be gleaned from the observations. It would be useful to know when galactic activity occurs during the life of a galaxy. We cannot examine the life cycle of a particular quasar or active galaxy, but we can see whether there is some kind of overall, cosmic evolution of the number of quasars in the universe.

When were these energy sources most plentiful? Look out into the universe and see. You can examine various epochs of the universe by examining objects at different distances from us. Light takes time to reach the earth from a distant object. The further we probe the depths of the universe, the further we are probing into the past. We see the sun as it was when sunlight left it eight minutes ago; we see the Andromeda galaxy by light that left it two million years ago; and we see the most distant quasar known, OQ 172, as it was 15 billion years ago, as it was when the universe was approximately 5 billion years old (these numbers assume a Hubble constant of 50; see Chapter 14). Thus you can see how many quasars were around at various epochs by counting quasars with various redshifts, or various distances.

Maarten Schmidt of Hale Observatories undertook quasar counting. To do anything like this, you need an unbiased sample of quasars: one for which you can argue that every quasar that is brighter than some limiting magnitude has been found. Limiting your search to such a well-defined sample has the unfortunate consequence of giving you a small sample: 33 quasars, in the case of quasars producing enough radio emission to be in the 3C catalogue and brighter than optical magnitude 18.5. Counting these quasars indicates that quasars were far more abundant in the early universe than they are now. If the universe contained now the same density of quasars that it did in its first two or three billion years, several hundred quasars would be as bright as 3C 273, instead of only one.

According to the current picture of cosmic evolution, all galaxies formed in the first few billion years after the Big Bang. Thus the prevalence of quasars in the young universe and their absence now indicates that the type of activity that produced extremely high luminosities is something that occurred shortly after galaxies first formed. This analysis does not put any tight, quantitative constraints on what the energy source should be like, but it does direct attention toward events happening in the cores of newly formed galaxies.

The problem of the energy source is totally unlike any other field of quasar investigation. In Chapters 7, 8, 9, and 10, the only theory involved was model fitting. Theorists have made observations and have been happy if a model more or less fitted the data. But here we have very little data bearing directly on the central energy source. We can't see it. We can only see what it *does*, and there isn't much to go on there. In summary, we need an energy source that (1) can produce 10^{47} ergs per second in the form of high-energy electrons, (2) is small, and (3) is the kind of object that you would reasonably expect to find at the center of a young galaxy. Little else is known about the energy source itself. With so few constraints on the model, theoreticians are free to roam into the far reaches of the model world in their search for a possible model for this energy source. The heart of the quasar is a mysterious place.

The models presented in this chapter are described in both philosophical and chronological order. Some people want to make the energy source as prosaic as possible, while others like to let the imagination run free and see what sort of weird object it can produce. The general starting point is some kind of massive object at the center of a galaxy: a supermassive cluster of more or less ordinary stars. Some investigators believe that these stars can do the job. Others envision the formation of a superstar, a Gargantua formed by coalescence of millions of stars, which in some way belches out the necessary clouds of high-energy electrons. Others do not see the quasar phenomenon occurring until this superstar has formed a gigantic black hole, of 10^8 solar masses or so, which will produce high-energy electrons as it swallows the clouds of matter surrounding it. I finish with a very brief look at some quite radical theories: white holes, antimatter, and quarks. Please remember that these are all theories.

Colliding stars

A galaxy forms somehow in the early universe. How is not known; perhaps the quasar event triggers the formation of galaxies. Anyway, it is reasonable to expect that as the protogalactic cloud of gas collapses, a concentration of matter forms in the middle. Computer calculations of the dynamics of collapsing objects indicate that the formation of a central concentration is inevitable. To start with, let's stay fairly close to real galaxies and suppose that this central condensation is a superdense star cluster, an extreme version of what the observations of the center of our galaxy show.

It is not clear whether the stellar density at the center of this star cluster will ever reach the point for which collisions between stars occur often enough to be interesting. Princeton's Lyman Spitzer calculates that one collision of stars per year is necessary if stellar collisions are to provide enough energy (ignoring, for the moment, the troublesome issue of whether you can get relativistic electrons out of such a scheme). The density of stars has to be 10^{11} per cubic parsec for collisions to occur this often. This density is 10^5 times the density at the galactic core today. Such a star density is tremendous. Think of what it would be like to live on a planet of such a star. The night sky would be as bright as the full moon. Of course, you would run the risk of having your sun collide with another star, wiping you out of existence; and the deadly radiation coming from the heart of the quasar would be lethal to life.

It is not at all clear how a collision between two stars could produce high-energy electrons, but supernova expert Stirling Colgate has an idea. Suppose that colliding stars rapidly evolve to the supernova stage. This idea is attractive because supernovae are known to produce high-energy electrons. Maybe you don't get the high-energy electrons until *after* the supernova. Thus the heart of a quasar may be millions of pulsars, ticking merrily away. (Martin Rees and Jeremiah P. Ostriker, among others, take the credit for this idea.) Perhaps when two stars smash into each other they blow off part of their mass as high-energy electrons. The slingshot model for the origin of double radio sources fits nicely into the colliding-star class of models for the quasar energy source. Here a close collision in a dense star cluster eventually results in the ejection of one or two objects into intergalactic space at high speeds.

The attractiveness of this particular class of energy-source models stems from the similarity of high-energy phenomena in quasars and in supernovae. The similarity of the radiation from the Crab Nebula and the radiation from quasars has been noted several times. To a large extent, these theories view the activity in galactic nuclei as the sum of activity in independent supernova events (or other violent phenomena) in the galactic nucleus.

Quasar superstar

When a galactic nucleus evolves to the point at which stellar collisions become frequent, it is not at all clear what should happen. Would the collisions themselves cause the required high-speed electrons to be produced either by accelerating stellar evolution to the supernova stage or by some other means? Is it possible that the colliding stars coalesce and merge? It seems that the result depends on the relative velocities of the stars. If they hit each other too hard, they blow each other up, but if the collisions are less violent, the two stars merge.

Robert H. Sanders, working at Princeton, has followed what happens in a galactic nucleus that has colliding, coalescing stars. The colliding of small stars gradually builds up larger stars. After a few million years, one very massive star forms and grows quite rapidly. It has such a strong gravitational force that it consumes all the smaller stars that get in its way. Such a Gargantua, of 10^6 to 10^8 solar masses, is known as a superstar. Although superstar evolution is not known with any certainty, it seems probable that most of the mass in the galactic nucleus would end up in such an object.

Now that the origin of this superstar has been accounted for, how do we explain the energy? Different investigators have different ideas. Remember that we are trying to generate about 10^{46} ergs/sec in the form of clouds of high-energy electrons, thrust out at intervals.

Flares

One way to approach the superstar theories is to suppose that superstar activity is like stellar activity, but scaled up by many orders of magnitude. Peter A. Sturrock hypothesizes that gigantic solar-flare types of eruptions burst forth from the surface of a superstar (or galaxoid, as he calls it). We see solar flares on the sun, but they are not particularly well understood, so it is hard to scale an unknown phenomenon to unknowable dimensions. Still, the picture is intriguing. At the surface of this vast superstar, containing 10^8 solar masses, the word goes out: *Flare!* The magnetic fields on the superstellar surface have merged, uncoupled, and merged again, but violently this time. They become entwined in each other, instabilities develop, and *Whoosh!* great streams of gas burst forth into space. (This fanciful idea is based on what happens during a solar flare. If you ever get a chance to see a movie of the sun in action, you'll get some idea of what is going on.) As these streams of gas burst forth, high-energy particles accompany them. In the case of the sun, these high-energy particles move toward the earth, causing auroras and such phenomena.

Nice, idea, isn't it? It seems to account for what we need. It would be better if we knew more about solar flares, however, before we become too enamored of the superflare hypothesis.

Super-supernovae

The presence of high-energy particles in supernovae has caused many people to examine whether there are any possibilities in a scaled-up supernova model for the quasar powerhouse. Suppose that a superstar explodes in the same way a supernova does, producing high-energy particles. The ejected gas then forms another superstar, which explodes again. Unfortunately, it is not easy to explain some source like 3C 120, which explodes fairly often, once a year or so. Let's see if there is any more promise in a supernova remnant.

Giant pulsars

One of two reasonably detailed models of the energy source is the giant pulsar—or *spinar*—theory originated by Philip Morrison. Rotation dominates many of the theories remaining for us to examine. As the superstar forms and contracts, it spins faster and faster in the same way that a skater spins faster and faster as the skater pulls his or her arms in toward the body. Rotation can store a tremendous amount of energy in a large object like a superstar, and any mechanism that can store tremendous amounts of energy is as attractive to quasar theorists as catnip is to cats. We want energy, and rotation can store it. If you can figure out a way to get this energy out, even in small fractions, the energy-source problem can be solved.

Once again attention falls on the smaller analogues of superstars, the pulsars, and on the Crab Nebula. The Crab Nebula produces many of the same things that quasars produce, on a much smaller scale: emission lines from filaments, radio, optical continuum, x rays, and high-energy electrons. Morrison views the quasars as giant cousins of the Crab, containing spinars, or giant pulsars. Pulsars rotate, and their rotational energy is converted into bursts of radio emission that are emitted in our direction once (or perhaps twice) in each rotation period. The mechanism for the conversion of the rotation into radio waves is far from understood, but it seems clear that large-scale electric or magnetic fields in the pulsar are somehow involved. Pulsars are also known to be the source of high-energy electrons in the Crab Nebula. All we need is magnetic fields to get synchrotron radiation, and we know that pulsars have magnetic fields, since the pulsing is electromagnetic in nature.

The spinar model has one distinct advantage over the other models in this chapter. You have probably noticed that my descriptions of each of these ideas have ended with some comment like, "Well, it's a nice idea. But so what? How can we show that it's correct?" A spinar is rotating, and anything rotating should have some sort of periodicity. Pulsars emit radiation in pulses, so spinars should, too. Early support for the spinar idea came from the quasiperiodic variations of the quasar 3C 345. It appeared, for a while, to brighten once a year, but the regularity has not persisted. Perhaps it changed phase, entering on a new sequence. When you seek to verify the existence of a period in an astronomical object, you have to observe it over many periods, for many years. More observations of 3C 345 are needed to show whether the quasiperiodic fluctuations are regularly spaced or random.

Seven other objects may also have quasiperiodic behavior, but the observations are so few that the existence of periodic variations has not been universally accepted. It is difficult to verify this periodic behavior because irregular changes in the brightness of these objects can obscure regular, periodic variations. But if the periods that have been reported really do exist, this observation of variability strongly supports the spinar model.

The superstar models—giant flares, super-supernovae, and spinars—have the advantage of representing scaled-up versions of well-known, if not well-understood, effects. As stars are tightly packed in galactic nuclei, it seems plausible that some massive object containing between 10^6 and 10^8 solar masses, or a superstar, would form. One type of superstar theory—the spinar or giant-pulsar idea—is the only quasar theory that is amenable to direct check with observations, since it predicts that quasar variations should be periodic. Some theorists, however, believe that the quasar phenomenon is involved with the next phase of evolution of a superstar, its collapse toward the black-hole stage.

Black-hole theories

As the collapse of a very massive object proceeds, the object rotates faster and faster. No one knows exactly what happens. Several poorly understood effects must be waved away as unimportant if theoretical calculations are to be practicable. There are no real supermassive objects around to enable us to confirm theory. The general consensus is that a very rapidly rotating disk would form, and that this disk could offer some promise as a quasar powerhouse. It is probable that a black hole would eventually form at the center of this disk, but it is not necessary that the hole form before the quasar phenomenon begins. It is rotation and infall that characterize this class of theories, since the material falling into and onto this disk is compressed drastically as the dimensions of the disk shrink toward the Schwarzschild radius, toward the event horizon.

The general picture is that the center of a quasar or active galaxy contains a huge disk or black hole of 10^8 solar masses or so. If the efficiency of energy generation is somewhere near the maximum theoretical efficiency, the infall of one solar mass of stuff per year would provide the necessary amount of energy. This object would lie at the center of a galaxy, eat the galaxy's core, and provide the energy for the quasar phenomenon.

Efforts to make the general black-hole picture a little more detailed have focused in three areas. Theorists seek to answer some more pointed questions regarding these models of black holes to see if they are a viable, or even a preferred, model for the energy source. How is the black hole fed with the one solar mass per year that it has to eat in order to power the quasar? How does this black hole produce high-speed electrons? How do the theories of black holes fit into the evolution of galaxies?

Feeding black holes: black tides?

A black hole powers a quasar by eating matter. Matter falling into a black hole travels faster and faster as it nears the black hole; gravitational energy is being converted into kinetic energy. Postponing the question of how this energy is converted to fast electrons for the moment, let us ask whether a typical galactic nucleus can feed a black hole fast enough to produce quasarlike luminosities of 10^{46} ergs/sec. Several possible sources for the one solar mass per year of black-hole fuel have been explored.

When a star evolves into and through the red-giant stage, it sheds mass. Low-mass stars shed planetary nebulae and become white-dwarf stars, and heavier stars shed supernova remnants like the Crab Nebula. If the stars in a galactic core have the same relative numbers of large and small stars as the stars in the solar neighborhood have, an entire galaxy of 10^{11} stars produces one solar mass per year of gas from the mass lost in late stages of stellar evolution. But can all this gas end up in a galactic nucleus and feed a black hole? Martin Rees of Cambridge University says it will; Gregory Shields and Craig Wheeler of the University of Texas argue that it won't. More work based on more detailed calculations of the dynamics of galaxies is needed to settle the issue.

A picturesque source of fuel for the black hole is the disruption of stars in the core by the strong tides near the central black hole. In this theory, a star passing too close to a central black hole is ripped apart by the tidal gravitational forces. These forces are the same forces that made life uncomfortable for the heroic astronaut of Chapter 4. Theorists disagree on whether enough stars approach the black hole closely enough to be eaten. This model has the wonderful name of the *black tide*.

And there is a very prosaic source for the mass that must fall down a black hole in order to power it. Galaxies form from gas, and some of the gas may well be left over after the central black hole has evolved. It could be that the black hole just eats this gas.

Making high-speed electrons

Material falling into a stellar black hole forms an accretion disk around the hole. In binary systems, this accretion disk heats up and produces x rays, not high-speed electrons or synchrotron radiation. How might a galactic black hole produce high-speed electrons? There are a couple of ideas, and both work.

One possibility comes from theoretical studies of just how hot the accretion disk in a galactic nucleus becomes. Some theoretical studies indicate that if the infall rate is high enough, temperatures in the accretion disk will rise, rise, and rise still more, to the point at which electrons start moving at speeds close to the speed of light. Donald Lynden-Bell, one of the original advocates of black-hole power in quasars, created a picture of a black hole eating so much matter that runaway temperatures resulted in the violent expulsion of matter from the accretion disk out to intergalactic space. In another scenario, proposed by Jonathan Katz of the Institute for Advanced Study in Princeton, charged particles in a hot gas can collide with photons, creating high-energy photons that might even be able to explain the entire quasar spectrum.

Still another class of black-hole models appeals to electrical and magnetic fields created by the swirling currents of charged particles in the region surrounding the black hole. In this model, moving particles create electrical forces that accelerate the particles to ever-higher velocities. Something similar seems to happen in the sun, where magnetic fields produced by the slow bubbling of gas near the solar surface eventually accelerates particles to extremely high velocities. No one has been able to produce detailed models that predict the acceleration of particles in solar flares, though the general ideas seem fairly successful. We know from direct observation that solar flares produce high-speed electrons. Similarly, the calculations of electromagnetic phenomena in accretion disks are fairly generalized. But, as you can see, there is no shortage of possible mechanisms for transforming the energy of particles falling toward a black hole into high-speed particles. These mechanisms all seem to work, in theory. No one knows whether any of them actually represents what is going on in the heart of a quasar.

Black power and the evolution of galaxy cores

Thus giant black holes seem to be able to provide the energy needed to power the quasar phenomenon. Donald Lynden-Bell, in a recent review, borrowed a phrase from the 1960s to describe this model: "black power." But do such objects tend to evolve in the nuclei of galaxies? What happens to them afterward?

One of the attractive features of the black-hole model, in the view of some theorists who tend to favor it, is that black holes are the natural end point of most of the theories considered in this chapter. The models can be grouped into three classes. One class involves densely packed collections of stars that can collide, whip around each other in slingshot near-collisions, produce myriads of pulsars, or do a variety of other things. A second class involves superstars—also called magnetoids, galaxoids, or *spinars*—which rotate, explode, flare up, or whatever. The third class involves black power. It is probable, though not certain, that in many cases a densely packed collection of stars produces a superstar, a massive object, as the result of collisions between the stars. A superstar is far more massive than the limiting mass for white-dwarf stars or neutron stars, and thus ends its life cycle as a black hole. So, in some ways, all these models deal with various plausible stages in the life cycle of a galactic nucleus. Theorists differ on just when the energy is produced.

Considering the evolution of a galactic nucleus brings up the question of what happens after all the excitement involved in quasar activity dies away. The massive black hole remains in the galactic nucleus, accreting matter at a slower rate. One speculation is that the more modest activity in radio galaxies, Seyfert galaxies, and other forms of active galaxies is powered by mass falling onto a black hole that at one time, in its youth, was a more energetic quasar. Was our Milky Way galaxy ever a quasar? Probably not. Investigations of the motions of gas near the galactic core shows that, within 0.4 parsec of the center, there is 5×10^6 solar masses of matter present in some form. Although it is possible that much of this mass is in the form of a black hole with a mass of 5 million solar masses (with a Schwarzschild radius of one-tenth of the distance from the earth to the sun), such a black hole could not produce the power necessary to produce the quasar phenomenon, even with a very high rate of accretion. The nearest dead quasar, in this picture, would be likely to be Centaurus A (Figure 10-4). The observed number of quasars indicates that at least 1% of all galaxies were quasars at one time.

How do these models agree or disagree with what we know about activity in quasars and galactic nuclei? The alignment of large and small radio sources and the existence of jets tends to argue against models based on stellar collisions or independent phenomena occurring in dense star clusters. The beam model for radio sources tends to support the idea that high-speed electrons are generated by some rotating body, an accretion disk or a superstar, and funneled in particular directions. If the observations supporting the presence of a massive object in the nucleus of M 87 (Chapter 10) stand up, there is direct evidence for the existence of a massive object in a galactic nucleus. This massive body need not be a black hole.

The black-hole model has become increasingly attractive in recent years. This popularity could be the result of a bandwagon effect. Scientists often appreciate the approval of their colleagues and it is very tempting to follow the fashions of the day. A recently quoted dialogue puts the black-hole models in perspective. Martin Rees, of Cambridge University in England, is a proponent of the black-hole models, and Geoffrey Burbidge, director of the Kitt Peak National Observatory, is noted as someone who resists jumping on scientific bandwagons. Both have contributed greatly to theoretical analyses of quasars and active galaxies.

REES: The energy source in these galaxies will be shown to be a black hole, I think, even though it may take 100 years before we have proven it.

BURBIDGE: I think it will take 1000 years and we may very well be on the wrong track. These [black hole] models are getting into the textbooks now, but there is never anything testable and people are working on smaller and smaller pieces of the problem.

REES: I agree, but I would argue that the way we are going about it is the most productive approach, even though the modelers may be getting the illusory satisfaction of a Ptolomean [sic] theorist who adds another epicycle.

BURBIDGE: I'm glad to hear you say that, Martin. The trouble is that so many people take these things more seriously than you do.[1]

Radical theories

From time to time some weird ideas surface in the literature. Since these ideas are spectacular, they often make their way into the popular press, with few caveats about their highly speculative nature. For example, white holes were first suggested in 1964 as possible sources of quasar energy. *White holes* are time-reversed black holes (see Chapter 6). Although it is theoretically possible that white holes exist, a prerequisite would be a singularity in just the right condition, ready to burst forth into the universe like a bubbling mountain spring. The idea about white holes was subsequently resurrected by the Russian astrophysicist Igor Novikov, and still later by an American, Robert M. Hjellming. Inevitably, a news article appeared, claiming that we had discovered white holes. Sorry, we were only thinking about them.

Another idea for power sources for quasars is *antimatter*. When antimatter and matter meet, both are annihilated and all the energy turns into gamma rays. But where are these gamma rays? Gamma-ray astronomers have been looking for them for years. Even though they recently found gamma rays from electrons annihilating their antiparticles at the galactic center, the intensity of the gamma rays is far too low if antimatter is to power the quasars. Some people speculate that quarks may power quasars. *Quarks* are the building blocks of matter. You put quarks together and you get protons, electrons, neutrons, and all the particles that high-energy physicists have observed. Yet people have searched for quark atoms in quasars and they have not found any.

Have you had fun with this chapter? It's really enjoyable to be able to let your mind run free, not constrained by many observations, to see how you can construct a quasar powerhouse while still staying within the laws of physics. There are many, many ideas for possible energy sources for quasars, and all of them work. We don't really know very much about these energy sources, so a search for them is an exploration of the model world only. There is not much sign of the real world here. It is groping in the dark. Even though there are many observations of quasars, these data from the real world provide little direct information about the energy source. There is no check between the model world and the real world, save for the spinar theories. And although it's fun to theorize, I feel a little uncomfortable that people have occasionally taken particular models too seriously. We know so little of what is happening in the nuclei of ordinary galaxies, to say nothing of the quasars.

To briefly review the models: One group of models has a vast concentration of stars at the center of galaxies or quasars, and these stars do interesting things—collide, become supernovas, become pulsars, and so on. The superstar theories see some large object, 10^8 solar masses or so, causing the ejection of fast electrons. Maybe it flares, maybe it becomes a supernova, or maybe it is a giant pulsar (and if so, we can check on it and see whether it's periodic). At third class of models sees a disk at the center, with a black hole forming sooner or later at the center of this disk. A fourth invokes alternative energy sources, not seen elsewhere in the universe: white holes, antimatter, or quarks.

Which is correct, if any? I do not know. More than one type of phenomenon may be involved. If I were asked to bet on one of these models right now, I should bet on something that involves rotation and magnetic fields—something like the spinar model or the black-hole model. These are all ideas. They are fun to play with, and who knows? One of them may be correct.

12 ALTERNATIVE INTERPRETATIONS OF THE REDSHIFT

One fundamental assumption underlies the interpretation of quasars and active galaxies developed in the last five chapters. The discussion has assumed that the redshifts of quasars are *cosmological;* that is, that they are a result of the expansion of the universe. The belief that quasars are very luminous objects located at vast distances is based on this assumption. These large luminosities require that violent, explosive events of unparalleled magnitude occur in the nuclei of galaxies. This fundamental assumption has been battle-tested. Soon after quasars were discovered, a number of alternative explanations of the redshift, based on known physical laws, were considered and rejected because the models conflicted with observations. A number of photographs that showed apparent links between quasars and nearby galaxies caused some people to advocate that quasar redshifts were noncosmological. In such a view, new physical laws would be required in order to explain the redshifts of quasars.

But supporters of the traditional view—that quasar redshifts are produced by the expanding universe—struck back. People looked for direct evidence that quasars are at cosmological distances. Two independent lines of work converged in the late 1970s to show fairly conclusively that at least some quasars are as far away as their redshifts indicate. The successful identification of the BL Lac objects with galaxies shows that objects with quasarlike luminosities do, in fact, exist. A series of searches for galaxies associated with quasars of the same redshift culminated in the discovery of eight such objects as part of a well-defined sample.

The present situation, then, is that at least some quasars are at the distances that their redshifts indicate. This conclusion is reasonably safe. But the photographs showing apparent links between quasarlike objects and galaxies must be explained. They may be just coincidental juxtapositions of objects in the sky, or they may represent a real phenomenon. The interpretation of these photographs, though, is now somewhat divorced from quasar research. It is clear that the objects described in the last five chapters—quasars and active galaxies—do exist in the real world.

Evidence for noncosmological redshifts

The first indication that the cosmological interpretation of redshifts might be in trouble appeared in 1966. Fred Hoyle of Cambridge University, collaborating with Geoffrey Burbidge and Wallace L. W. Sargent, then at the University of California, pointed out problems posed by the rapid variations of brightness of some quasars. These rapid variations indicate that these quasars are very small. It is difficult to understand how the high luminosity in quasars can come from such a small object. The problems pointed out in 1966 can be avoided with more complex models, as was shown subsequently. But some doubt had been cast on the cosmological-redshift theory. Two of the authors of the paper, Hoyle and Burbidge, have since been prominent in the challenge to the traditional explanation of quasar redshifts.

But how can the challengers prove that quasars are closer than the redshifts indicate? The centerpiece of their evidence is a group of photographs taken by H. C. Arp of the Hale Observatories. Arp's photographs seem to show connections between several quasars and galaxies with much lower redshifts. These connections, if real, imply that the quasar and the galaxy in each case are at the same distance from the earth. Since the redshift of the galaxy is much smaller than the redshift of the quasar, the inevitable conclusion is that some of the quasar's redshift is "extra," or—to use the current term—discordant, and the cosmological-redshift theory is thereby demolished.

Yet two objects close together in the sky may or may not be real companions, since we can see the universe only from one vantage point—the earth. For example, Mizar, the second star from the end of the handle of the Big Dipper, appears to have two companions (Figure 12-1). One of these companions, Alcor, is distant, whereas Mizar B is so close to Mizar A that you need a telescope to split them. When the system is analyzed, it turns out that only Mizar A and B are true companions, while Alcor is a background star, much farther away.

Thus if we seek to show that quasars are associated with lower-redshift galaxies, we must show a real physical connection between the objects. One connection has been investigated more thoroughly than others, and I present it as an example.

Markarian 205 and NGC 4319

The remarkable pair of objects Markarian 205 and NGC 4319 is shown in Figure 12-2. Galaxy NGC 4319 is the large spiral galaxy at the top. It has a redshift of 1800 km/sec, or $z = 0.006$. Markarian 205 is the starlike object below the galaxy. It has a quasar spectrum with a redshift of 21,000 km/sec, or $z = 0.07$. If these two objects are true companions, most

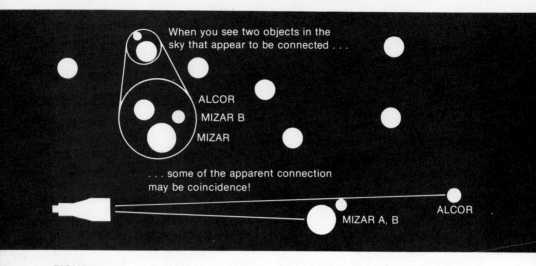

When you see two objects in the sky that appear to be connected . . .

ALCOR
MIZAR B
MIZAR

. . . some of the apparent connection may be coincidence!

MIZAR A, B

ALCOR

FIGURE 12-1   Two stars, close together in the sky, may be far apart in space, as Alcor and Mizar in the Big Dipper are.

of the redshift of Markarian 205 is extra, or discordant, coming from some physical cause now unknown to science.

But are these two objects true companions? Two possible views are shown in Figure 12-3. How are we to decide? Look at Figure 12-2 again, the photograph of the objects, and concentrate on the area between the galaxy and the starlike object. There seems to be a bridge of luminous material connecting it and the galaxy. This photograph can be most straightforwardly interpreted by supposing that this luminous matter connects the two objects. Such an interpretation indicates that the starlike object and the galaxy are at the same distance, and that the starlike object has a discordant redshift.

Yet many of us hesitate to call for new laws of physics until all attempts to explain the observations by the old laws have failed. Other explanations of the observations have appeared.

1. Maybe it isn't there after all. Some investigators have failed to photograph this bridge when they have photographed these objects.

2. Perhaps it is just the overlap of the images of the galaxy and the starlike object, since images can blend together in the photographic process.

3. It is a short, stubby spiral arm of NGC 4319 that just happens to point right at Markarian 205.

4. It is a background galaxy that happens to be there, fooling us by looking like a connection between the two objects.

Thus the photographic evidence shown in Figure 12-2 allows two interpretations. In one, the juxtaposition of these objects is just coincidence

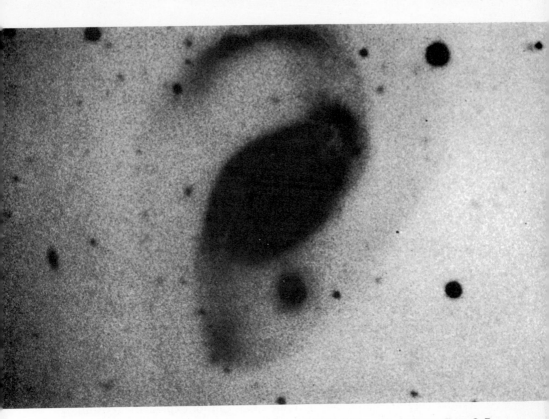

FIGURE 12-2   Markarian 205, a quasar, and NGC 4319, a spiral galaxy. (From C. R. Lynds and A. G. Millikan, *Astrophysical Journal Letters,* vol. 176, page L5, 1972, published by the University of Chicago Press. Copyright © by the American Astronomical Society. All rights reserved.)

and the apparent connection is an illusion. In the other, the apparent connection is real and part of the redshift of Markarian 205 has to be explained by new physics. Consider another example before evaluating the evidence.

A high-redshift object in front of a galaxy?

Figure 12-4 shows another pair of objects in which anomalous redshifts may be involved. At first glance the photograph looks like a collection of dots and blobs. Its significance is that one of the blobs—the one marked with an arrow—has a quasarlike spectrum and a high redshift, a redshift corresponding to a velocity of 13,300 km/sec. The blob above it is an overexposed image of a galaxy called NGC 1199. The galaxy has a redshift of 2700 km/sec.

So far, big deal. But look at the photograph more closely. This is a negative photograph, printed as a negative so that the contrast between

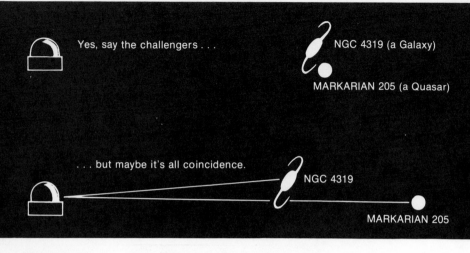

Yes, say the challengers . . .

NGC 4319 (a Galaxy)

MARKARIAN 205 (a Quasar)

. . . but maybe it's all coincidence.

NGC 4319

MARKARIAN 205

FIGURE 12-3   Is Markarian 205 really a companion of NGC 4319?

objects of differing brightness will be enhanced as much as it can be. There is a ring around the (as-yet-unnamed) high-redshift object. Halton Arp interprets this ring as a dust ring around the high-redshift object that absorbs some of the light from the background galaxy. If this dust ring absorbs light from the galaxy, the high-redshift object must be between us and the galaxy. In such a case, Hubble's Law cannot explain the redshift of the high-redshift object, and we need to appeal to new physics.

The investigation of this high-redshift object, which may or may not be associated with the galaxy NGC 1199, has not been confined to taking photographs. (Photographic images of objects are subject to a wide variety of distorting effects. For example, a photograph of a sharp-edged object may not, in fact, show a sharp edge.) Detectors that resemble the detectors used in television cameras have been used to increase the sensitivity and precision of astronomical observations. These detectors have been used in investigations of the apparent dust lane around NGC 1199. I do not find that the measurements show the dust lane as definitively as the photograph seems to. This is a personal judgment. At this writing, Figure 12-4 has not been critically evaluated in published astronomical literature.

Photographs of apparent associations between galaxies and objects with higher redshifts are not limited to the two presented in Figures 12-2 and 12-4. In a recently published list, Arp cited two dozen cases that he and other astronomers had pointed out. L. Bottinelli and L. Gouguenheim, two French astronomers, as a result of statistical analyses, claim that smaller, companion galaxies have redshifts that are systematically larger than the redshifts of the larger galaxies. Other astronomers dispute whether these analyses show a real discrepancy or whether the results could be chance

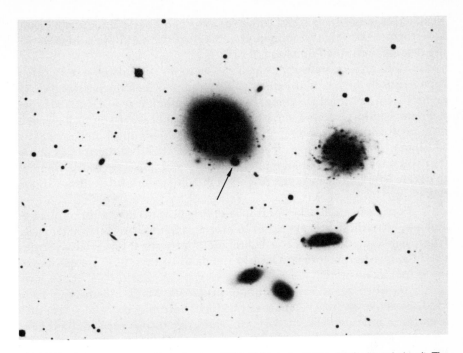

FIGURE 12-4 The elliptical galaxy NGC 1199 and a high-redshift object below it. The high-redshift object has a higher redshift than the galaxy. The bright ring around the high-redshift object, in this negative photograph, has been interpreted to indicate that the high-redshift object is absorbing light from the galaxy and is therefore nearer than the galaxy. (From "A Compact, High-Redshift Object Silhouetted in Front of the E Galaxy NGC 1199," by H. C. Arp, *Astrophysical Journal* 220 (1978), 401. Copyright © 1978 by the American Astronomical Society. All rights reserved.)

effects. Because statistical analyses have played a part in the redshift controversy, the role of statistics in settling questions like this one is worth considering in more detail.

Statistics: big ideas from small numbers[1]

The heart of the disputes about peculiar objects like the Markarian 205–NGC 4319 pair is the question, Are these two objects really associated with each other or is the line-up just due to chance? You might think that the tools of statistics would be good to use to attack this question, since statistics do not involve the judgment of individuals. Statistical equations have no axe to grind. Yet the investigators who use the equations do have an axe to grind, so that statistical arguments have not proved of much assistance in settling the controversy. There are two fundamental weaknesses in applying statistical arguments to quasars: You are not trying to test hypotheses that were formulated before the data appeared, and you are dealing with small numbers of objects.

You can use statistics to answer questions only if the question has been precisely formulated before you go out and seek the data. It is erroneous to say, as some have done, that there is one chance in a thousand that a quasar would be found close to a particular galaxy. If you make such a statement, you are reasoning in circles. You take a photograph (or measure redshifts), notice a strange phenomenon, and then use statistics to prove that you have found a strange phenomenon. You must use your judgment to decide whether this strange phenomenon is meaningful; statistics are no help.

Statistics can also be used in a different way. Take a list of quasars and ask, "Are these quasars closer to bright galaxies than you would expect from chance alone?" This is a legitimate approach, for you have formulated the question before you look at the data. You can now use statistics to try to find an answer.

Yet once again you encounter problems. Many investigators have used the above approach, and they produce contradictory results. Whether a statistical investigation supports the challenge or not depends on the list of quasars used. The reason that statistics are ambiguous here is that the number of quasar-galaxy associations is so small. For example, one heralded investigation showed that 5 out of 50 quasars in a sample chosen from the 3C catalogue were quite close to bright galaxies, a result that had a chance probability of 1 in 100,000.[2] Yet five quasars is a very small number on which to base a conclusion. Other investigations have failed to find any statistical correlations.[3]

Somewhat similar in nature are claims that quasars are found with particular redshifts. In the 1960s, it was claimed that there were far more quasars with redshifts of 1.95 than with other redshifts near 1.95, for example. But again small numbers of objects were involved, and none of these claimed correlations has stood the test of time.

Support for cosmological redshifts

Thus the challenge to the idea that the redshifts of all quasars are cosmological is based on several independent pieces of evidence. This evidence has appeared over a fair period of time. The first photographs were published in the late 1960s, and more data accumulated during the 1970s. The growing body of evidence prompted other astronomers to seek independent evidence that quasars were, in fact, at the distances that their redshifts indicate. Such evidence has appeared. Therefore it seems fairly safe to conclude that at least some quasars are at the distances indicated by their redshifts.

Associations between galaxies and quasars

Beginning in 1971, some astronomers who sought clusters of galaxies that were associated with quasars had a viewpoint somewhat different from that adopted by supporters of noncosmological redshifts. In their view, faint galaxies near a quasar might be associated with the quasar itself. Were quasars and galaxies to be at the same distance, the quasar would be far brighter, since it has a hyperactive nucleus, while the galaxies have quiescent, less powerful nuclei. Measurement of the redshifts of the galaxies would provide a secure estimate of their distance, since there are a number of indications that Hubble's Law is valid for galaxies. If the redshifts of galaxies were the same as the redshift of a given quasar, one could conclude that the quasar was a member of that cluster of galaxies and that its redshift was a valid indicator of its distance.

The first association between a galaxy and a quasar involved the quasar PKS 2251+11. Caltech astronomer James E. Gunn measured the redshift of one of the four galaxies surrounding this quasar and found that it was the same as the redshift of the quasar. Over the 1970s, more and more similar associations appeared, one at a time. Although the data kept accumulating, a definitive test was yet to come. Galaxy-redshift measurements often involved only one of the galaxies in a cluster, and the spectrum of the faint galaxy was not as well defined as one would like. Supporters of the noncosmological-redshift view advanced counterarguments: Maybe the associations were coincidental, maybe the fuzzy images were spurious, and so on.

A definitive investigation thus seemed necessary. Alan Stockton, of the Institute for Astronomy in Hawaii, examined the neighborhoods of each of 27 quasars selected from a comprehensive catalogue. He selected these quasars because they were the nearest and most powerful objects, because problems of understanding their tremendous power would be most severe, and because they were at the same time near enough so that if there were normal galaxies associated with them, one would be able to see those galaxies. Stockton found that there were 29 galaxies that were apparently associated with 17 of the quasars. For 8 of the quasars, the redshifts of one or more galaxies near it agreed with the redshift of the quasar, considering the rate at which galaxies move randomly through space. In the other 9 cases, the galaxies turned out to be foreground galaxies. Figure 12-5 shows photographs of some of these quasars and their associated galaxies.

Stockton's investigation was well planned; he formulated the question to which he wanted the answer precisely before he obtained the data. The question was: Do quasars have galaxies with similar redshifts associated with them? He sought such galaxies in an unbiased manner, and found many. In 8 out of 27 cases, he found galaxies that were associated with quasars whose redshifts were the same. Since he formulated the question first, he was able to use statistics to test the strangeness of what he found.

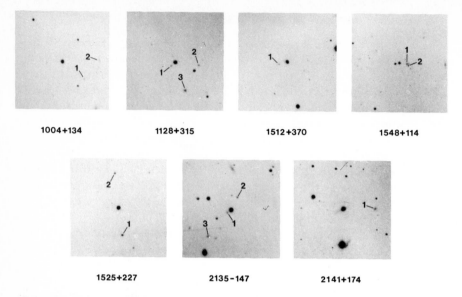

1004+134   1128+315   1512+370   1548+114

1525+227   2135-147   2141+174

FIGURE 12-5   Photographs of a number of quasars with galaxies of similar redshifts near them. The fuzzy spots with numbers identifying them are the galaxies, and the central object in each photo is the quasar. The numbers below each photograph identify the quasar by its position in the sky. (From A. M. Stockton, "The Nature of QSO Redshifts," *Astrophysical Journal* 223 (1978), p. 747. Copyright © by the American Astronomical Society. All rights reserved.)

The probability that 8 of 27 quasars would have at least one galaxy within 45 arc-seconds of them (1/40th of the diameter of the full moon and less than the size of the image of Mars seen through a telescope) is 1.5×10^{-6}. Thus there is a little more than one chance in a million that Stockton's result could be coincidence.[4]

The reaction of much of the astronomical community to Stockton's work is summarized in a report on a workshop at the Lick Observatory held in the summer of 1978. Previously the redshift question had been a hot topic of controversy, widely debated at these conferences. John Faulkner and Martin Gaskell, summarizing the Lick Observatory workshop, state, "No doubts at all were raised over the cosmological interpretation of emission-line redshifts. Indeed, almost the last nail in the noncosmological redshift's coffin was hammered in by the acclaimed work of A. Stockton (Hawaii), who gave strong evidence for the association of galaxies with bright QSO's."[5] I do not know whether the astronomers who would have continued to raise doubts were present at the workshop to raise them, but I think they would find it hard to disagree with Stockton's conclusion that there definitely are some quasars at the distances indicated by their redshifts.

New and old physical laws

One reason that there has been such intense interest in the redshift question is that — if the redshifts of quasars are not produced by the expansion of the universe — there is need for new physical laws to explain them. When quasars were first discovered, many theorists explored various alternative causes of the redshift. One idea was that redshifts are gravitational. This idea fails, for all spectrum lines from a quasar have the same redshift, and if gravitational redshifts are to be the same at all points in the cloud, the emitting gas cloud would have to be unacceptably small. Another idea was that quasars were objects ejected at high velocities from our galaxy. We should expect that in such a case other galaxies would eject quasars too, some of which would be traveling toward us and would show a blueshift. No such objects have been observed. The failure of these two attempts to find an alternative cause for the redshift points to one inescapable conclusion: If quasar redshifts are due to some physical law now known, they are cosmological.

The need for a new physical law to explain a noncosmological quasar redshift is much greater than the need for a model to explain the high luminosity and the small apparent size of some particular objects. We mentioned earlier the problem of generating high luminosities in small volumes of space. The pioneering work of Hoyle, Burbidge, and Sargent showed that a measurement of the time that it takes a quasar to brighten, along with a measurement of the quasar's luminosity, could tell us the size of the region that the luminosity was coming from. In a few cases it is difficult to see just how so much power could come from such a small space. But there is still room for adjusting the models within the current scientific paradigm, within the currently accepted laws describing the way that matter behaves. Complex models have been proposed that can account for the large power and small size of the objects, but they are so complex that not all astronomers believe them. However, it is one thing to disbelieve a model and another to disbelieve a physical law.

The BL Lac objects also serve to divorce the problem of modeling the high-powered, rapidly varying sources from the cosmological-redshift question. In these objects, the redshifts of the spectra that come from the starry fuzz around the central nuclei provide estimates of distances that astronomers use to derive luminosities or total power. At this stage of the game, no one has found extremely large noncosmological redshift effects that are associated with stars. (Some small effects in galaxies have been found, but their significance is disputed.) Thus an astronomer can apply Hubble's Law to the redshift of the starlight and derive a distance for the BL Lacs. These objects produce some of the most serious problems regarding high power and small size. But here again, new laws of physics — the hallmark of a scientific revolution — are not involved.

The redshift controversy and scientific revolutions

The challenge to the accepted idea that the redshifts of all extragalactic objects are cosmological and the astronomical community's response to that challenge fit the pattern of scientific revolutions described by Kuhn and paraphrased in Chapter 1. That is, a scientific paradigm, or accepted viewpoint, is always beset by anomalies in which the model based on the paradigm does not quite match the observations. When normal science fails to bring model and observation closer together by refining the model or reexamining the observations, a crisis results, in which people question the underpinnings of the model. If the model and the observational results turn out to be irreconcilable, scientists who are not too attached to the original paradigm (generally a somewhat irreverent group) search for new scientific laws, which—if models based on the new paradigm can explain the anomaly and conform with existing observations—may form the basis for a new paradigm.

Normal science

The pattern of normal science Kuhn described corresponds well with the way that the challenge was posed and with the way that science responded. Once the paradigm of cosmological redshifts was established, most astronomers expected that the redshifts of all quasars would be cosmological, in accordance with the paradigm, or pattern. The most prestigious journals rejected the first articles containing evidence for noncosmological, discordant redshifts because astronomers demanded a higher standard of proof for such paradigm-shattering ideas. Yet the anomaly persisted.

Crisis

There was a point, in the mid-1970s, at which the anomaly of noncosmological redshifts reached the crisis stage. A few papers appeared in which theorists began to search at random for new physical laws to explain the redshifts. Supporters of the cosmological view became genuinely worried, and spent a great deal of time rebutting the case for noncosmological redshifts. When I was writing the first edition of this book (in 1974), it seemed as though we could be on the brink of a scientific revolution.

But now the crisis has passed, as far as quasar research is concerned. Stockton's work and the confirmation of the nature of the BL Lacs shows that there are some objects with tremendously powerful sources of energy in galactic nuclei. Anomalies persist: Despite the apparent completeness of the models of quasars and active galaxies, described in Chapters 7–11, questions remain. Where does the energy come from? How is it transferred to the radio-emitting clouds? How can we explain the rapid variations of brightness and the high luminosities? Where are the absorption lines formed?

A personal perspective

Stockton's work on associations between bright quasars and galaxies of similar redshifts—along with the confirmation of the nature of the BL Lac objects—appeared so recently that the supporters of noncosmological redshift have not had a chance to react thoroughly. Elsewhere in this book I present the views of the astronomical community on these frontier areas of research. Where there is controversy, I present both sides of the question. But here, astronomers favoring noncosmological redshifts have been struck with twin hammer blows and have not had a chance to respond in detail. I can't speak for them, but I can provide my own views of the likely impact of recent work. I have hesitated to evaluate the arguments about the cases discussed in this chapter so that I could leave my own views until the end. Here they are.

First: I find it very, very hard to believe that no quasars and BL Lac objects are at the distances their redshifts indicate. Certainly those 8 quasars for which Stockton found associated galaxies with similar redshifts are at such distances. Stockton also found, near 3C 273—the most famous of all quasars—a galaxy which has a redshift similar to that of 3C 273. Geoffrey Burbidge, an astronomer who is well known for his ability to withstand the lure of bandwagons, agrees that the BL Lac objects are also as far away as their redshifts indicate.

As a result, I suspect that quasar researchers from now on will spend less time on the redshift controversy. Since some objects like those described in Chapters 7–11 exist, an astronomer can spend time investigating these things with confidence that he or she is producing models of real objects. Someone seeking to explain x-ray emission, for example, can model the x rays and gamma rays from 3C 273 (recall Figure 9-1), secure in the knowledge that this quasar is in fact 820 megaparsecs away.

But does this mean that all quasars are at the distances that their redshifts indicate, that associations of the type illustrated in Figures 12-2 and 12-4 are just coincidental? One possibility is that there are two types of starlike, high-redshift objects in the universe. One type is represented by the quasars, BL Lacs, and active galactic nuclei whose redshifts are clearly cosmological. Another type could include Markarian 205, the high-redshift object associated with NGC 1199, and other high-redshift objects associated with low-redshift galaxies. New physical laws would be needed to explain the redshifts of this second kind of object. This idea has been around for years and no one has really liked it. The concept of two apparently similar objects being different in a very basic way—with one needing new physics and the other not—violates the universal scientific prejudice that nature is fundamentally simple. There is no obvious difference between the properties of these two classes of objects, though some finer distinctions have been claimed. I suspect that supporters of noncosmological redshifts will have to take the view that there are two types of starlike object: one with a cosmological redshift and one with a noncosmological redshift. Stockton's work is very hard to contradict.

But the photographs—pictures like Figures 12-2 and 12-4—are still there. Either there are two classes of high-redshift, starlike objects or these pictures are just coincidences. I'm impressed by the photographs and certainly leave my mind open to the possibility of noncosmological redshifts in some objects. More research is needed to show whether the apparent connections shown in Figures 12-2 and 12-4 are real. I have a suspicion, really no more than an intuitive feeling, that they will be shown to be illusions— that all quasars and high-redshift starlike objects are as far away as their redshifts say they are. But I could be wrong, and I hope that further analysis of Figures 12-2, 12-4, and similar pictures is not ignored just because most astronomers believe that quasar redshifts are cosmological.

I have tried to present the case for noncosmological redshifts fairly. For more on the controversy, look at the recent reviews by Halton C. Arp, listed in the Suggestions for Further Reading section; they are eloquent and readable expressions of the case for noncosmological redshifts. These reviews are written at the same general level as this book.

The redshift controversy is—to me—a vindication of Kuhn's conception of the way that science progresses. The centerpiece of Kuhn's thesis is the paradigm—an accepted view of the world. In his view, most of science consists of exploring the consequences and implications of the existing paradigm before venturing forth to seek a new one. In agreement with this model of the scientific process, most work by astronomers has involved deciding whether the anomaly of noncosmological redshifts is real, not exploring new laws of physics. Kuhn's model of a scientific revolution imparts a certain timeliness to different phases of the scientific process. This timeliness has been apparent in the kinds of work that people have done in connection with the redshift controversy.

Since the controversy is now somewhat divorced from the analysis of quasars, BL Lacs, and active galaxies (or at least that subclass of those objects in which the redshift is known to be cosmological), you may be wondering why I spent a chapter—even a short chapter—on this topic. Well, this book focuses on the cutting edge of research. Exploration of one particular controversy in some depth illustrates just how controversy proceeds in the scientific world. Scientists do not just ignore new ideas, but they don't necessarily embrace them wholeheartedly either. It's easy to believe that all new ideas lead to new insights, for it is the successful new ideas that find their way into the textbooks. The Galileos, Darwins, Newtons, and Einsteins are remembered, and the numerous false starts of the less-well-known scientists are forgotten. Most new ideas and new experimental discoveries have failed to lead to new insights. When you look at the history of science, you find that many new ideas ended up in blind alleys. As a result, scientists follow an anomaly in the current paradigm to the crisis stage before they seek a new paradigm. Further, the evidence supporting the anomaly must be stronger than the evidence supporting the existing paradigm, since the existing paradigm is supported by a large number of observations that do not specifically bear on the anomaly itself.

Astronomers have critically examined the fundamental assumption that the redshifts of quasars are valid indicators of their distances. Thanks to the discovery of galaxies and quasars in clusters in which the objects have similar redshifts, this assumption seems reasonably safe as far as it affects at least some quasars. Analysis of the nature of the BL Lac objects further shows that violent explosions, with a power in the 10^{46} erg/sec range, occur in the nuclei of some galaxies.

But photographs of quasarlike objects that are apparently connected to galaxies of much lower redshifts still present puzzles. Either these photographs are coincidences, or new laws of physics are needed to explain the high redshifts of the radiation from the quasarlike objects. The interpretation of these photographs, along with the whole question of the existence of noncosmological redshifts, is now somewhat divorced from the rest of quasar research. But even if the redshifts of some quasars are accepted as cosmological, these pictures remain as a scientific anomaly.

SUMMARY OF
PART 2

The completion of the 200-inch Hale reflector at the end of World War II marked the beginning of the intensive study of extragalactic objects that has paid great dividends for twentieth-century astronomy. Quasars, the most luminous objects in the universe, were discovered in 1963. Since then, we have realized that there are numerous objects similar to the quasars: the active galaxies. Many mysteries remain in our work on these objects.

The following display summarizes the essential results of this section, again differentiating among fact, concrete theory, informed opinion, and speculation. Here the observers are in the forefront of research, so that most theory consists of models that are very closely tied to the real world, in marked contrast to the black-hole business where we are not even sure black holes exist.

But some questions do remain. To a certain extent, these are questions of a more detailed nature, but some of them may be rather fundamental. The new generation of satellite observatories, to be launched in the 1980s, will provide a deeper view and perhaps some answers to these questions. And it is possible that some new physical laws will emerge from further research.

OBSERVATIONAL FACT	Quasars and active galaxies exist, with emission lines, radio-frequency radiation, infrared excesses, optical variations, and x rays
	Very-Long-Baseline observations of the radio structure of these objects show both compact and extended sources which are aligned with each other
	There is activity at the center of the Milky Way galaxy
CONCRETE THEORY	Radio-frequency radiation from quasars and active galaxies is synchrotron radiation
	The emission-line spectra of these objects come from clouds of gas subject to ultraviolet, ionizing radiation
WORKING MODEL	Variations in the intensity of quasar continuum radiation come from the production and expansion of clouds of high-speed electrons
	The infrared continuum comes from dust in some cases and synchrotron radiation in others
	The faster-than-light expansion of radio galaxies is an illusion
	BL Lac objects are quasars without emission lines
	Some quasars, BL Lac objects, and active galaxies are at the distances indicated by their redshifts; thus they have high luminosities
	Galactic activity—from the core of the Milky Way galaxy through the active galaxies, the BL Lac objécts, and the quasars—is basically the same phenomenon

OPEN	Are continuum x rays due to synchrotron radiation or to collisions
QUESTIONS	between high-speed electrons and photons?

Are high-speed electrons in radio sources produced by a beam of fast particles, by the slingshot ejection of a massive object from the core of a galaxy, or do they come from an expanding, turbulent cloud?

Does the energy in quasars come from collisions of stars, from the rotation of a massive object, or from a black hole?

Are all quasar redshifts cosmological?

3   THE
UNIVERSE

robably all astronomers—from the first prehistoric man or woman to follow the days, nights, and phases of the moon by making notches in a bone fragment to the twentieth-century astronomer using the latest gadgetry on the 200-inch telescope—became astronomers in order to determine our place in the universe. Yet to understand the universe, we have to understand galaxies, and to understand galaxies, we have to understand stars, somewhat at least. Furthermore, the evolution of the universe is governed by gravity, so an under-standing of the way gravity works is prerequisite to accepting the challenge of cosmology, the study of the evolution of the universe.

Until the twentieth century, cosmology was largely the province of philosophers. People continued to learn more about the nearby planets and stars, but really knew nothing about the vast expanses of the universe. Observational cos-mology is a product of the last 70 years. We have finally developed the tools to pick up some observational facts bearing on the evolution of the entire universe. Cosmology is no longer limited to the model world. Real data can be brought into the picture and various models can be confirmed or disproved.

Here, as in Part 1 (Black Holes), the story begins with the model world. Chapter 13 describes the cosmological model that almost all astronomers believe in these days—the Big Bang theory. Subsequent chapters show why this theory enjoys such widespread support, as we compare the model world with the real world. Chapter 14 examines the age of the universe, the time interval that has passed since the Big

Bang. In Chapter 15, we examine the observational facts that confirm the Big Bang picture, and discuss the evidence bearing on the Big Bang theory's one-time rival, the Steady State theory. Chapter 16 looks to the future: Will the expanding universe expand forever, or will it slow down, stop, and start contracting? At this time, we do not know what it will do. The evidence at hand seems to indicate that it will expand forever. But there are ambiguities in the interpretation of observations.

13 LIFE CYCLE OF THE UNIVERSE: A MODEL

To study the life cycle of the universe, we begin with theory, or a model history of the cosmos. The universe is expanding now and has been expanding for a long time. Any hypothetical history of the universe must account for this expansion, and also for the observed fact that the universe looks generally the same in every direction. If we follow the expansion backward in time, we see that the universe must have been smaller in the past. Following it back still further, we find that there was a time when all the matter in the universe was packed tightly together. The explosion of this cosmic egg provided impetus for the expansion that we now observe. This theory is the Big Bang theory, which is currently accepted by all but a handful of astronomers. (Chapter 15 contains a brief discussion of rival theories, including the Steady State theory, and a discussion of the evidence against these theories.) Between the Big Bang explosion, some 10 to 20 billion years ago, and the present, some interesting events have occurred. These occurrences have had observable consequences that enable us to verify the model.

This chapter outlines our model of cosmic evolution. Chapters 14, 15, and 16 describe the observational foundations of our tantalizingly complete picture of the life cycle of the universe.

The evolving universe

The discovery, in the 1920s, that the universe was expanding provided the foundation for modern cosmology. Follow the expansion backward, and we come to the Big Bang, the explosion that marked the beginning of the expansion. Follow it forward, and maybe the expansion slows to a stop. Or maybe it goes on forever.

Historical background

The idea of an expanding, or evolving, universe is new in Western thought. Most European philosophers have generally believed that the universe is static, unchanging. "One generation goeth, and another generation cometh, but the earth abideth forever. . . . That which has been is that which shall be, and that which has been done is that which shall be done;

and there is no new thing under the sun" (Eccles. 1:4, 9). Others through history have expressed this same thought, but none so eloquently. The discoveries of the 1920s overturned this idea. They showed that the universe does indeed evolve as it expands. The Steady State theory, now in conflict with the evidence, was an attempt to create an expanding but unchanging universe.

The evolution of the expanding-universe theory illustrates the interplay between the model world of the theorists and the real world of the observers. In 1917, two years after the presentation of the General Theory of Relativity, Albert Einstein and the Dutch physicist Willem de Sitter independently applied the new theory to the universe as a whole. Their research showed that a universe that obeyed Einstein's theory of gravitation must either expand or contract; it could not remain static. Because of the Western predilection for an unchanging universe, this idea was quite alien. Einstein even tried to modify his theory so that the universe could remain static, making what he later called "the greatest blunder of my life."[1] The Russian mathematician Alexander Friedmann ignored Einstein's modification and, as a mathematical exercise, calculated some model universes. These models have endured; at the current level of sophistication, they are the best models of the universe that we have.

The observers were not so far behind. In the early 1920s, American astronomers in California were extending the frontiers of observational astronomy beyond the Milky Way. It became clear that the wispy spiral nebulae were galaxies, each one containing billions of stars. Looking at one of these galaxies can give you some perspective on the vastness of the universe, as you realize that this fuzzy little patch is a galaxy equal to our Milky Way. If you can, go somewhere out in the countryside on a dark autumn night and try to find the Andromeda galaxy, the nearest large spiral. (Directions for finding it are given in Figure 13-1.) You do need to be some distance away from the glare of city lights to see it. As you look at this little fuzzy patch of light, think how many stars you are seeing: 100,000,000,000 of them, each one somewhat like the sun. Binoculars will enable you to see the galaxy a little better.

The evolution of the real universe was revealed later in the 1920s, when Edwin P. Hubble of the Mount Wilson Observatory extended earlier measurements, by Vesto M. Slipher, of the spectra of distant galaxies and found that all but a few of the nearest galaxies were speeding away from the earth. Hubble and his colleague Milton Humason discovered the law that bears Hubble's name: Velocity of recession equals Hubble's constant times the distance of the galaxy. If Einstein had not modified his original theory in 1917, he would have made one of the most stupendous predictions in the history of science: the prediction of the expanding universe. As it was, the expanding universe was discovered in due course, so it is unclear whether the history of science would have been substantially affected if Einstein had not erred.

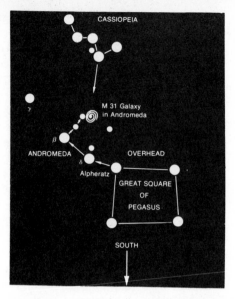

FIGURE 13-1   How to find the Andromeda galaxy. At about 9 p.m. in late October, look high in the southern sky, almost overhead, and find the Great Square of Pegasus, marked by four moderately bright stars and about 15 degrees on a side. Messier 31, the Andromeda galaxy, is northeast of the Square, toward the W- (or M-) shaped constellation of Cassiopeia. From the northeast corner of the Square, find δ (delta) and β (beta) Andromedae; let your eyes make a right turn at β and go toward Cassiopeia. You will find a fuzzy star, the Andromeda galaxy, which you can see if the sky is dark.

The nature of the expanding universe

Two questions about the nature of the expanding universe immediately arise. How do we explain Hubble's law, that the expansion rate *increases* as we inspect areas farther and farther from the earth? Since the expansion is uniform in all directions, aren't we at the center of the expansion? These two questions are naturally answered if we realize that it is the entire universe that is expanding. You have to think about it for a little while.

Writers on cosmology have concocted different analogies to make the expansion of the universe seem more concrete; I shall mention two of them and hope that you will find that at least one helps you understand how each galaxy, in an expanding universe, sees all the others recede, following Hubble's Law. In Figure 13-2, the universe is seen to be analogous to a raisin cake, with the raisins representing the galaxies. As the universe expands, or as the cake rises in the pan, the galaxies (or raisins) get farther apart. The analogy that seemed most appropriate to me, when I first saw it, is George Gamow's jungle-gym analogy (Figure 13-3). Gamow likened the universe to a gigantic jungle gym made of telescoping pipe.

When a raisin cake rises in a pan, each raisin moves away
from its neighbors as the dough expands . . .

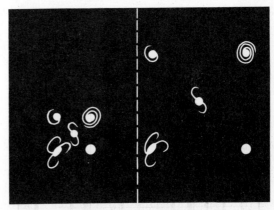

FIGURE 13-2
The expanding universe,
seen analogous to a raisin
cake.

. . . just as, in the expanding universe, every galaxy moves
away from its neighbor galaxies.

The galaxies are represented by children (or in the figure, Escher's cubes) sitting at intersections of the jungle gym. As the pipe expands, the whole framework expands, moving each cube away from all the others, since there is more pipe between it and each of its neighbors as time goes on. Every cube seems to be in the center of the expansion, but there really is no center because the entire framework is expanding. The idea that it is the framework that is expanding is the key to the concept of the expanding universe.

A little more thought with one of these analogies will produce the Hubble expansion law: that the rate of recession increases with distance (see Figure 13-4). Here we follow the expansion over the time that it takes the universe to double in size. The farther apart the two galaxies at the beginning of the picture, the more space between them to expand and the more the change in separation. Since the velocity is just a measure of the rate at which the two galaxies separate, the velocity is higher for the galaxies that were more distant from each other at the beginning. Figure 13-4 contains some numerical examples.

But where did it all begin? A Belgian cleric, the Abbé Georges Lemaitre, made a suggestion in the 1920s that has subsequently been accepted by most astronomers, as it has been shown to agree with observations made in the 1960s. He proposed that in the beginning all the galaxies in the universe were concentrated in a single lump, which he called the primeval atom. This primeval atom then exploded, flinging the galaxies off into space. Although the term *primeval atom* is no longer used very often, the basic idea is still the same. In the beginning the Big Bang sent all elements in the universe moving away from one another, and this motion is still visible at the expansion of the universe.

Yet until the observations of the 1960s, this picture was not entirely satisfactory. It was just a model, and there were other ways to explain the expanding universe. People could get together and argue about which model of the universe they preferred, but all they could talk about was the esthetics of each model—which one was more philosophically pleasing. It was the introduction of observational data bearing on the early stages of the evolution of the universe that made cosmology a science. Two Big Bang remnants were discovered: the processing of one-quarter of the universe into helium, and radiation left over from this primeval explosion.

But how can we be sure that these two observations do indeed tell us about the Big Bang? Recall that the essence of the scientific process is model-building and fitting models to observations of the real world (recall Chapter 1). In cosmology, the models came first, so I turn to a detailed description of the Big Bang model, the one almost all astronomers believe in. The basic elements of this model were developed by George Gamow and Chushiro Hayashi, along with coworkers, in the 1940s and 1950s.

The beginning: the Big Bang

The story starts with a homogeneous glob of matter containing all the substance of the universe. This glob was hot and dense, with a temperature of 10^{12} to 10^{13} K and a density of 10^{14} g/cm^3. This matter was in the form of a number of particles with exotic names: pions, hyperons, mesons, muons, neutrinos, along with the more familiar protons, neutrons, electrons, and photons. (If you feel a bit at sea with protons, electrons, neutrons, and photons, you might wish to refer to the Preliminary section or to the Glossary.) When the story begins, in frame A of Figure 13-5, all these particles are in equilibrium, as they are being continually transformed through various reactions. Billions of times a second, electrons collide with

FIGURE 13-3   The analogy between the expanding universe and an expanding three-dimensional framework is well illustrated by M. C. Escher's "Kubische ruimteverdeling" (Cubic Space Division). (Courtesy of the Escher Foundation—Haags Gemeentemuseum—The Hague.)

their antimatter counterparts, positrons, and disappear in a flash of gamma rays. Similarly, gamma rays collide, producing electron-position pairs. Every reaction is balanced by an inverse reaction.

What preceded the hot, dense state of frame A, the microsecond-old universe? We really don't know, and it's not likely that we ever will. Before this point, the expansion might have been smooth and uniform, or chaotic. One of the chaotically expanding models is given the picturesque name of the "Mixmaster universe." It is possible that we may someday discover tiny primeval black holes that were made at this very early stage of cosmic evolution. But until and unless we discover them, we have no observational evidence whatever that pertains to this very early stage. You can speculate

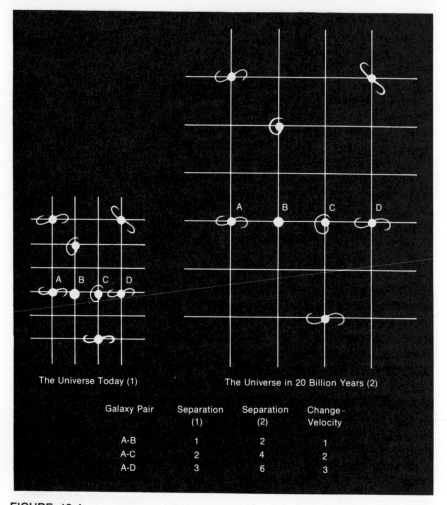

The Universe Today (1)   The Universe in 20 Billion Years (2)

Galaxy Pair	Separation (1)	Separation (2)	Change-Velocity
A-B	1	2	1
A-C	2	4	2
A-D	3	6	3

FIGURE 13-4   As the universe expands, the separation of more distant galaxies increases faster than the separation of nearby galaxies. The expansion velocity follows Hubble's Law: Velocity is proportional to the distance.

that our expanding universe was the result of a previous stage of cosmic evolution, in which a predecessor universe expanded and collapsed. In such a model the universe oscillates—it could oscillate for as many cycles as you might like to think of. But this sort of speculation can neither be proved nor shot down by the key ingredient of scientific cosmology—experimental fact. Therefore I leave this early stage of cosmic evolution alone.

While the universe evolves, it cools and expands. Reactions no longer occur fast enough so that inverse reactions can balance them, as is

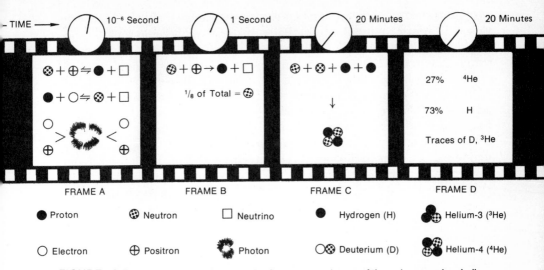

FRAME A   FRAME B   FRAME C   FRAME D

● Proton   ⊗ Neutron   ▢ Neutrino   ● Hydrogen (H)   Helium-3 (³He)

○ Electron   ⊕ Positron   Photon   ○⊗ Deuterium (D)   Helium-4 (⁴He)

FIGURE 13-5   A movie depicting the first twenty minutes of the universe, when helium was formed. See text for details.

shown in frame B. The positrons, being in an antimatter minority, disappear from the scene, since gamma rays cannot collide and make them in appreciable quantities. The density is just too low. Once the positrons go, neutrons become stable because they no longer become protons by combining with positrons. At this one-second mark, with a density of 10^5 g/cm³ and a temperature of 10^{10} K, the number of neutrons was frozen at one-seventh of the number of protons.

Helium production

Neutrons are important because they are responsible for the transformation of roughly one-quarter of the universe into helium. Between the one-second mark (frame B) and the 20-minute mark (frame D), the series of reactions summarized in frame C make helium. A neutron and a proton collide, stick together, and from deuterium, or heavy hydrogen. Deuterium nuclei almost always pick up another proton and neutron to become helium-4: two protons, two neutrons. A very small fraction of the deuterium does not react further and remains deuterium. The sequence stops there, because a helium nucleus cannot go anywhere by picking up another particle, proton or neutron. It would have a mass of 5, and there is no stable nucleus of mass 5. As George Gamow put it, it cannot leap over the mass 5 crevasse, so a helium nucleus finds itself at the end of the primeval element-production line.

The amount of helium produced in a hot Big Bang does not, fortunately, depend on the details of the cosmological model. Complex theoretical calculations, which carefully follow each reaction taking place in the early universe, confirm the qualitative picture sketched above: For a wide variety of initial conditions, a Big Bang universe will become roughly one-quarter helium in the first 20 minutes of its evolution. The observational confirmation of the fact that the universe is one-quarter helium thus supports the Big Bang model. (I return to the helium question in Chapter 15.)

Once the helium was made, nothing very interesting happened for a long time. The universe cooled and expanded, but no transformations occurred. The next interesting thing occurred much later, when the universe was some hundreds of thousands of years old. For the sake of definiteness, I refer to the age of the universe at this next milestone as one million years, although the exact figure depends on the precise figures you choose for the age of the universe and the details of the cosmological model. The one-million-year figure results from assuming a Hubble constant of 50 km/sec per Mpc (Chapter 14) and an open, ever-expanding universe.

Radiation

While the universe was between one second and one million years old, the photons kept colliding with the matter in the universe. The principal result of these collisons was that the photons remained in balance with the matter, even though nothing changed as a result of these collisions. At the million-year mark, however, the temperature of the universe had dropped to 3000 K. At this point, the universe was cool enough so that the electrons recombined with the protons in the universe, forming hydrogen atoms. (Recall from Chapter 9 the discussion of recombination.) There were no longer any electrons with which the photons could collide, so these photons simply traveled unimpeded through the universe.

The million-year point (Figure 13-6) is an important milestone because it provides another relation between cosmological models and the real world of observations. We can observe photons that last interacted with matter when they scattered off electrons just before recombination. Now redshifted to become radio photons, they have been detected by radio astronomers. If radio astronomers have really detected Big Bang relics and not something else, then when we look at this radiation we are looking back to the million-year-old universe.

Origins: galaxies, stars, planets, and life

At the million-year mark, then, the universe contains hydrogen and helium gas at a temperature of 3000 K, expanding uniformly. Twenty billion years later, it is a far more interesting and complex place. It contains galaxies like the Milky Way, stars like the sun, planets like Earth, and people

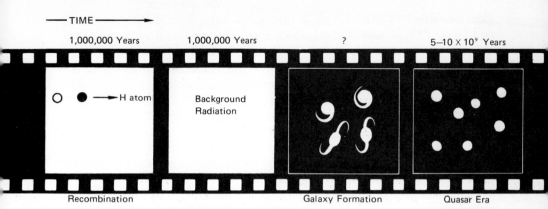

FIGURE 13-6 A movie depicting cosmic evolution from the million-year mark through the first five or ten billion years. A continuation of Figure 13-5; to be continued on Figure 13-8, 13-9, and 13-10.

like you and me. There are other fascinating objects around too: a black hole in Cygnus X-1 (probably), supernova remnants like the Crab Nebula, giant radio galaxies, quasars, pulsars, and so on. But all these things are related to the life cycle of the four basic constituents of the universe: galaxies, stars, planets, and living organisms.

How did the universe evolve from a smooth mixture of hydrogen and helium gas to this fascinating collection of complex structures? Parts of this question fall within the domain of cosmology, but a full answer to it covers the entire field of astronomy, along with a fair piece of biology and geology as well. Thanks to progress in all these fields, a rather important conclusion has been reached: We can understand how these structures came to be as the result of natural processes. We are a long way from unraveling the details of this complex web, and the material in this section cannot be supported by observations as completely as the model of the first million years can be. Many of the steps in the evolutionary chain leading from hydrogen and helium at the million-year mark to now have not been elucidated. But we think we can visualize the general picture.

Galaxies

The next cosmically significant event after the million-year mark occurred a few billion years afterward when stars and galaxies formed. These condensed objects emerged, somehow, from the primeval mist of expanding hydrogen and helium gas. Stars are concentrated in galaxies. Galaxies are scattered through space in groups that vary from twos and threes to enormous clusters of galaxies. P. James E. Peebles and his co-workers at Princeton have spent years mapping and analyzing the distribution of galaxies. The Princeton group works from a number of surveys of galaxies made by California astronomers, including Fritz Zwicky at Caltech and Donald Shane and Carl Wirtanen at the Lick Observatory. Figure 13-7 shows a picture of the universe derived from the Lick Observatory data. The smoothness of the early stages of cosmic evolution is preserved only over very long scales. Otherwise galaxies in the universe break up into clusters of ever-increasing size—a lacy network of galaxies. The largest clusters seen are 20 megaparsecs across.

When and how did galaxies form? Take these questions one at a time. When? That one we can answer, at least approximately. Look outward into space and you are looking backward in time, since light traveling toward earth from a faraway object takes a while to get here. The quasar OQ 172 has stood for five years as the holder of the record for the highest redshift of any quasar. Light that we see now left OQ 172 five billion years after the Big Bang, and took fifteen billion years to reach us. The most distant bona fide galaxy is the radio galaxy 3C 343.1. Light was emitted from this object 11½ billion years after the Big Bang. Thus we can place the formation of galaxies sometime between the formation of the microwave background at the million-year mark and the appearance of quasars like OQ 172 five billion years later.

But now to the second question. How did galaxies form? We really don't know. Did swirling gas streams coalesce, become slightly denser, and then coalesce with other gas streams, leading to the formation of eddies that became protogalaxies and subsequently galaxies? Or did stars form first, coming together through gravitational interactions to form star clusters and then larger galaxies in the vast wilderness of intergalactic space? Were primeval black holes the seeds around which star clusters or galaxies grew? These questions are some of the most important ones in theoretical cosmological research. Since galaxies exist, it is embarrassing that we can't make galaxies in a hot, Big Bang cosmology.

Stars and planets

We can't say exactly how galaxies formed, but we can visualize a couple of ways in which they might form by turbulence or by gravitational interaction. Our understanding of the formation of stars is slightly better, because it is going on right now in our own Milky Way galaxy. The Milky Way contains a fair quantity of gas, located in between the stars. This gas is often associated with the dust that produces the picturesque dust lanes shown in Figure 7-1. We can see some of the stages in star formation, and so we have some confidence that our general model is correct. I focus on the formation of one particular star, our sun, about five billion years ago, fifteen billion years after the Big-Bang.

Five billion years ago, one of the outer arms of the Milky Way galaxy contained a smallish cloud of gas. Something caused this cloud of gas to

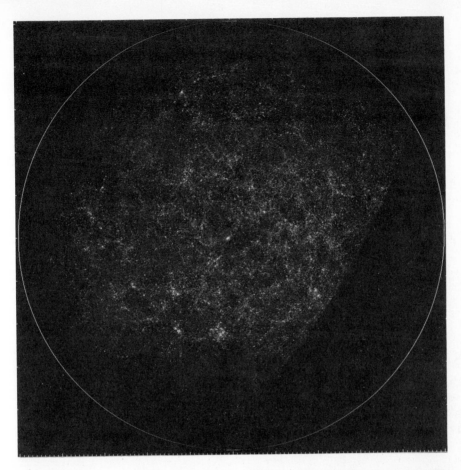

FIGURE 13-7   A map of the universe, including a million galaxies brighter than 19th magnitude that were recorded by C. Donald Shane and Carl A. Wirtanen at the Lick Observatory. Individual galaxies are not shown; the dots represent different gray shadings showing the number of galaxies in small parts of the universe. The dark part represents the southern hemisphere sky that is invisible from the Lick Observatory. Fewer galaxies are visible toward the edge of this map because absorption of light in our own galaxy obscures them. The general, statistical properties of the distribution of galaxies are determined from maps like these, the best large-scale views of the universe that we have. (Courtesy P. James E. Peebles.)

start becoming smaller. The initial trigger for the formation of the solar system could have been the effect of the gravitational field of a spiral arm of the galaxy, or it could have been the blast wave from a nearby supernova explosion. Recent evidence indicates that the supernova idea may well be the correct one, for we see other newly formed stars at the outer edges of supernova remnants. Once it became small enough, gravitational forces within the cloud pulled its atoms toward the center, continuing the collapse.

Because the cloud was rotating slightly, the collapse eventually formed a disk of gas and dust surrounding the central object, which can be called the *protosun*. The gas cloud cooled, and solid particles condensed from it. These solid particles eventually coalesced to form some rockballs that orbited the sun, and, farther out in the cloud, the nebular gas coalesced to form some larger planets that contained more gas and less rock.

Evolution of life

Now focus on one of these planets around the sun, the third one out (Figure 13-8). This rockball was just far enough from the sun so that liquid water in the form of oceans could exist on its surface. In those oceans, hydrogen, carbon, oxygen, and nitrogen atoms collided, stuck together, and changed places in a great chemical dance. The energy needed for some of these chemical reactions could have come from many places—from lightning or from ultraviolet radiation from the sun, for example. After many, many chemical reactions, the first living cell was produced.

During the next several billion years, living cells became more and more complicated. The more complicated cells were better suited for survival in this environment. They were able to get to the point of reproducing and making still more cells. A few hundred million years ago, some living organisms left the friendly ocean that spawned them and struggled to exist on land. A few million years ago, these same evolutionary pressures led to the development of intelligent creatures, human beings. A few thousand years ago, a few of these creatures, the first astronomers, looked up at the sky and started to ask questions, started to wonder just how the universe came to be the way it is.

Origin of the chemical elements

But these creatures trying to understand cosmic evolution are not made up of hydrogen and helium, the two elements present in the million-year-old universe. The chemical reactions that make life possible simply cannot occur with such simple atoms. The panorama of cosmic evolution is missing one essential story: the genesis of the atoms that this book is made of. Parts of this story have been mentioned earlier, and so a short summary can suffice to complete the picture.

In the beginning the universe was hydrogen and helium. A few other minor constituents like *deuterium,* a form of hydrogen in which the nucleus contains a neutron in addition to the proton that forms the nucleus of most hydrogen atoms, were present, but they play a small role in the production of the elements that you and I are made of. The first stars formed from this hydrogen and helium and survived by cooking hydrogen and helium in their interiors, fusing these nuclei to make heavier elements.

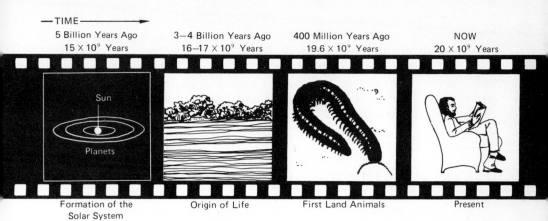

—TIME——————————▶

| 5 Billion Years Ago | 3—4 Billion Years Ago | 400 Million Years Ago | NOW |
| 15×10^9 Years | $16-17 \times 10^9$ Years | 19.6×10^9 Years | 20×10^9 Years |

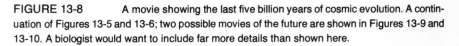

| Formation of the Solar System | Origin of Life | First Land Animals | Present |

FIGURE 13-8 A movie showing the last five billion years of cosmic evolution. A continuation of Figures 13-5 and 13-6; two possible movies of the future are shown in Figures 13-9 and 13-10. A biologist would want to include far more details than shown here.

They then exploded as supernovae, ejecting these heavy elements back into the galaxy, so that stars and planets that formed subsequently contained a larger fraction of heavy elements than the first generation of stars did. Still more stars formed, had fusion reactions in their interiors, and ejected the results into the interstellar medium. Our sun formed five billion years ago, something like ten billion years after the galaxy first formed, and so it is a later-generation star. Thus the carbon atoms that this book is made of were synthesized in the bowels of a massive star something like ten billion years ago.

This skeletal picture looks reasonable, and in its broadest outline it is probably correct. But there are many, many unanswered questions. Confirmation of this picture constitutes an active area of research on galaxies. The last few years of work have made it clear that this simple model leaves many, many questions unanswered. Ten years ago we didn't even know enough to ask these questions that presently baffle us. Ivan King of the University of California, in Berkeley, asked many of them in his opening paper delivered to a conference on the evolution of galaxies held at Yale in 1977.[2] Some of King's questions deal directly with the story outlined here. Where are the first-generation stars, stars made of pure hydrogen and helium? Even very old stars contain some heavy elements. Galaxies interact with the surrounding mass and with other galaxies; how do these interactions affect the evolution of galaxies? Are heavy elements produced at the same rate in all parts of a galaxy?

But all these parts of the story of cosmic evolution are sideshows in the grand cosmological panorama. The field of cosmology is concerned with the bigger picture, the evolution of the entire universe. The smaller fragments of the story of cosmic evolution shown in Figure 13-8 are an important part of scientific knowledge, since they play a key role in our understanding of the way that we, the human race, came into being. But strictly speaking, they are not part of cosmology.

We thus understand the past evolution of the universe reasonably well. We have a good model for the first million years of cosmic evolution, and this model provides a natural explanation for the expanding universe. But what of the future? What eventually happens to the universe?

The future evolution of the universe

The expansion of the universe is being continually slowed, since every galaxy in the universe is affected by the gravity of every other galaxy. Gravitational attraction is attempting to pull these galaxies together, opposing the universal expansion and causing the expansion rate to decrease. However, the expanding universe is causing the galaxies to move farther and farther apart, decreasing the gravitational attraction between them. Two alternatives seem ultimately possible. Either gravity is strong enough so that the universal expansion will slow down, stop, and become a contraction; or the expansion is rapid enough so that the rapidly increasing distances between the galaxies will overcome gravity, and the universe will go on expanding forever. The universe that stops expanding is called a *closed* universe, and the ever-expanding one is an *open* universe. There is a borderline case, in which gravity is just strong enough so that the distance between any two galaxies increases toward a limiting value, as the expansion rate is slowed toward zero by gravity, but gravity is not strong enough to bring the universe back together again. But if this borderline case is to represent the real universe, the universe must contain exactly the right amount of mass per cubic megaparsec, an unlikely state of affairs.

What will future universes look like? As galaxies evolve, they grow old. Stars within the galaxy are evolving, becoming red giants, and dying. Some of the mass present in stars is recycled, as gas is ejected from stars in the form of planetary nebulae or supernova remnants (recall Chapters 2 and 3). Some of this ejected gas forms new stars. But the recycling is not perfect. A star forms from the interstellar gas. When it dies, it returns some of its mass to the interstellar gas, but some of its mass is interred in the stellar graveyard as a white dwarf, neutron star, or black hole. Stars in a galaxy evolve as the galaxy ages; bright blue stars disappear, returning some but not all of their mass to the interstellar medium. As the eons pass, more and more of the interstellar gas is processed through stellar evolution, finally becoming stellar corpses. There is progressively less gas to make young new stars. All that is left are the low-mass, low-luminosity stars, which burn their nuclear fuel very slowly. Eventually these stars die, too. In the end, a galaxy consists of white-dwarf stars, neutron stars, black holes, and smaller masses like planets. The white dwarfs and neutron stars cool and cease to radiate very much visible light. In the long run, any galaxy will consist of nothing but stellar corpses.

This rather depressing picture of the evolution of galaxies is brightened a little when we consider what happens in a closed universe, depicted schematically (and perhaps somewhat fancifully) in Figure 13-9. Individual galaxies still evolve toward darkness, but the evolution of the universe itself provides some hope for rejuvenation. The expansion will slow down, stop, and reverse itself, turning into a contraction. The galaxies will start approaching each other again, will coalesce and will approach a time when the universe looks as it did at the beginning, a dense glob.

What happens next? Our view is limited. We cannot guess. Theory says that the universe will form a singular state, just like the star that made a black hole. But clearly, the theory is inadequate to describe what happens. We scientists can speculate about what will happen, but when we speculate in this way we are leaving the territory of science and are trespassing on the territory of the philosopher and the theologian.

The future of a closed universe is a mirror image of the past. As the closed universe coalesces, it becomes much hotter. The sequence described earlier will reverse itself; instead of recombination at the million-year mark, there will be ionization. Helium atoms will be disrupted because the universe will become too hot to allow them to exist as nuclei; neutrons and protons make their appearance. The quasi-equilibrium state of frame A of Figure 13-5 follows shortly, and all kinds of subnuclear particles appear. But then what happens? No one really knows. You can speculate that the collapsing universe will bounce and that the cycle of cosmic evolution will start all over again. But such speculation goes beyond the known laws of physics and reaches into the untestable domain of metaphysics.

However, the evidence shows that the universe may well go on expanding forever. In this view, cosmic evolution follows an inexorable path toward a dark, dreary, and cold final state in the very distant future. Galaxies will separate more and more, and the stars in them will fade away. The cosmic clock has been wound up once only, and it keeps on ticking as the spring winds down. Like everything in it, the universe will follow its one-way life cycle: born in a blaze of glory in the Big Bang, living while the galaxies form and evolve, producing stars, planets, and living organisms in them, and dying when all the stars die and the galaxies fly farther and farther apart.

It is amusing to stray from the hard-nosed, testable cosmological models of what happened in the past and consider what might eventually happen in an open universe (Figure 13-10). The stars in a typical galaxy will have finished their evolutionary cycles in about 100 billion years. At this point a galaxy will contain stellar corpses and planets. These objects slowly, slowly lose energy by gravitational radiation. Galaxies in clusters also radiate gravitationally when they orbit each other. This loss of energy

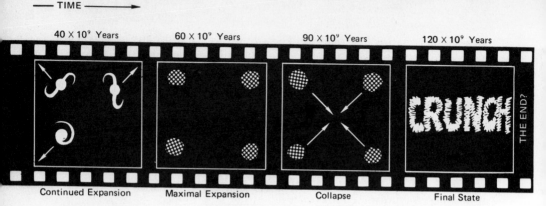

40 × 10⁹ Years   60 × 10⁹ Years   90 × 10⁹ Years   120 × 10⁹ Years

Continued Expansion   Maximal Expansion   Collapse   Final State

FIGURE 13-9 Future evolution of a closed universe. After 60 billion years have passed, the galaxies are shown as faded blobs because stars in them will be much dimmer. A hundred billion years from now, the universe will collapse catastrophically and the end of the present cycle will occur. What happens next is anybody's guess. (The time of maximum expansion depends on how much mass there is in the universe. The figure given is for a universe with twice as much mass as is needed to cause the expansion to come to a stop.)

eventually causes many of the small particles to spiral into the center of a galaxy. After 10^{29} to 10^{34} years, the universe will contain galactic and supergalactic black holes. When the galaxies collapse, some smaller bodies are tossed out into intergalactic space by stellar collisions, and so some white-dwarf stars, neutron stars, black holes, and planets will be around too.

One can stray still further from the hard-nosed cosmological models and look deep, deep into the future. After 10^{100} years, the galactic black holes will evaporate. F. J. Dyson, of the Institute for Advanced Study at Princeton, has pointed out that eventually the solid bodies too will evaporate.[3] After a long, long time, $(10^{10})^{77}$ years, quantum-mechanical processes will allow white-dwarf stars to tunnel through the barrier of degeneracy pressure and collapse to form black holes, which will evaporate after another 10^{66} years. (If you want to amuse yourself with names for these long time intervals, refer back to the Preliminary section.) Thus eventually the universe will contain nothing but electromagnetic radiation in the form of very, very low-energy photons. These photons will be redshifted to lower, lower, and still lower energies while the universe continues to expand indefinitely.

Thus we have two cosmological models for the future evolution of the universe. In the closed-universe model, the expansion stops, reverses itself, and ends in a final cosmic collapse. If you buy the theory that the universe is open, the expansion goes on forever. Although the story of just what happens in each of these models is a purely theoretical one, observations may enable us to decide which of these futures correctly represents what will happen to the real universe. But before we proceed to discussing observational verification of our models for the past (Chapters 14 and 15)

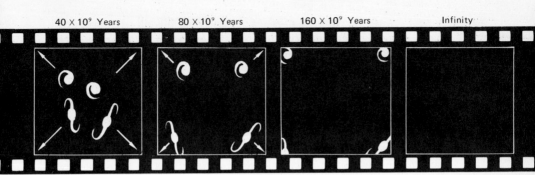

Continued Expansion

FIGURE 13-10   Future evolution of an open, ever-expanding universe. The expansion continues forever and galaxies fade away. (The detailed representations of the universe in Figures 13-8, 13-9, and 13-10 are not meant to be taken too seriously.)

and future (Chapter 16) evolution of the universe, there is one more subtle feature of the theoretical cosmological model that is worth exploring. The geometry of the universe depends on whether it is closed or open. But we have not been able to use the geometry of the universe as an observational cosmological test, and so the rest of this chapter is not background material for anything else that comes later.

The edge of the universe

"What is the universe expanding into? Where did the Big Bang explode? Does the term Edge of the Universe have any meaning?" Students and readers often ask me such questions. In this last section of the chapter, let us briefly deal with them. Unfortunately these apparently innocent questions cannot be neatly and completely answered in a short space. Complete answers to them would require another chapter, a chapter written at a pretty abstract level. Readers interested in deeper answers to these questions might enjoy puzzling over a recent *Scientific American* article on the geometry of space-time in evolving universes.[4] Take the questions one at a time.

What is the universe expanding into? This question presumes that there is a finite, limited amount of matter in the universe, and that it exists in an infinite sea of space. Such a mental conception is perfectly natural, but it happens to misrepresent the true nature of space on the cosmic scale. Concepts of space defined by our intuitive ideas of what space is like on a human scale just don't make sense when applied to very large or very small scales. On the cosmic scale, space is defined by the matter in the universe. The expanding universe grows larger because space itself is expanding.

Space exists because it contains mass. The universe fills all of space, and when space grows larger, galaxies in the universe run away from each other. Remember the jungle gym made out of telescoping pipe (Figure 13-3). It's like that.

A somewhat related question is, "Where did the Big Bang explode?" The answer is: Everywhere. At that time, the universe was far smaller than it is now. The giant elliptical galaxy M 87, now 20 megaparsecs away, was much, much closer to us. Its particles were only 6 kiloparsecs away at the time of recombination and 0.02 parsec away during helium synthesis. Every particle in the universe participated in the explosion known as the Big Bang. Since the universe includes the location of all particles in it, the Big Bang occurred everywhere.

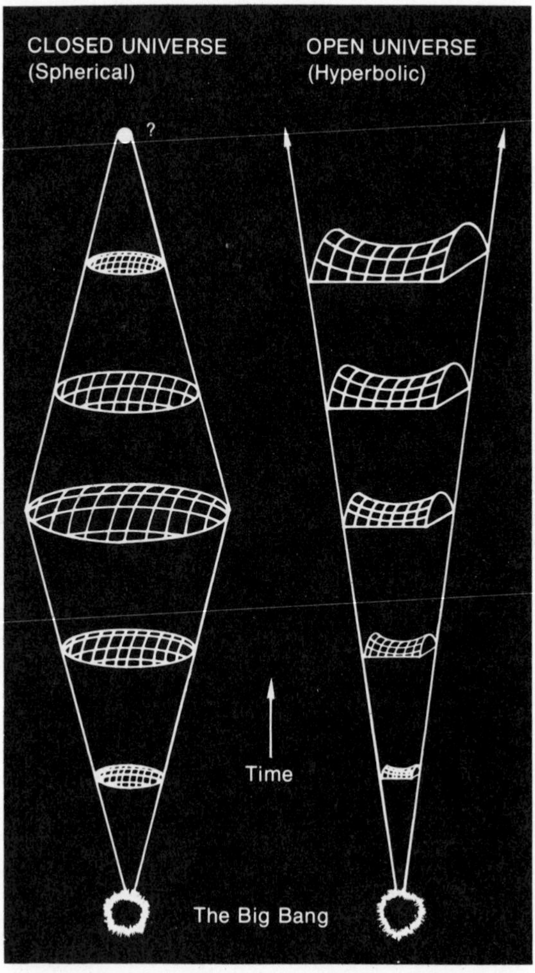

FIGURE 13-11

Open and closed universes. As the universe expands, it becomes flatter. A closed or spherical universe will stop expanding at the point of maximum size and then recontract. An open universe will go on expanding forever. The drawing shows two-dimensional analogues of three-dimensional surfaces; three-dimensional surfaces would have to be drawn on four-dimensional paper.

The question of edges is a bit more complex, since an answer to it requires some consideration of the shape of the universe, or cosmic geometry. The shape of the universe is hard to comprehend; one tends to look at shapes from the outside, and we are living *within* the universe. Furthermore, you cannot visualize the shape of a three-dimensional space, the universe, unless you can think in four dimensions. I cannot think in four dimensions, much less draw in four dimensions, so you will have to be content with two-dimensional analogues of the three-dimensional universe, drawn in Figure 13-11. When you look at these things, think like a Flatlander or an ant-crawling on the surface of these universes.

The closed universe is a hyperspherical universe. A hypersphere is a three-dimensional analogue of the two-dimensional surface of a sphere. We live in galaxies on the surface of this sphere and are constrained to move along its surface. The open universe is an even stranger one; it is shaped like a saddle or hyperboloid. Although different shapes of the universe can, in principle, result in observable effects, the universe is so large that it is impossible to detect these effects with current techniques.

Neither the open universe nor the closed universe has an edge. Consider the closed, spherical universe: The surface of a sphere has no edge. The open universe is infinite and has no edge either. When people refer to the "edge" of the universe, they are referring to the edge of the *observable* universe. If you take the figure of 20 billion years for the age of the universe, anything more distant than 20 billion light-years could not be seen by us, for it would take light longer than the 20-billion-year age of the universe to get from there to here. The age of the universe limits the extent of the *observable* universe, but does not limit the extent of the universe itself, since there are galaxies beyond the observable universe that we just cannot see because they are too far away. The term *edge of the universe* is spectacular-sounding but misleading; a better term would be *limits of the observable universe*.

This chapter described the Big Bang model of the universe. Concocting myths of creation has been a favorite preoccupation of poets, storytellers, and philosophers since ancient times. The detailed pictures of the future of the Big Bang model are a bit similar to the stories of philosophers, since they cannot be verified observationally, even though they are based on physical laws that have been verified observationally. The part of the story that deals with the origin of galaxies, stars, planets, living organisms, and chemical elements is on somewhat firmer ground. In this part of the story, we don't know all the detailed steps, but most scientists are convinced that the creation of all the wonderfully complex and beautiful structures and creatures in the universe can be explained on the basis of natural forces.

If you view the universe on a large scale, the best-understood phase of cosmic evolution is the first million years. Three key aspects of the Big Bang model that relate to that time interval have been verified by observation. Chapter 14 illustrates that the age of objects in the universe and the age of the universe itself are the same, as required by the Big Bang model. Chapter 15 discusses measurements of the abundance of helium in astronomical bodies. These measurements show that one-quarter of the universe is helium, as the Big Bang model says it should be. At the million-year mark, the hot, dense universe left a cosmic footprint in the form of radiation. This radiation was observed in 1965. As a result of these observations, astronomers and physicists have a good deal of confidence in the Big Bang model.

Now that we have a cosmological model, we need observational evidence that gives us some reason to believe that the model is more than just philosophy. The theorists were first on the scene in the history of cosmology, and the cosmological models used today are essentially those that Alexander Friedmann first concocted in 1922. Most present-day work on cosmology seeks to match these models to the real world of observations.

The most obvious feature of a Big Bang cosmological model is its statement that the universe began at some definite past time. Is there any evidence for this? Objects in the universe are being formed continuously, so an examination of any one object cannot tell us how old the universe is. We can, however, seek out the oldest objects in the universe and see whether their ages point to a definite beginning of cosmic evolution. We can then compare the ages of objects in the universe with the age of the universe itself and see whether the two ages correspond.

Ages of objects in the universe

Stars

Stars are continually being born in spiral galaxies like ours. There are some clusters of stars, however, that date from the early days. These globular clusters are found in the halo of our galaxy, far from the activity in the spiral arms. We believe that the globular clusters were formed at some time in the early stages of the evolution of our galaxy; therefore if their age can be measured, we can obtain an approximate figure for the age of the galaxy.

As a star cluster ages, stars of steadily decreasing mass reach the stage at which they exhaust their central hydrogen fuel and become red giants. Many astronomers have obtained some observations of these clusters that indicate that stars like the sun are just beginning to reach the red-giant stage. The astronomers then compare these observations with theoretical calculations of the evolution of stars to determine how long it takes such stars to exhaust their hydrogen. You have to allow for the scant abundance of metals in these stars. These stars evolve at a rate that is slightly different from the sun's rate of evolution. About ten years ago, Allan Sandage of the Hale Observatories found that the age of the globular clusters was between 10 and 15 billion years. Subsequent work refined this figure somewhat. Two recent estimates are 14 to 20 billion years and 14 to 16 billion years for the age of these oldest star clusters. Assumptions made about the compositions of these very old stars and difficulties in estimating their temperatures can affect estimates of the age of the universe.

Atoms

Some atoms are radioactive. As radioactive atoms age, they decay slowly. Nuclear reactions cause a radioactive atom to change its nature and turn into the atom of some other element. Uranium-238 eventually becomes lead-206 over billions of years. We can measure the ages of atoms by checking the progress of this decay. The procedures involved here are extremely complex; a *Scientific American* article by David Schramm describes all the details.[1] Current research indicates that the age of the oldest atom is between 7 and 20 billion years. This age refers to the age of the heavy elements produced in supernova explosions since the Big Bang. It is an average age; since these elements were produced over a period of time, it is not possible to say exactly when the first supernova exploded in the Milky Way galaxy. It is not possible to determine the age of the atoms produced in the Big Bang, since they are not radioactive.

Although stars and atoms of heavy elements are a little younger than the universe as a whole, their ages fall into the same general range of 10 to 20 billion years. The absence of any objects significantly older than this figure points to some finite time in the past as the beginning of our galaxy, at least. Yet our galaxy is not the universe. How does the age of the universe compare with the 10–20-billion-year age of our galaxy? If the ages are similar, we have evidence that the universe began at some definite time in the past—the time of the Big Bang.

The age of the universe

The rate of expansion determines the age of the universe. Follow the expansion backward, and you come to the Big Bang—the time when all matter was aggregated. The faster the expansion, the less the time from the Big Bang, because the universe has needed less time to progress from a dense state to its present dispersed state.

Hubble's constant measures how fast the universe is expanding. This constant H is involved in Hubble's Law describing the expansion of the universe. The velocity of recession v of an extragalactic object equals Hubble's constant H times the object's distance d; thus $v = Hd$. The larger the Hubble constant, the faster the rate of expansion.

The Hubble constant is a measure of the age of the universe. If it is large, the expansion proceeds rapidly and the universe is young. If it is small, the expansion is slow and the universe old. Such a statement presumes that there have been no changes in the expansion rate. The age of the universe derived in this way is called a *Hubble time*. Since the expansion was faster in the past, the universe is in general younger than the age the Hubble time indicates: somewhere between 60 and 100 percent of the Hubble time for reasonable values for the change in the rate of expansion.

To evaluate the age of the universe numerically, we must cast the above arguments into a more precise form. For a specific example, consider the Hydra cluster of galaxies, whose recession velocity is 61,000 km/sec. Hubble's Law states that their distance is 61,000 divided by H, or 1220 megaparsecs, with H equal to 50 km/sec per megaparsec. Following this expansion backward in time, you find that it takes 6.17×10^{17} sec, or 19.6 billion years, for the Hydra cluster to cover this distance if the expansion rate has been constant. A larger Hubble constant produces a faster rate of expansion and a shorter Hubble time. Numerically, the Hubble constant is usually expressed in units of kilometers per second per megaparsec, and the Hubble time equals 19.6 billion years times 50 divided by H.

The Hubble constant is also used in many other areas of extragalactic astronomy. It is used, for example, to estimate the distances to quasars. Yet quasars are strange enough so that few if any aspects of the quasar story would change materially if the Hubble constant were readjusted. Its exact value is more important for cosmology.

How do you measure the Hubble constant? In principle, it is quite easy. Go and find a galaxy, measure its recessional velocity by looking for a Doppler shift in its spectrum, measure its distance, divide, and out pops the Hubble constant, as the velocity divided by the distance.

Yet there are some problems with this method. You can only measure accurate distances for nearby galaxies where stars can be seen. But the random motions of these galaxies through space swamp the expansion of the universe. These random motions exceed a few hundred kilometers per second. Take M 81, a spiral galaxy some three megaparsecs away. At this distance, the expansion of the universe amounts to 150 kilometers per second, less than the random motions. Clearly you cannot use M 81 to measure the Hubble constant. To measure Hubble's constant, we need to examine galaxies that are tens or —better still— hundreds of megaparsecs distant, since at 100 megaparsecs the universal expansion velocity of 5000 km/sec will overcome any local motions.

Yet at such great distances we cannot distinguish individual stars in galaxies. How, then, are we to measure distances to those galaxies? We must use galaxies in the Local Group and other nearby galaxies as stepping stones. But how do you even measure the distance to a galaxy in the Local Group? To accomplish this basic activity, we need to know how astronomical distances are measured.

Techniques of measurement of distance

The most common way of determining distances in astronomy is by measuring the apparent brightness of familiar objects. As a star moves farther and farther from the earth, it becomes fainter and fainter according to an exact mathematical law. The amount of light energy that crosses the lens of your eyeball (F) falls off in the amount 1 divided by the square of the distance d to the star, so long as there is nothing but empty space between you and the star. (Thus F varies as $1/d^2$.) Therefore, if an astronomer is looking at a star that he or she recognizes as some familiar type of star, he or she knows how bright the star would be if it were some given distance from the earth. The astronomer then measures how bright the star actually is and thus determines how far away it is.

This easy-sounding procedure contains one critical assumption: that the space between us and the subject star is empty. When you measure distances in the Milky Way, star clouds, which absorb light, may lie between the earth and the star under scrutiny. These star clouds absorb light and complicate the procedure. (The clouds do redden the light, as was described in Chapter 5 in connection with the distance measurement of Cygnus X-1, so we can estimate the absorption from the color change and compensate for it approximately.) Since the Milky Way obscures our view of all but the closest galaxies, extragalactic work generally involves galaxies situated where our view is not obscured by the Milky Way and where interstellar absorption of starlight is not so important as it is in stellar astronomy.

An example of the application of this procedure is as follows. Suppose you want to measure the distance to the nearest star, Alpha Centauri. Begin by assuming that Alpha Centauri is identical to the sun. (Its spectrum is quite similar, so that this assumption seems reasonable.) The sun is some 5.2×10^{10} times brighter, as seen from the earth, than Alpha Centauri, because it is so much closer. Using the formula that light from an object varies as $1/d^2$, where d is the distance, you can find that the sun is 2.3×10^5 times closer [$(2.3 \times 10^5)^2 = 5.2 \times 10^{10}$]. Thus the distance to Alpha Centauri is 2.3×10^5 astronomical units, or a little more than one parsec. (Not bad for a crude approximation; the actual distance to the Alpha Centauri system is 1.3 parsecs.)

However, applying this method is awkward. Most stars are unlike the sun, so using the sun as a standard star, as in the preceding calculation, gives the wrong answer in most cases. The principle is still valid nevertheless, and partly for historical reasons, astronomers have adopted a different formalism to make such distance measurements. The brightness of a star, as seen in the sky, is measured by the *apparent magnitude* of the star. The brightness that that star would have if it were ten parsecs away is its *absolute magnitude*. The difference between the apparent and absolute magnitudes, suitably corrected for the absorption of light by interstellar clouds, is called the *distance modulus*. The distance modulus gives the distance of the star when we apply the law relating brightness to distance. (When you cast the law into this formalism, it becomes: Distance modulus $= -5 + 5 \log d$, where d is in parsecs.)

Returning to the example, we find that the apparent magnitude of Alpha Centauri is 0.0 and its absolute magnitude is 4.4. In other words, it is substantially brighter than it would be if it were 10 parsecs distant (remember, large magnitudes mean faint stars; see the Preliminary section if you're confused). The distance modulus is simply the difference between apparent and absolute magnitudes: $0.0 - (+4.4) = -4.4$, which corresponds to a distance of 1.3 parsecs, as before.

The use of absolute and apparent magnitudes casts this method of distance measurements into a much more convenient form. In summary: If we know the absolute magnitude of a star, then we know how bright it would be if it were ten parsecs away. It is easy to measure how bright it actually is. The difference between these two brightnesses, usually expressed in magnitudes, shows how much nearer or farther away the star is than the standard distance of ten parsecs.

The hard part of this method is determining the absolute magnitude of a given star. It would be nice if starlight came with a little flag on it telling what kind of star the light came from. We could then look that star up in tables and determine its absolute magnitude, for the tables would tell us the absolute magnitudes of nearby stars. (The distances to nearby stars can be determined by triangulation; see the Preliminary section.) In many cases, the spectrum of the star can act as a flag; similar stars have similar spectra. But for our purposes in this investigation, spectra are no help. Stars in the Andromeda galaxy are simply too faint to have their spectra measured. Where do we go from here?

The Local Group of galaxies

To measure the distances to relatively nearby galaxies, those within the Local Group, we need to observe objects in these galaxies whose absolute magnitudes can be determined. Fortunately there is a type of star whose light, properly interpreted, signals its absolute magnitude as clearly as if its photons carried a little sign. These stars, called Cepheid variables, do not maintain constant brightness. Over a period that can be as short as a few days or as long as a few months, they first increase in brightness and then decline, only to increase again in a very regular periodic fashion. Henrietta Leavitt in 1912 discovered that the longer the period, the brighter the Cepheid (in general). This was a milestone of early-twentieth-century astronomy and the key step that enabled astronomers to measure the distances to galaxies. Subsequent analysis of Cepheid variables in our galaxy, whose distances can be determined by other methods, has produced a very well-defined relation between the period of a Cepheid and its absolute magnitude, provided that the location of the Cepheid in the galaxy is known. Spiral-arm Cepheids and halo Cepheids with equal periods have different absolute magnitudes, and this difference must be allowed for.

Thanks to the Cepheids, it is not too hard to determine the distance to a relatively nearby galaxy, in which these stars can be seen. We obtain a dozen or two nights' time on a large telescope, take many photographs of the galaxy of interest, and then look for variable stars among the stars in the galaxy. We then scrutinize the photographs with a blink microscope, which allows us to look alternately at two photographs of the same object taken at different times. Any star that does not maintain constant brightness stands out, since its image changes from one photograph to the other while the other stars in the sky remain constant. Once we find a variable, we can determine its period if we measure its brightness on all the other photographs and see how long it takes for the star to go through a complete cycle. In principle, this process is easy, though it is time-consuming and sometimes exasperating. It is the inspiration that led to the original proposal to observe and the knowledge that the results are worth while that makes this drudgery rewarding.

Once the hard work is done, the astronomer has only to interpret the cosmic signposts that the Cepheids are displaying. The period of pulsation of a Cepheid variable is like a label on the star, telling absolute magnitude. Apparent magnitude can be measured quite rapidly, and we now know the distance to the galaxy.

It is really quite remarkable that the study of a certain relatively obscure type of star—a Cepheid variable—should prove to be the key to determining the distances to the galaxies. Henrietta Leavitt had no idea that this discovery was to be the result of her research; she was just determined to know what the variables in the Magellanic Clouds were doing. This story demonstrates how unpredictable and interrelated scientific inquiry is. You would not normally believe that something prosaic like variable-star work would turn out to be a critical step in our understanding of distances in the universe. But science works that way. We cannot understand how the universe works without understanding its parts. Sometimes parts of the universe turn out to be relatively uninteresting, but some parts, like the Cepheids, can turn out to be the key items in solving pieces of the cosmic puzzle.

But we cannot use the distances to galaxies within the Local Group—galaxies like the Andromeda galaxy—to measure the Hubble constant. Their random motions are much larger than their movement away from us, owing to the expansion of the universe. Attempts to extend the Cepheid method to more distant galaxies are futile, because Cepheids are too faint to be seen in galaxies more distant than four megaparsecs in photographs taken by ground-based telescopes. When the Large Space Telescope, a 90-incher, is built and launched into orbit around the earth, the Cepheid range will be extended to 40 megaparsecs. The orbiting telescope will not have to look through the obscuration and turbulence of the earth's atmosphere. But how can we extend our vision far enough to measure the Hubble constant now?

Far beyond the Local Group

Although we cannot use Cepheids to measure directly the distances of the more distant galaxies, the distances that Cepheids provide within the Local Group do enable us to use the nearby galaxies as stepping stones. Edwin Hubble, when he first determined the Hubble constant, noticed that the brightest stars in Local Group galaxies all seemed to have the same absolute magnitude; that is, the same luminosity. Since, as he thought, these stars could be seen in galaxies outside the nearby ones, he could use the Local Group galaxies, in which Cepheids could provide distances, as an intermediate link. He determined the distances to Local Group galaxies by using the Cepheids, determined the absolute magnitudes of the brightest stars in those galaxies from these distances, and then determined the distances to more distant galaxies by using these absolute magnitudes of brightest stars. His results, announced in 1936, indicated that the Hubble constant was equal to 526 km/sec per megaparsec.

This value for the Hubble constant was annoying, since the Hubble time for such a constant is 1.86 billion years, and thus Hubble's result indicated that the universe was less than 2 billion years old. There are fossils on the earth that are 3.5 billion years old, and the rocks containing these fossils were known to be this old in 1936. Subsequently, we have found moon rocks that are 4 billion years old. Clearly something must be wrong with Hubble's result. The earth cannot be older than the universe.

Hubble had made two mistakes. In 1950, Walter Baade of the Hale Observatories discovered that there were two types of Cepheids. Hubble had been comparing halo Cepheids in the Andromeda galaxy with spiral-arm Cepheids of our own galaxy. These two types have different intrinsic brightnesses, and as a result Hubble thought that the nearby galaxies were 2.5 times closer than they really are. This error caused him to underestimate the luminosity of the brightest stars in the Local Group galaxies. In addition, what Hubble thought were bright stars in the more distant galaxies were in fact clouds of ionized hydrogen gas, which are considerably more powerful than the bright stars. Allan Sandage, in rectifying these two mistakes in 1958, deduced that the Hubble constant was somewhere between 50 and 100, between one-half and one-quarter of Baade's value and one-fifth to one-tenth as large as Hubble's value.

Since 1958, scientists in numerous observatories have tried to refine this value for the Hubble constant by different methods. The astronomer must be able, through a number of intermediate links, to reach beyond the nearby galaxies to more distant ones. For nearby galaxies, random motions overwhelm the expansion of the universe. Even for galaxies as far away as the Virgo cluster, the nearest rich cluster of galaxies, the motions of galaxies may not be entirely uniform. You must explore the areas several tens or even hundreds of megaparsecs distant from the earth before you can be sure that you are looking at the expanding universe and not just at some local motion.

The Virgo cluster: a stepping stone

The Virgo cluster of galaxies is the nearest large cluster of galaxies, so let us see if we can use it to measure H (the Hubble constant). Rene Racine, a collaborator of Sandage's at Hale Observatories, measured the magnitudes of some 2000 globular clusters near M 87, the dominant galaxy of the Virgo cluster. Assuming that the brightest of these clusters is as bright as the brightest globular clusters in the Andromeda galaxy and the Milky Way, Racine and Sandage find that the Virgo cluster is 14.8 megaparsecs distant. Combining this with the redshift of the Virgo cluster, 1136 km/sec, yields a Hubble constant of 77. Unfortunately there are two reasons for distrusting this estimate of H. It is not clear that you have pushed far enough away from the earth to overcome local peculiar motions, and comparing the globular clusters around a peculiar elliptical galaxy like M 87 (recall Chapter 10) with those near a spiral galaxy is treacherous.

Subsequent work, published early in 1975, indicates that this value of 14.8 megaparsecs may be too low. Sandage and G. A. Tammann compared spiral galaxies in the Virgo cluster with other closer spirals and found that the probable distance of the Virgo cluster is about 20 megaparsecs. Other investigators disagree, believing that the value is nearer 15. In any case, it would seem premature to use the Virgo cluster as a means for measuring the Hubble constant, since random velocities may interfere with measurements of the expansion of the universe, or the pure Hubble flow. (Earlier in this book I used the old value for the distance to the Virgo cluster.)

Yet the Virgo cluster can still be used as a stepping stone. Astronomers George Abell and James Eastmond at UCLA compared the Virgo cluster with two similar clusters at greater distances: the Coma cluster and the Corona Borealis cluster. Using the 14.8-megaparsec distance to the Virgo cluster, we can then determine the absolute magnitudes of the galaxies themselves. Since all three of these clusters are of the same type, it is reasonable to assume that the absolute magnitudes of galaxies in the two more distant clusters would be the same as the absolute magnitudes of comparable galaxies in the Virgo cluster. The brightest galaxy in Coma should have the same absolute magnitude as the brightest galaxy in Virgo, the next-brightest galaxies should be equally comparable, and so on. Abell and Eastmond then found distances of 130 megaparsecs to the Coma cluster and 410 megaparsecs to Corona Borealis. The redshifts of these two clusters produce a Hubble constant of 53 km/sec per megaparsec in each case. Is this the answer?

At the present time, I think a value of 50 is a "best-buy" value for the Hubble constant, agreeing essentially with the Abell-Eastmond number. An enormous number of determinations of the Hubble constant have appeared in the last few years. I describe one of these determinations in detail to show you how complex the determination of the Hubble constant is, and to review the basic steps outlined in previous sections. Recall that the determination of H described in the next section is only one of many recent determinations. I think it is one of the better ones, but not all astronomers would agree with that assessment. Other determinations are equally complex, in most cases.

Eight steps to the Hubble constant

Sandage and Tammann use a unified approach to determining the Hubble constant, and the description of the method serves to review the basic steps on which the method depends. It goes all the way back to the beginning. The eight steps are as follows.

1. Determine the distances to nearby stars using triangulation, and the distance to the Hyades star cluster using other geometrical methods.

2. Determine the distances to Cepheids in our own galaxy by finding clusters of stars that contain Cepheids, compare those clusters with nearby stars and the Hyades, and determine the distance to those clusters. In this way you can determine the absolute magnitude of a Cepheid of a given period.

3. Determine the distances to nearby spiral galaxies in the Local Group, the M 81 group, and the South Polar Cap group by determining the absolute magnitudes of Cepheids in these galaxies. (These last two groups are similar to the Local Group and a few megaparsecs away.)

We discussed these steps previously. Now Sandage manages to avoid using the Virgo cluster, confining his investigation to spiral galaxies.

4. From the distances to these nearby galaxies, determine the average size of the largest clouds of ionized gas in spiral galaxies. For example, the biggest such cloud in class Sc spiral galaxies measures 245 parsecs across.

5. As you look off into the distance, objects of the same size appear to become smaller and smaller. By comparing the angular size of gas clouds in galaxies some tens of megaparsecs distant with the sizes of gas clouds in

nearby galaxies, measured in Step 4, you can then measure the distances to spiral galaxies in the 10–50 megaparsec range.

6. Now you have measured the distance to a sufficient number of spirals, so that you can determine the average absolute magnitude of a spiral galaxy of a given shape classification, such as the Sc's.

7. You now obtain spectra and redshifts of faint spiral galaxies hundreds of megaparsecs away. You can determine their distances because you know their absolute magnitudes from Step 6 and you can measure their apparent magnitudes.

8. From the redshifts and distances of many distant spiral galaxies, determine the Hubble constant H.

The original investigation produced a value of 55 ± 7 for the Hubble constant H. The distance to the Hyades star cluster (Step 1) has been remeasured, and applying this change brings the Hubble constant down to 50 ± 5.

The value of the Hubble constant

There are many, many astronomers who have joined Hubble, Baade, Sandage, and Tammann in the quest for the Hubble constant—a number of great cosmological significance. Most follow the essential procedure used in previous sections, using the Local Group galaxies as stepping stones. For example, Sidney van den Bergh of the University of Toronto and the Dominion Astrophysical Observatory in British Columbia uses supernova luminosity as the connecting link between the Local Group and more distant galaxies. He obtained values of 95 ± 20 (in 1968), later refined to 93 ± 15 (in 1977). But there are others. Some recent other values, for example, are 99, 82, 120, 86, 40, and 76![2] It looks as though we are no better off than we were in 1958, with the Hubble constant being somewhere between 50 and 100. It is distressing that estimated uncertainties in the individual measurements of H are in the 10-to-20-percent range, while different measurements differ by 100 percent of the smaller ones. Presumably the uncertainties are somewhat larger than the investigators think they are. But whose Hubble constant do you pick as *the* Hubble constant?

A few people have tried to avoid the complex procedures of the sort illustrated here. You can, if you are clever, make some assumptions about objects that you can see in distant galaxies, make an end run around the complexities, and determine the Hubble constant more directly. R. P. Kirshner and John Kwan, then at Caltech, used an ingenious geometrical method, based on the expansion of supernovae, to obtain $H = 60 \pm 15$. Later refinements of this method produced $H = 49 \pm 9$. Donald Lynden-

Bell, of Cambridge University in England, based his determination on the apparent expansion of radio galaxies at speeds exceeding the speed of light. He made some assumptions about the precise nature of the illusion that led to these fast expansions (recall Chapter 8) and obtained $H = 110 \pm 10$. Here again, the claimed uncertainties are 10 to 20 percent and the difference between the two determinations is a factor of 2. These discrepancies are a bit easier to understand because we need some assumptions about the fundamental nature of supernovae or radio galaxies in order to do the determination at all, and it's hard to estimate just how correct those assumptions are.

It seems to me, though, that the evidence taken as a whole argues for a value of H that is closer to 50 than it is to 100. Recall that the universe must be younger than the Hubble time. The Hubble time is 19.6 billion years for $H = 50$ and 9.8 billion years for $H = 100$. If you adopt $H = 100$, the time it takes the universe to expand becomes uncomfortably close to the time it takes stars to evolve and atoms to be produced. Two conventions have emerged in the research literature. Some workers use $H = 50$, others use $H = 75$, and everybody states explicitly which value of H is being used in a particular investigation.

In addition, any scientist must realize how sensitive his or her conclusions are to the assumptions that are made regarding the value of particular numbers like the Hubble constant. To calculate ages, distances, luminosities, and so on that are quoted in this book, I have used $H = 50$ km/sec per megaparsec, and I have assumed that the universe expands at a constant rate. In most cases, the particular value of H that is used does not affect the fundamental conclusions here. For example, if H is 100 rather than 50, all the quasar and active-galaxy luminosities quoted in Part 2 will be one-fourth of the values quoted there. The exact value of H does become somewhat important in deciding whether the universe will expand forever. Table 14-1 lists the determinations of the Hubble constant discussed in some detail in this chapter. (This table is not meant to be an exhaustive list.) If you want to know exactly how the numbers cited in this book will change if the Hubble constant changes, the Appendix to this chapter gives you some prescriptions. I certainly hope that the value I choose, $H = 50$, is more or less correct. The wide range of values in Table 14-1 (next page) makes me more than a little cautious.

This chapter has examined several different methods for determining the age of the universe. Stellar evolution, radioactive decay, and the Hubble constant all point to an age between 10 and 20 billion years. Each of these methods has some uncertainty—the logical chains leading from the observations to an age for the universe are long. The coincidence of all these results lends some support to the Big Bang idea that the universe began at a definite point in the past, some 10 to 20 billion years ago. In any case, the Big Bang theory has survived a test, for the existence of an object with an age considerably greater than the Hubble time would make it difficult to sustain a belief in the creation of the entire universe at one time in the past.

The Big Bang picture does more than just postulate a single date for the creation. It describes what happened in the very early universe. Now that I have shown that the idea of a universe created at a definite time in the past is compatible with observations, I shall direct your attention to other ways in which the theoretical picture can be confirmed.

TABLE 14-1   DETERMINATIONS OF THE HUBBLE CONSTANT

INVESTIGATOR	H (km/sec/Mpc)	REMARKS
Hubble, 1936	526	Original determination
Baade, 1950	200	Corrected Hubble's value for the difference between halo and spiral-arm Cepheids
Sandage, 1958	50–100	What Hubble thought were stars were really clouds of ionized gas
Racine and Sandage, 1968	77	Globular clusters in the Virgo cluster of galaxies
Abell and Eastmond, 1968	53	Coma and Corona Borealis clusters of galaxies
Van den Bergh, 1968	95 ± 20	Supernova luminosities
Sandage and Tammann, 1975	50 ± 5	Eight Steps to H (see text)
Kirshner and Kwan, 1976	60 ± 15	Supernova expansion
Lynden-Bell, 1978	110 ± 10	Faster-than-light expansion of radio sources

APPENDIX

What if the accepted value of the Hubble constant changes again? It may be that by the time you read this book, the value of 50 km/sec per megaparsec that I have adopted for the value of the Hubble constant will no longer be correct. I hope this will not be the case, but I wish to hedge my bets. Changes in the Hubble constant by a factor of 2 or so will not affect the basic picture of quasars presented in Part 2, but many of the numbers in this book will need to be adjusted. The precepts below enable an energetic reader to make these adjustments, which apply only to the distances of objects more distant than the Virgo cluster of galaxies. In the formulas below, subscripts refer to the value of the Hubble constant and $h = H/50$.

$$\text{Absolute magnitudes: } M_H = M_{50} + 5 \log h$$
$$M_{75} = M_{50} + 0.87$$
$$M_{100} = M_{50} + 1.5$$

$$\text{Luminosities: } L_H = \frac{L_{50}}{h^2}$$
$$L_{75} = \frac{L_{50}}{2.25}$$
$$L_{100} = \frac{L_{50}}{4}$$

$$\text{Distances: } D_H = \frac{D_{50}}{h}$$

Ages: Ages of objects are proportional to the Hubble time, which scales as $1/h$.

The ages of stars, atoms, and the entire universe are in the same range — 10 to 20 billion years. This age coincidence points to a definite time as the beginning of the evolution of the universe, but this evidence in favor of the Big Bang model is very indirect. Is there no more direct way of confirming that our universe was once in a hot, dense state?

There is. Recall the Big Bang history (Chapter 13). For all reasonable Big Bang models, roughly one-quarter of the universe is transformed from hydrogen into helium in the first 20 minutes of cosmic evolution. In addition, there was radiation in the early universe, and this radiation has not disappeared. If we can observe this primeval fireball radiation, we can confirm the Big Bang theory. The Big Bang theory makes two unequivocal statements about the present condition of the universe: It should be roughly one-quarter helium and it should be filled with the primeval fireball radiation. As this chapter will show, these theoretical ideas are confirmed.

Another cosmological model, the Steady State theory, was quite popular in the 1950s and is still found in the literature. This theory can produce an expanding universe, but if the theory is to be sustained it must find another source for the helium in the universe and for the primeval fireball radiation

Other cosmological theories have appeared from time to time. Yet it is the Big Bang theory that is consistent with the observations, for the other theories generally run into problems of one sort or another when they confront the experimental evidence.

Helium in the universe

In the first 20 minutes of the evolution of a Big Bang universe, almost all the neutrons that were around at the one-second mark were incorporated in helium nuclei, producing a universe that was roughly one-quarter helium. These neutrons were around at the one-second mark because the early universe was hot enough so that protons and electrons collided, forming neutrons. It is a remarkable result of Big Bang models that the amount of helium produced is insensitive to the initial conditions, since the fractional abundance of helium is between 23 and 29 percent for all reasonable Big Bang models. We can now check the Big Bang models by asking: Is this helium really there?

The helium-abundance problem has engaged the attention of many astronomers working in widely different fields. Some of us look for helium directly; we focus our instruments on some object that produces helium lines as helium atoms emit or absorb radiation. Others use more indirect methods. The evolution of stars containing 25 percent helium is quite different from the evolution of stars containing no helium, and comparing real stars with model stars enables us to determine the helium abundance of these stars.

Table 15-1 lists a number of different determinations of the cosmic abundance of helium. This list is not meant to be exhaustive; there must be almost a hundred different helium-abundance determinations in the literature. But it is representative in one important respect: Of all the determinations of helium abundance made, there is not one piece of evidence that indicates a helium abundance of less than 25 percent. Although there are a few stars whose visible surface layers contain little or no helium, the current consensus is that the abundance of surface helium in these stars is not representative of the abundance of helium in their interiors. Thus the observations of helium abundance are consistent with the Big Bang theory, since theory says that about 25 percent of the universe became helium in the hot Big Bang.

Cosmological helium?

So far, it seems that the Big Bang theory has passed the helium test with ease. The theory says that the universe should be roughly one-quarter helium, and a wide variety of observations show that it is indeed one-quarter helium. But how meaningful is the helium test?

The helium-abundance test is more like a gate in an obstacle course than an accolade of approval. If we found that some part of the universe contained very little helium, the Big Bang theory would be disproved, for it states that helium abundance of about 25 percent should be universal. Yet

TABLE 15-1   DETERMINATIONS OF HELIUM ABUNDANCE

TYPE OF OBJECT	HELIUM ABUNDANCE, PERCENTAGE
Gaseous Nebulae	
Milky Way	29
Small Magellanic Cloud	25
Messier 33	34
Milky Way (Radio Observations)	26
Young Stars	30–33
Globular Clusters (Indirect Methods)	30

the existence of a high abundance of helium means only that the theory is *consistent with* the observations, not *proved by* them. The Big Bang theory could be wrong and the helium could have been made somewhere else. Suppose, for example, that a supermassive star formed at the galactic center early in the galaxy's evolution and then exploded, spreading helium all over the galaxy. Then there would be a high abundance of helium that did not come from the Big Bang. Thus the Big Bang theory has passed a test, but the test may not be as significant as we should like. There is a less equivocal Big Bang test.

The primeval fireball

The hot, early universe contained photons. At the million-year mark, these photons interacted with matter for the last time as they bounced off the electrons that were then just about to recombine with atomic nuclei. As we see these photons now, they have been redshifted and are visible as radio waves and far-infrared (or microwave) radiation. The universe is filled with these photons. Two terms are used to describe these photons. The *microwave background* is what is observed, and *primeval fireball* refers to the radiation that should be there according to the model. The identification of the microwave background, an observed phenomenon, with the theorists' primeval fireball was a milestone of twentieth-century cosmology. Arno A. Penzias and Robert W. Wilson, the Bell Laboratories physicists who made the first observations of this background, won half of the 1978 Nobel Prize as the reward for their discovery.

Discovery of the microwave background

Many discoveries in astronomy were accidental. Penzias and Wilson, in 1964, were not looking for background radiation. It could even be argued that they were not doing radio astronomy at the time. Bell Laboratories had a particular type of radio antenna that was very useful for determining exactly how much radio energy was coming from a particular radio source. A person could calculate the sensitivity of this horn antenna from the basic laws of physics. When communications satellites were being launched in the early 1960s, Penzias and Wilson were asked to get the horn antenna working. It would first be used for communications satellite work; radio astronomy might come later.

The Telstar satellite was launched in 1964. Penzias and Wilson then had to calibrate the antenna. They had to make a precise measurement of the amount of radio noise that the amplifier would produce even when the antenna wasn't looking at anything. Such a calibration was desirable for the communications-satellite people and essential for the next project Penzias and Wilson had in mind. The Holmdel horn was one of the few radio antennas in the world whose sensitivity could be calculated precisely, and so they wanted to measure the brightness of the Milky Way galaxy with it. (Other antennas could determine, say, that source A was 1.30 times brighter than source B, but could not determine the brightness of either source in absolute terms as well as the Bell antenna could.)

So Penzias and Wilson set to work. Earlier workers had found that the antenna was producing a little more noise than it should. Penzias and Wilson taped the seams in the antenna, cleaned its interior surfaces, evicted two birds that had found it a nice place to nest, pointed it in different directions to see if there were some unexpected radio sources in the beam, and made all the other checks that careful experimenters do. Excess radiation was still coming into the antenna. But what was it? Penzias and Wilson had no idea. In an essay written long before the Nobel Prize was awarded, Penzias wrote, "Having explored and discarded a host of terrestrial explanations, and 'knowing,' at the time, that no astronomical explanation was possible, we frankly did not know what to do with our result."[1]

The theoretical background

The theoretical story of background radiation starts with George Gamow and the small group of Big Bang theorists who worked with him in the late 1940s. Gamow, Ralph Alpher, and Robert Herman were primarily interested in explaining the abundance of the various elements in the universe. Their initial vision was that all the elements in the universe were produced in the Big Bang during the first half hour or so of cosmic evolution. They struggled to get the nuclear reactions to go past helium, since helium cannot add a proton or neutron and produce a stable atomic nucleus. Try as they might, they were unable to make all the elements in the universe, though they were able to make helium in the Big Bang.

Gamow, Alpher, and Herman realized that one of the by-products of the hot, dense phase of the Big Bang universe was some leftover radiation. Alpher and Herman produced the first numerical prediction of the intensity of this background radiation. Expressed as a temperature, they put it at 5 K in their 1949 paper. Four years later Gamow set it at 7 K. For a variety of reasons, their work was forgotten. In these early days, the Big Bang theory ran into the difficulty that the age of the universe, determined from the Hubble time, was thought to be between 2 and 4 billion years, and star clusters were known to be older than that.

Subsequent theorists rediscovered Gamow's earlier work. In the late 1950s, the Soviet physicist Igor Novikov began working on cosmology. It is variously reported that he rediscovered Gamow's result or found it in the literature. P. James E. Peebles, who became interested in Big Bang cosmology in the early 1960s, began to do some calculations on the abundance of helium produced in a hot Big Bang universe. He, too, rediscovered Gamow's result, wrote a paper on it, and gave a colloquium (scientific talk) at Johns Hopkins University in early 1965. Stimulated by Peebles' theoretical work, Robert Dicke and a group at Princeton started to build some equipment to look for this background radiation.

There are various stories about how Penzias and Wilson heard of the work being done at Princeton. One story has it that Penzias called the radio astronomer Bernard Burke to discuss another matter, and casually mentioned the Bell Laboratories results. Burke said that he had seen a preprint of Peebles' work. Another story has it that Burke went to the Johns Hopkins colloquium and then sat next to Penzias on a plane trip, where the link was made. Certainly the link was made during one of those informal conversations that often stimulate ideas for scientific research.

Once Penzias and Wilson heard of the activity at Princeton, they obtained a copy of Peebles' paper, invited the Princeton group to visit Holmdel, and agreed to publish the results jointly. The Bell Laboratories paper announced the experimental results, and paper by Dicke, Peebles, Philip Roll, and David Wilkinson of Princeton provided the interpretation. What had begun as work on telecommunications satellites had turned into a cosmological gold mine.

Confirmation

But was this radiation really the primeval fireball? Its photons did not carry name tags with them while they traveled from the million-year-old universe to the waiting radio telescopes. To be sure, the excess radiation discovered by Penzias and Wilson was about as intense as theoretical predictions said it should be. Such a discovery requires more detailed confirmation than the rough agreement of one measurement with theory, though. Could Penzias and Wilson be observing something else?

If this radiation were the primeval fireball, it would have a characteristic type of spectrum called a *blackbody spectrum.* This type of spectrum is emitted by an object without intrinsic color. One million years after its beginning, the universe was filled with radiation that had this type of spectrum. The corresponding temperature of the radiation was 3000 K. Since then, this radiation has been traveling through the universe. At the million-year mark, electrons, protons, and helium nuclei recombined, and so the radiation no longer collided with charged particles. Now 20 billion years later, the radiation is redshifted. It still has a blackbody spectrum, but this spectrum is characterized by a much lower temperature of 2.9 K, and is found in the radio part of the electromagnetic spectrum. The solid line in Figure 15-1 shows this type of spectrum.

As soon as the background radiation was discovered, several radio astronomers measured its intensity at different wavelengths. Figure 15-1 shows the results; the radio measurements from the ground are shown as

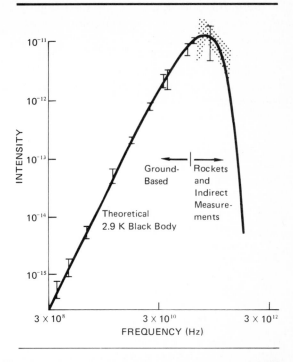

FIGURE 15-1   Observations and theoretical expectations of the microwave background. Vertical lines show measurements at one frequency; the length of the line indicates estimated error of the measurement. The shaded area is a recent rocket measurement at several frequencies. [Rocket measurement by D. P. Woody, J. C. Mather, N. S. Nishioka, and P. L. Richards, "Measurement of the spectrum of the Submillimeter Cosmic Background," *Physical Review Letters,* 34 (1975), pp. 1036–1039. Other data from P. Thaddeus, "The Short-Wavelength Spectrum of the Microwave Background," *Annual Review of Astronomy and Astrophysics,* 10 (1972) pp. 305–334.]

open circles. The observations fit the blackbody curve expected from the Big Bang theory quite well. Yet the best confirmation of the blackbody character of the radiation would be to see the turnover in the spectrum at wavelengths of 0.1 centimeter, or 1 millimeter. Such observations cannot be made from the ground, because the earth's atmosphere absorbs radiation in just the region in which the spectrum starts to turn over. (Recall Figure P-4.) Instruments have to be flown above the atmosphere in a rocket or balloon.

In the late 1960s, a number of scientists built telescopes to detect the short-wavelength portion of the background radiation. The U. S. government maintains facilities for launching rockets or balloons for scientific purposes. These scientists sought to use these facilities to confirm that the background radiation did indeed follow a blackbody spectrum. Their requests for balloon and rocket time were granted. Once again, astronomers had to go above the atmosphere to perform an astronomical experiment.

The first results from the rocket flights were a bit disturbing. A 1968 flight found that the background radiation seemed to be 100 times as intense as the fireball would be in the 0.01–0.06 cm wavelength range. Earlier indirect measurements of the background intensity in this wavelength range showed no such excess radiation. What was wrong? Was the background radiation coming from a bunch of distant radio galaxies, rather than from the million-year-old universe?

This first rocket experiment did not cause the astronomical community to abandon its primeval-fireball interpretation of background radiation. Everyone involved, including the scientists who flew the first rocket experiment, was quite cautious. Many things can go wrong in a rocket experiment. You build your telescope, take it to New Mexico, put it on top of a rocket, push the launch button and watch the rocket go up, and hope that the equipment and rocket do what they are supposed to. It turned out later that the original experiment was indeed flawed. Some radiation from the earth's atmosphere had entered the telescope and contaminated the measurement of background radiation. Subsequent balloon flights showed that the spectrum of the background radiation did indeed turn over as the fireball model says it should (Figure 15-1). In late 1974 and early 1975 a group of astronomers from Queen Mary College in England and one from the University of California at Berkeley independently measured the intensity of the background radiation and showed that their measurements fit the blackbody curve quite nicely.

It took nearly a dozen years to confirm that Penzias and Wilson had indeed discovered the background radiation. If background radiation were being emitted by a collection of distant radio galaxies, its intensity would keep rising to the upper-right-hand corner of Figure 15-1, and not turn over. Now the data points all follow the blackbody curve, severely restricting possible interpretations of the background radiation. The primeval-fireball model fits the data; few other explanations do. Now the blackbody nature of this radiation and its virtually certain interpretation as the primeval fireball are secure, and Penzias and Wilson have won the Nobel Prize for their efforts.

Analysis

Any new phenomenon plays a succession of roles in the grand drama of science. First it has to be discovered, then it has to be confirmed. Following the initial discovery, there is often debate about the nature of the discovery (if it is an observation of something unexpected) or the correctness of the theory (if it is a new theory). This new result then enters the scientific mainstream. The basic nature of the phenomenon is accepted, and it is used as a source of additional information. Measurements of particular properties are pushed to the limits of precision, and progressively more detailed models are developed to match the observations. Finally, once all the insights have been milked from it, it becomes old textbook knowledge, and becomes less important in research. The primeval fireball is no longer a new character on the stage of cosmology. Current research is directed toward analysis, the use of fireball radiation as a probe of the early universe.

When we look at background radiation, we are looking at the universe as it was one million years after creation. Every photon in the microwave background last interacted with matter 20 billion years ago, at these very early times. Astronomers have spent many, many hours of radiotelescope time looking at the million-year-old universe by carefully measuring the intensity of this radiation in different places. These astronomers search for hot spots in the background, hot spots that might be the seeds of primeval galaxies. They have found no hot spots. The million-year-old universe presents a uniformly dull picture, with no structure whatsoever. This uniformity is a crucial foundation for the cosmological models discussed in Chapter 13, which presume that on a large scale the universe is quite uniform. Furthermore, these measurements fix the time that galaxies formed as being after the million-year mark when the background radiation was formed.

Astronomers can also use primeval-fireball radiation to probe the earth's motion relative to the rest of the universe, as shown in Figure 15-2. The radiation should be more intense in the direction that the earth is moving, since it will be blueshifted. The variation in intensity is small—a tenth of one percent—but nevertheless present. For a long time, people tried to measure this large-scale variation in the intensity of background radiation. A few positive results, apparent detections of the earth's motion, were announced from time to time, but until 1977 none of these measurements was unambiguous enough to provide positive evidence of motion.

In 1977, a team from the Lawrence Laboratory, Berkeley, designed, built, and flew an instrument on board a U-2 aircraft operated by NASA. These aircraft were at one time used for spy work (one of them was shot down over the Soviet Union in a celebrated incident in the late 1950s). Now that satellites can do some of the spy work, the U-2 planes are available for other tasks—such as carrying telescopes to high altitudes, far above the interfering radiation of the earth's atmosphere. This Lawrence Laboratory team finally discovered the long-sought drift of the earth relative to the rest of the universe.

The motion of the earth in space is relatively complex. Earth orbits the sun at a speed of 30 km/sec. The direction changes depending on the season of the year. Figure 15-3 shows the direction of the earth's motion in December and in June. The sun is carried around the Milky Way galaxy at a speed of 250 km/sec in the direction of the constellation Cygnus, as it

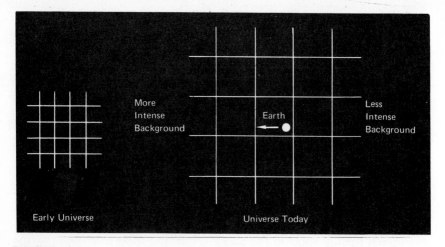

More Intense Background

Earth

Less Intense Background

Early Universe

Universe Today

FIGURE 15-2   The earth's motion relative to the frame of reference of the rest of the universe produces variations in the intensity of the background radiation. Suppose that the earth moves toward the left, as in the above picture. Because of the Doppler shift, radiation from the million-year-old universe will be more intense in the direction that the earth is moving.

whirls around the galactic center in an almost circular orbit. All this was known before the 1977 flights. The Berkeley experimenters found that the earth moves at a speed of 400 km/sec in the direction of the constellation Leo. This motion is the sum of the sun's motion around the galaxy and the galaxy's motion through the universe. Allowing for galactic rotation, the Berkeley group determined that the Milky Way is moving through the universe with a velocity of 600 km/sec in the direction of the constellation Hydra.

This speed is surprisingly high; the astronomical community is only now beginning to digest its implications. Measurements of the redshifts of galaxies show that, in general, any galaxy moves randomly relative to the universe with a speed of a few hundred kilometers per second. Why should our own galaxy be so speedy? No one knows, yet.

Detection and confirmation of the primeval fireball was one of the most important discoveries of twentieth-century cosmology. As a Big Bang relic, the primeval fireball radiation has the advantage of being unequivocal. Although the discovery of an apparently cosmic abundance of helium is only indirect confirmation of the Big Bang picture, it is very difficult to explain the primeval fireball as anything except a relic of the Big Bang. It enables us to look back at the universe as it was at recombination, roughly a million years after the Big Bang, in a way that we cannot do if we look at galaxies or even quasars.

It is even more important that the primeval fireball was predicted. Because astronomy is an observational rather than an experimental science, theoretical work generally consists of explaining what the observers

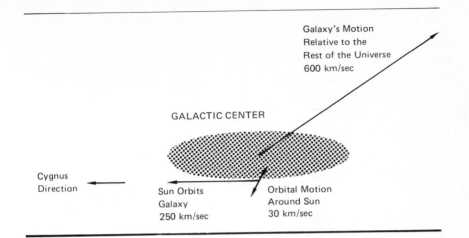

Galaxy's Motion
Relative to the
Rest of the Universe
600 km/sec

GALACTIC CENTER

Cygnus
Direction

Sun Orbits
Galaxy
250 km/sec

Orbital Motion
Around Sun
30 km/sec

FIGURE 15-3   The earth moves around the sun at a speed of 30 km/sec in directions that change depending on the time of year. The sun moves around the galaxy at a speed of 250 km/sec in the direction of the constellation Cygnus. Measurements of variations in the intensity of the background radiation show that the galaxy moves at a speed of 600 km/sec in the direction of Hydra.

have discovered. It is much more impressive and much more important for a theoretical prediction to be verified; this happened in the case of the primeval fireball. The impressiveness of the prediction is not diminished by the fact that the radio astronomers who discovered the microwave background did not realize that it had been predicted. This discovery discredited a number of other theories of cosmology, including the Steady State theory that was widely discussed in the 1950s.

Demise of the Steady State theory

From the evidence presented in the last few pages, it is apparent that the Big Bang theory fits the data. But the theory is not necessarily the correct one; there may be other theories that also fit the data. Before the mid-1960s, one such theory was widely discussed: the Steady State theory. This theory still makes its appearance in the literature, and a few of its proponents have not given up on it. Yet the Big Bang theory has much more experimental support. If the Steady State theory had survived, it would have meant a scientific revolution, but the revolution failed, as far as we can tell now. There are still a few believers, and who knows? They may turn out to be right and the rest of us all wrong.

The Steady State theory and other rival theories have been good for Big Bang cosmology. Without them, the Big Bang theory would have been just accepted, not tested. Thus there are two reasons for examining the reasons that most astronomers do not believe in the Steady State theory. You may run across this theory in some cosmology books, and it is is a good foil to the Big Bang scheme. To show why the Steady State theory does not fit the data, let us turn from the real world of observations, back to the model world of cosmological theories.

The theory

The basis of the Steady State theory, which Fred Hoyle, Hermann Bondi, and Thomas Gold formulated in 1948, is the old philosophical principle first stated in Ecclesiastes: "There is nothing new under the sun." (Eccles. 1:9) This idea prevails in Western philosophy. In its mathematical form, it is termed the Perfect Cosmological Principle: The overall appearance of the universe does not change. Whenever you look at any part of the universe at any time, you will see, for example, the same number of galaxies per cubic megaparsec.

Yet the universe is expanding. How does the Steady State theory explain the expanding universe? Herein lies the theory's revolutionary character. Suppose that matter is being continuously created somewhere in space. Creation of this matter would cause a pressure that would force galaxies to move away from each other. This continuous-creation scheme violates some of the conservation laws of nineteeth-century physics, but no matter. The theory is revolutionary and thus exciting.

Figure 15-4 explains the difference between the Steady State and Big Bang views of a small part of the universe. Focus on the middle panel of the Big Bang picture first. At the present time, our small sample of space shows four galaxies, labeled a, b, c, and d, with a fifth, e, halfway out of the picture. In the past (lower panel), these five galaxies would have been closer together. Since this compaction applies to all galaxies, the lower panel contains many galaxies that cannot be seen in the middle panel, for they have expanded out of our small volume of space. In the upper panel, the galaxies have moved toward the edges of our picture; the universe has expanded. Galaxy e has left the picture completely. The Big Bang picture is that of an evolutionary universe; the universe looks quite different in each of the three panels. The distance between galaxies changes as the universe ages.

The Steady State view of the same volume of space is shown in the right-hand side of the figure. At the present time, the Steady State universe looks exactly like the Big Bang universe. But as the Steady State universe sees time pass, the galaxies that exist now, a, b, c, and d, do move away from each other, just as they did in the Big Bang view. However, the average distance between galaxies remains the same as a new galaxy, galaxy f, appears. Galaxy f is formed from this newly created matter that is forcing the other galaxies to move away from each other. Follow the Steady State

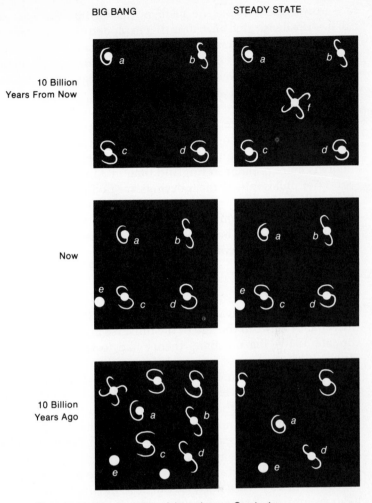

BIG BANG   STEADY STATE

10 Billion
Years From Now

Now

10 Billion
Years Ago

FIGURE 15-4   Two views of the evolution of the universe. See text.

universe backward, and you see that the average distance between galaxies is still the same, since galaxies b and c did not exist ten billion years ago. The Perfect Cosmological Principle holds: On the average, the distance between galaxies does not change in time. Small differences exist, but the overall picture remains constant. "There is nothing new under the sun." It may not be the same sun, or the same galaxies, but there is nothing new about the appearance of the universe.

The Steady State theory has two philosophical consequences that, while they cannot be used to test the theory scientifically, may well have been responsible for some of its popularity. The Big Bang theory has to

RELICS OF THE BIG BANG   291

face a problem: Who put the Ylem, or Cosmic Egg, or whatever you wish to call it — the dense glob of matter that exploded — there in the first place? In the Big Bang theory, creation took place at some definite time in the past. (This question can be circumvented by invoking an oscillating universe, so that the Big Bang followed the collapse of a predecessor universe, but such a model is conjectural at the present time.) The Steady State theory is also called the continuous creation theory, for matter is being created all the time. The Steady State theory thus avoids the touchy problem of origins. Furthermore, it does not produce a universe that is continually running down. (These philosophical implications may explain why the Steady State theory is more popular in England than in America.)

Yet philosophical meanderings are no help when it comes to testing the theory. Let us now return to the real world to show why the Steady State theory is no longer compatible with observed data. It was once compatible with the real world, but one now has to stretch both observations and theory to make the theory fit the data.

Cosmic ages

An initial motivation for the Steady State theory was a one-time discrepancy between the age of the universe and the ages of objects in it. In the late 1940s, the Big Bang theory had to face a serious problem. The Hubble time, the upper limit to the age of the universe, was then 1.8 billion years, as derived from Hubble's (1936) value of the Hubble constant. Yet the oldest rocks on the earth are known to be 3 billion years old, and people had begun to analyze the ages of stars and had come up with numbers like 10 to 20 billion years (not too different from the ages described in Chapter 14). The idea that the earth, the galaxy, and the atoms were older than the entire universe sounds absurd.

The Steady State theory avoids the age difficulty quite nicely. Any individual galaxy in the universe can be as old or as young as you like. Some galaxies are very, very old, as shown by Figure 15-3. When it was initially proposed, it was a neat way around the embarrassing problem of ages.

The Big Bang theory has managed, with another 20 years of research on extragalactic objects, to cope with the age problem. Hubble's constant is now one-tenth as big as it was in 1950, and the universe has become ten times older. The Hubble time, the upper limit to the age of the universe, is now 19.6 billion years, and there is no longer any age embarrassment.

We have not managed to knock down the Steady State theory, though. I have just shown that the ages of objects in the universe are consistent with both the Big Bang and the Steady State theories. We seek some evidence that the universe was at one time quite different from what it is now. Counts of radio sources do show the evolution of the universe, and historically they were the first pieces of evidence that cast doubts on the

Steady State theory. I have not discussed this evidence until now because the interpretation of the source-count data is difficult. About all that they prove, at the present time, is that the universe does change.

Counting radio sources

The strict, straightforward form of the Steady State theory is easy to refute, since it makes an unequivocal prediction: The universe looks the same at all times. We can see the universe at different times, in a sense, as we look farther and farther out into space. Quasars afford us a view of the universe as it was long, long ago, since they are very distant objects and their light takes a long time to reach us. Study of the evolution of quasars shows that there were more quasars around in the early universe. This is an observed fact, as long as you agree that the redshifts of quasars are cosmological. Advocates of the Steady State theory extricate themselves from this conflict by advocating noncosmological redshifts for quasars, which have been proposed on other grounds.

There is another way, independent of quasars, to test the strict Steady State theory. This test was the one that hit the Steady State theory with a blow in the 1960s. Radio astronomers counted the number of sources of cosmic radio-frequency radiation. It was assumed initially that the nearby sources were the bright ones and the distant sources the faint. There were more faint sources than there should have been for a universe that was always the same.

It soon became apparent that the interpretation of the radio-source counts was more complex than it seemed at first. We need to assume some sort of model for numerous different types of radio sources in the sky. Cosmological effects become entangled with the nature of the radio sources themselves. At one time it was hoped that radio-source counts would provide substantial information about the evolution of the universe, but these hopes have not been realized.

It is still true that the strict Steady State theory is incompatible with the existing data on the number of radio sources, as shown in Figure 15-5. The lines are the results of a Steady State theory, indicating how many sources there should be relative to the number of sources in a flat, nonexpanding universe, with different radio brightness. It is impossible to fit any of the curves to the data. There are either too few bright sources or too few faint ones. The strict Steady State theory is simply not compatible with the data.

Present advocates of the Steady State cosmos have thus modified the theory. Both the radio-source counts and the quasar data refer only to evolution in the past 10 billion years or so. Suppose that the universe goes through cycles lasting tens of billions of years, varying from epochs with many quasars and radio sources to epochs like the present one, with few quasars. These cycles do not have to be even 10 billion years long, if you go

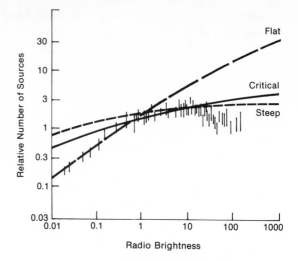

FIGURE 15-5   Source counts and the Steady State. The vertical lines are the data, the actual number of radio sources with a given brightness relative to the number in a flat, static universe. The radio brightness is in flux units (10^{-26} watt per square meter per hertz) at a frequency of 178 megahertz. The curves are theoretical results from a Steady State universe; each curve corresponds to an assumed relation between the number of radio sources and their luminosity. It is evident that no curve fits the data. (Adapted from S. von Hoerner, *Astrophysical Journal,* vol. 186, p. 750, 1973, published by the University of Chicago. Copyright by the American Astronomical Society. All rights reserved.)

along with the idea that quasars are not cosmological and hypothesize that the radio sources are nearby too. Such ideas are reasonable, even if there is no proof of them. All that the quasar and radio-source-count data prove is that the universe evolves wherever the radio sources are. The initial philosophical thrust of the Steady State theory is not so strong as it was, but the general idea persists.

The only way to rule out this revision of the Steady State theory, named the "Fluctuating Steady State," is to show that at one time the universe existed in a hot, dense state, radically different from the universe of today. The Big Bang is such a hot, dense state, and two strong pieces of evidence in this direction can be summarized.

Background radiation

The Big Bang theory states that the universe should be filled with primeval fireball radiation having a blackbody spectrum. As far as we can tell, this is exactly the case. The Steady State theory must find another source for this radiation. Of course, if we postulate that many galaxies are putting out radiation that just happens to mimic the blackbody background, we can save the Steady State theory. Unfortunately most radio galaxies do not radiate like blackbodies at a temperature of 3 K. Maybe we could hypothesize that there are dust grains around at just the right temperature. They would probably evaporate, unless they were big enough. At this date there is no simple way to make the background radiation in the context of the Steady State model. There are complicated ways, to be sure.

Cosmic abundance of helium

The Big Bang theory states that the universe should be between 23 and 29 percent helium. It is, so far as we can tell. The Steady State theory must provide some other source for this helium. It is somewhat easier to make helium than background radiation, but the consensus is that the helium is cosmological.

Table 15-2 shows a summary of the arguments. We could argue that the Steady State theory is not dead yet, that it can be saved in different ways. However, it is certainly losing, and losing badly. In science, particularly in astronomy, it is often possible to save a theory by patching it up to cope with the data. However, when a theory becomes more patches than original cloth, it is difficult to sustain your belief in it. (Recall Occam's Razor.) For this reason, few astronomers still support the Steady State theory. The Big Bang model explains the data in a much more straightforward way.

TABLE 15-2   EVIDENCE BEARING ON THE STEADY STATE THEORY

		BIG BANG	STRICT STEADY STATE	HOW TO SAVE THE STEADY STATE
Source	Counts			
	Quasars, cosmological redshifts	✓	X	Allow fluctuations
	Quasars, noncosmological redshifts	✓	✓	
	Radio	✓	X	Allow fluctuations
Background radiation		✓	X	Find another source
Cosmological helium		✓	X	Make helium somewhere else

Other cosmological theories

Cosmology—the study of the entire universe—is one of the most fascinating areas of science. Many people have tried to find cosmological theories that might explain the evolution of the universe. The two cosmological facts discussed here, the abundance of helium and—most important—the background radiation, were not discovered until the mid-1960s. So before the 1960s a fair number of theories were proposed to explain the only known cosmological fact—the expanding universe. The Steady State theory was the most serious competitor to the Big Bang model. In this section I shall briefly dispose of some others.

Variations in the gravitational constant

Perhaps the most persistent class of alternative cosmological models is one in which the gravitational constant varies with time. In the 1930s, P. A. M. Dirac, one of the pioneers of quantum mechanics, noticed that there are some rather curious numerical coincidences that involve cosmology. The ratio of the *electrical* force between a proton and an electron and the *gravitational* force between a proton and an electron is 10^{40}. The ratio of the Hubble time to the time it takes light to cross an atom is the same number, 10^{40}. The number of atoms in a closed universe is 10^{80}, or $(10^{40})^2$. It seems strange that the same large number, 10^{40}, should appear in three fundamental aspects of the physical universe. Dirac's hypothesis was that the equality of the ratio of forces, the ratio of times, and the square of the number of particles in the universe was no coincidence. This idea is often referred to as the *large-number hypothesis*.

But since the universe is closed in this cosmology, the magical large number of 10^{40} must change when the Hubble constant changes. Since the number of particles in the closed universe, 10^{80}, is equal to the square of the magic large number of 10^{40}, continuous creation is required, but at a rate that is so small as to be unobservable. However, consider that 10^{40} equals the ratio of electrical to gravitational forces. That is, electrical force is 10^{40} times as strong as gravitational force. When this number changes, either the electrical or the gravitational force must change if the large-number hypothesis is to be correct. There are other cosmological models that produce a changing gravitational constant, but none has the esthetic attraction that the large-number hypothesis has.

Observational evidence is now able to rule out these cosmologies almost—but not quite—definitively. In a widely publicized study, an astronomer used timings of the moon's passage in front of certain stars, or *lunar occultations*, to measure the variation of the gravitational constant. He did observe an apparent change, but the uncertainties turned out to be nearly as large as the proposed change. This result was not accepted within the scientific community as a demonstration that the gravitational constant was in fact changing.[2] Radar measurements of the moon's distance from the earth rule out a change as large as the change supposedly determined from lunar occultations. A changing gravitational constant would have other consequences. If gravity had been stronger in the past, the sun would have been more luminous and the primeval oceans would have boiled. The primeval fireball radiation might not have been produced in a large-number cosmology. Put all the observations together, and you see that astronomers are almost able to definitively rule out the change in the gravitational constant called for by the large-number hypothesis.

This measurement of the constancy of gravity has been accompanied by some other measurements of the constancy of other constants of Nature. People have examined spectral lines in distant quasars to demonstrate the constancy of atomic structure over the billions of years that it took the light to travel from the quasar to us. It turns out that Planck's constant, the number that relates photon energy to photon frequency, must have been the same in the Big Bang as it is now in order to produce the correct amount of helium in the primeval nuclear cooker. These types of cosmological tests give us confidence that the laws of physics as we know them are correct. They can, for example, rule out alternatives to Einstein's theory of gravitation.

Antimatter in the universe

Another alternative to the conventional Big Bang picture involves the possible presence of antimatter in the universe. People seeking to understand the ultimate makeup of matter spend a great deal of time, money, and electricity in accelerating subnuclear particles to high speeds and whacking other particles with them. These experiments have demonstrated the existence of *antimatter,* a type of stuff in which particles have electrical charges opposite to their matter counterparts. When a particle and its antiparticle meet, they annihilate each other in a blast of gamma rays. Antimatter was mentioned earlier as one of the more speculative, far-out theories to explain the luminosity of quasars and active galaxies.

The symmetry between a particle and its antiparticle, observed on a small scale, leads us to ask whether a similar symmetry exists in the universe. It is esthetically satisfying to suppose that the Creator had the same sense of symmetry that we do and produced a universe with equal numbers of particles and antiparticles. It is possible to produce models of a Big Bang in such a universe that do not lead to complete annihilation of matter and antimatter. There is only one problem with these theories—their conflict with observation. Matter and antimatter produce a blast of gamma rays when they meet. If every other galaxy were an antigalaxy, there would be many strong gamma-ray sources in the universe. These gamma rays aren't there, and so the antimatter isn't there—at least in the amounts that the models predict. None of the antimatter cosmologies proposed so far agrees with observation.[3]

These alternative cosmological models have their attractions, then, but do not agree with the observations. More cosmological models may be proposed in the future. The Big Bang cosmological model lacks symmetry, and symmetry has always been something that has attracted physicists seeking to model the universe. There is only matter, no naturally occurring antimatter. The evolution of the universe is one-way in Nature, from primeval explosion to continued expansion. If the universe is open and will expand forever, the one-way, antisymmetric nature of the Big Bang theory is particularly poignant. The universe begins with a tremendous bang and gradually whimpers away, to paraphrase T. S. Eliot. But will the universe expand forever? The next chapter shows how we astronomers try to answer that question.

The Big Bang left several fossils or footprints that allow us to verify that it actually happened. An indirect test of Big Bang cosmologies is provided by the observation that approximately one-quarter of the mass of the universe is helium. Nuclear reactions in the first 20 minutes of cosmic evolution made this helium. A more direct test is provided by microwave background radiation. The detection and confirmation of this Big Bang relic provide direct evidence that the universe was hot one million years after creation. The microwave background radiation is now being observed carefully to provide more information about the million-year-old universe and our motion relative to it.

A number of alternative cosmological models have come and gone since the Big Bang theory was proposed. The most persistent of these was the Steady State theory. However, this theory cannot explain the fireball radiation, cannot explain cosmological helium, and cannot explain the relative abundance of bright and faint radio sources. But even though these theories turned out to be wrong, they have been good for cosmology. Without them, the Big Bang, expanding-universe cosmology of the 1930s would have remained alone on center stage, and it would have been accepted whole rather than tested. The radio-source counts would never have been made. The large telescope used for the source counts might not even have been constructed, and the telescope at Cambridge University might not have gone on to discover a number of other exciting phenomena, such as the radio galaxies. People might have just accepted the background radiation, and not measured its spectrum in detail. Observational cosmology would be a much poorer science if alternative models like the Steady State theory and the less renowned large-number hypothesis had not been developed.

16 THE FUTURE OF THE UNIVERSE

It is much easier to investigate the past than to predict the future. The theory that the universe began as a dense glob that exploded, making helium and background radiation in the process, is strongly supported by the evidence. The present expansion is a result of this past explosion. But what happens now? Will the expansion continue forever, or will it slow down, stop, and reverse itself? It seems foolish to try to predict the future, but scientists have an advantage over historians, for scientists know the forces that govern the future evolution of the universe. Gravity is the force that will cause the expansion to slow down. Is gravity strong enough to stop the expansion?

There are some philosophical reasons for hoping that the universe will be closed or open, depending on your point of view. Some people find the idea of an oscillating, recycling universe more appealing than the idea of a universe expanding forever, with galaxies getting farther and farther apart. Some find the oscillating-universe idea less acceptable than the open one. But see what rational investigation has to say.

We can approach the decision between the alternative futures directly or indirectly. The direct way is to count up all the mass in the universe and see whether its gravitational attraction is sufficient to halt the expansion. Alternatively, we can look backward in time by looking at far-distant galaxies whose light has taken a long time to reach us and see whether we can detect any changes in the expansion rate. A third approach, also indirect, is to look for additional effects caused by the amount of matter in the universe. Other approaches have been tried, too.

The verdict is still open. The literature on research in astronomy reflects changing trends. A few years ago, the preponderance of opinion favored a closed universe, in which expansion slows down, stops, and turns into a contraction. In the mid-1970s, the trend reversed, and today the evidence seems to favor an open universe, one that expands forever. What these changing trends really mean is that the evidence is not at all complete. This chapter provides a framework for understanding how astronomers are attempting to answer the question: How will the universe end?

I first consider the direct method of attacking this question: counting up all the mass in the universe and ascertaining whether there is enough mass there to close the universe. First, we must know how much mass is needed.

Escape velocity

Gravitation is the force that causes a closed universe to stop expanding and start contracting. Any pair of galaxies that are flying apart from each other feel a gravitational drag on expansion, because gravity is trying to pull the galaxies together. As the galaxies separate more and more, however, this gravitational drag weakens, because the force of gravity decreases with increasing distance. Whether gravity or expansion will win depends on the speed of the expanding galaxies. If the galaxies are traveling fast enough, they will overcome gravity and continue to separate from each other, forever. If the gravitational force is too strong and the galaxies are traveling too slowly, their speed will be insufficient and the moving galaxies will slow down, stop, and come together again.

A familiar example of the conflict between gravity and motion comes from the space program. Suppose we want to launch a rocket that will pass by Saturn and keep on traveling, as the Voyager probe will be doing. Try to do this job with a slow rocket—say one that accelerated the Voyager probe to only 3 km/sec—and we would fall short of our goal. To escape the earth's gravitational field, pass beyond Saturn, and keep going indefinitely, we need to exceed the earth's escape velocity of 11 km/sec (about 25,000 miles per hour). (See Figure 16.1.) Thus the cosmological problem can be restated: Are the galaxies traveling at escape velocity?

The magnitude of the escape velocity for any object or any collection of objects is determined by the strength of the gravitational forces that the objects produce. The gravitational force between any two objects is proportional to the product of their masses and inversely proportional to the square of the distance between them. Thus the more massive these two objects, the stronger the gravitational force and the larger the escape velocity. The farther apart the two objects, the weaker the gravitational force and the smaller the escape velocity. Hence, to determine whether the universe is open or closed, we must measure its mass, its speed, and its size.

Unfortunately, it is not easy to measure directly the mass, speed, and size of the universe. However, the problem can be cast into an alternative framework.[1] The Hubble constant is a rough measure of how fast the universe is expanding relative to its size. We can combine the values for the mass and size of the universe to give its density, or mass per unit volume. We can measure the density simply by counting, by looking at a few cubic megaparsecs of space and determining what is there. The cosmological question, What will happen to the universe? can be phrased in an alternative way that is easier to answer quantitatively: Is the mean density of matter in the universe sufficient to provide a large enough gravitational force to cause the universe to stop expanding? This critical density is generally referred to as the *closure density*, since it is the density required to make a closed universe, in which expansion will stop and turn into contraction. Numerically the closure density is $4.7 \times 10^{-30} \; h^2$ g/cm^3, where h is the

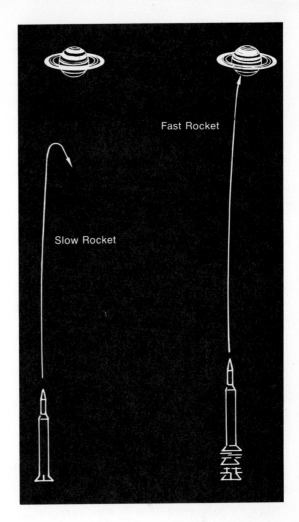

FIGURE 16-1
An interplanetary probe, such as the Voyager craft that will reach Saturn in the 1980s, must in practice escape the earth's gravitational field to reach its destination. Its speed must exceed the earth's escape velocity of 11 km/sec; if not, it will fall short. (In principle, you could be barely short of escape velocity, by 0.02 km/sec, and just make it to Saturn before you started falling back.)

Hubble constant divided by 50 km/sec per megaparsec. Since the current best value of the Hubble constant is 50, h equals 1 and the closure density is 4.7×10^{-30}, or one particle per cubic meter—one atom in a volume of space the size of a typical desk. It doesn't take much mass to close the universe, but the universe is quite empty.

The density method of deciding whether the universe is closed or open can really only show that the universe is closed. If we find enough mass to close the universe, fine; the universe is closed. If we fail to find this mass, we cannot be sure that the universe is open, for we may have overlooked some mass, or it may be invisible. Let's start counting to see whether we can reach the magic figure: the closure density of 4.7×10^{-30} g/cm³.

The mean mass density of the universe

The universe contains many types of objects: galaxies, clusters of galaxies, and probably intergalactic stars. There may also be intergalactic gas, and possibly other, stranger constituents like black holes and planet-sized objects in intergalactic space. In this section I shall consider each class of object individually, and determine the fraction of the closure density present in each form. (If you're anxious to find out what the answer is, skip ahead and look at Figure 16-5, which summarizes the results.) In addition to providing information about the future of the universe, this section should be a useful summary of just what is out there.

Visible galaxies

Galaxies are the most obvious components of the universe. It is easy, in principle, to determine how much mass is contained in galaxies. You just look at wide-field photographs of the sky, such as those taken with the Palomar 48-inch Schmidt telescope, and count galaxies. Tedious, but not difficult. You determine the brightness of each galaxy and roughly estimate its distance, so you can determine the number of galaxies per unit volume. It turns out that there are between one and three large galaxies per 100 cubic megaparsecs, so the distance to the nearest large galaxy is, on the average, between three and five megaparsecs.

We then need to convert our number counts into a mass density, or mass per unit volume, by using standard values for the masses of galaxies. It is this step that is least certain. Allowing for a reasonable margin of error, between 1 percent and 4 percent of the closure density exists in the form of visible galaxies. Some investigators have argued that invisible, massive halos surround spiral galaxies, producing an underestimate of the mass. If these investigators are right, counting galaxies is like counting tips of icebergs: we're missing most of what is there. Tiny galaxies, galaxies too dim to stand out against the night-sky background, may also be missed.

And so counting galaxies is one way to start counting the amount of matter in the universe, but it is quite possible that in using such a procedure we may miss a lot of mass. So far, we have only found 1 to 4 percent of the mass needed to close the universe. Where else could this mass be? We can search for it by trying to detect the light it emits, or we can seek it by looking for the gravitational forces it produces. Let's take them one by one.

The cosmic light

Many people have spent a lot of time and effort measuring the light of the night sky. It is a frustrating business, for once we eliminate the light of terrestrial activities we have a long way to go before finding anything of cosmological interest. Much of the background light in the night sky comes from light sources that illuminate dust grains, a ubiquitous part of the universal landscape. What happens is that some source of light shines on a dust grain that then reflects this light back to the telescope. A large contribution to the light of the night sky comes from illumination of dust in the earth's atmosphere by city lights. This source of sky brightness is becoming increasingly annoying to astronomers. More interesting sources of night-sky emission are zodiacal light, which comes from sunlight illuminating interplanetary dust grains, and diffuse galactic light, starlight reflected by interstellar dust grains. Take out the light from the Milky Way too, and what remains is the cosmic light — light from galaxies, clusters of galaxies, and the baby galaxies that might have been overlooked in counting the mass density in the form of galaxies. The massive galactic halos would also contribute to this light.

The cosmic light has not been seen. But such negative results can often contain some information. Observations indicate that the cosmic light ranges from less than 3 to 10 times as intense as the light from visible galaxies, the galaxies that contribute from 1 to 4 percent of the critical density. Could we fill the universe with stars, have a stellar density equal to the closure density, and not exceed the limits to the intensity of cosmic light? These stars could be in galaxy halos or they could be more or less uniformly distributed through the universe.

If these hypothetical stars were the right kind of stars, they could easily close the universe and not be seen. Astronomers can see something only when it is emitting light, so any type of star that doesn't emit much light could do the job. A commonly used measure of the light-emitting power of a star is its mass-to-light ratio, measured in solar units. The sun, for example, has a luminosity of 1 and a mass of 1 in these units, so the mass-to-light ratio is 1/1 or just 1. A faint, low-mass star with a mass of 0.1 of a solar mass and a luminosity of 0.002 of the luminosity of the sun has a mass-to-light ratio of $[0.1/(2 \times 10^{-3})]$ or 50. Black holes emit no light (unless they accrete matter). Thus they have an infinite mass-to-light ratio. It turns out that this unseen matter must have a mass-to-light ratio that exceeds 50 if it is to close the universe. Low-mass stars on the main sequence could do the job as long as their average mass was less than 0.1 of a solar mass. Another possible type of star is the stellar corpse: white dwarf star, neutron star, or black hole. And so it is possible to close the universe without seeing light from the mass that will eventually cause the universe to collapse (if it's there at all). Is there any other way that we can determine whether that mass is there or not?

Clusters of galaxies

The disadvantage of trying to derive cosmologically interesting information from the cosmic light is that you observe light, not mass. A more direct measurement of mass can be made by studying clusters of galaxies. If a cluster of galaxies is to be stable, the energy of motion of its component galaxies must be balanced by their gravitational energy. If the cluster is to stay together, the galaxies must be traveling at less than escape velocity from the cluster. It is generally believed that clusters of galaxies were formed only a few billion years after the Big Bang, when the galaxies were formed, so that their continued existence from then until now is good evidence that these clusters are bound together. Since the escape velocity is determined by the cluster's total mass (among other things), we can measure the mass of a cluster by observing the velocities of the galaxies in it. It turns out that if the cluster is stable, the galaxies should be traveling at 71 percent of the escape velocity on the average; so we can measure the escape velocity of the cluster directly. Knowing the size of the cluster enables us to measure its total mass. Comparing the mass of the cluster with the masses of the individual galaxies tells us whether we are missing something or not.

Several clusters have been studied in this way. The most thoroughly investigated is the Coma cluster, named for the constellation Coma Berenices, which contains this cluster of galaxies (Figure 16-2). If we attribute normal masses to the galaxies in the Coma cluster, the individual galaxies are traveling at two or three times escape velocity (Figure 16-3). If the cluster really *were* flying apart, it would no longer be visible as a cluster. We must be missing some mass.

Astronomers have searched for this missing mass in the Coma cluster, by looking for its light, and have not found the mass. It may well be in the form of massive halos of low-mass, low-luminosity stars surrounding the individual galaxies, or it may be in some other form. But suppose it's really there. It may be that the dynamics of the motions of galaxies in clusters of galaxies are not what we think. Even if we accept the Coma results at face value, they indicate that galaxies are six times as massive as previously thought, or that some form of mass exists within clusters of galaxies that is six times the mass of the galaxies in the cluster. Even if we extend these results to all galaxies, the mass density reaches only about 15 percent of closure density.

Rich clusters of galaxies like the Coma cluster are strong sources of x rays. The most logical explanation of this x-ray emission relies on an *intracluster medium,* a cloud of hot gas spread within the cluster of galaxies. Collisions between speeding electrons and protons in this gas produce the x rays that satellite observatories have detected. But can this tenuous gas close the universe? Most models indicate that the amount of mass contained in clouds of gas like these is only a small fraction of the mass needed to hold the cluster together, and a still smaller fraction of the mass needed to close the universe.

Gravitational effects

Since there are galaxies and clusters of galaxies considerably nearer to us than Coma is, you might think that we could determine the motion of

FIGURE 16-2
The Coma cluster of galaxies. (Hale Observatories photograph.)

these nearby galaxies more accurately and therefore determine whether there were any invisible mass exerting considerable gravitational effects. Many efforts along this line have been made. The result is that there does seem to be a great deal of invisible mass around, but no one has found enough of it to close the universe.

One indication of large quantities of invisible mass comes from the analysis of the motion of stars in individual galaxies. A number of workers at Princeton—including Jeremiah Ostriker, P. James E. Peebles, and Amos Yahlil—have argued that spiral galaxies must have massive halos around them or the disk would not be stable. Measurements of the orbital speeds of stars in disk galaxies keep showing more and more mass in the galaxies as these measurements are carried farther out away from the nucleus. Two independent investigations have revealed a spherical haze of very faint light surrounding a couple of bright disk galaxies.

All these bits of evidence point to one conclusion: The visible parts of galaxies are indeed like the tips of icebergs. There is a great deal of hidden mass associated with these galaxies. But how much mass lies beneath the surface of the dark ocean of intergalactic space? How much hidden mass is there? There are no positive indications that these halos contain enough mass to close the universe. However, a careful reading of the scientific literature indicates that our knowledge of the mass of these halos is sufficiently incomplete that there could be enough mass there to close the universe. The mass needed to close the universe could also be hidden in a uniformly distributed intergalactic medium; its gravitational effects would be undetectable. How else might the intergalactic medium be found?

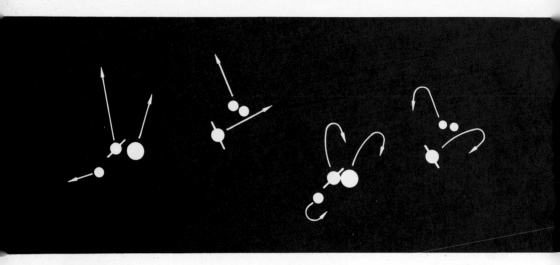

FIGURE 16-3   If the galaxies in the Coma cluster have mass normally attributed to galaxies of their type, the cluster will fly apart *(left panel)*; but if there is extra mass in the cluster, it will stay together *(right panel)*.

The intergalactic medium

Within our galaxy, the space between the stars is filled with a tenuous gas called the interstellar medium. Sometimes this gas tends to clump in clouds. You can see these clouds as the dark dust lanes in the Milky Way photograph, Figure 7-1. It is conceivable that the space between galaxies is filled with gas. Such gas has been called the intergalactic medium, even though its existence is still hypothetical. Perhaps it is here that people who want to see a closed universe will find the necessary mass. In fact, it would be quite surprising if the process of formation of galaxies were so efficient that all the matter in the universe found its way into galaxies, with none left over. An astronomer searching for this gas must imagine what form it might be in and then seek it observationally. If the intergalactic medium is there, it is likely to be mostly hydrogen.

Absence of neutral hydrogen

The intergalactic medium is not in the form of neutral hydrogen. Neutral hydrogen would absorb radiation from quasars in a part of the spectrum that the quasars' high redshift enables us to photograph from the ground (Figure 16-4). James Gunn and Bruce Peterson, then graduate students at Caltech, were the first to apply this test to a quasar spectrum. They found that there was no detectable atomic hydrogen in the interga-

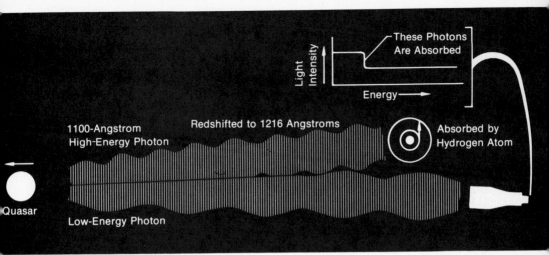

FIGURE 16-4   High-energy photons emitted by a quasar would be redshifted to 1216 angstroms and absorbed by hydrogen atoms in the intergalactic gas if the gas were there in the form of neutral hydrogen.

lactic medium. This test is remarkably sensitive; if there were only 3×10^{-7} of the closure density in the form of atomic hydrogen, they would have seen it. Our search for enough matter to close the universe does not end here.

Intergalactic ionized hydrogen

If the intergalactic medium is not neutral, it must be ionized if it is there at all. How could we find such a medium? In a hot ionized gas, electrons and protons fly around at high speeds. As an electron zips by a proton, its path is altered slightly as the proton's electrical force acts on it. The electron is slowed by the proton, losing some energy of motion. This energy is emitted in the form of a high-energy photon, and it is these photons that we hope to find. X-ray astronomers may have detected this radiation in the form of an x-ray background radiation that fills the universe. The background is definitely there, but its source is unknown.

Although the x-ray background could come from the intergalactic medium, recent work has provided little encouragement for the closed-universe model. If the intergalactic medium is massive enough to close the universe, this medium must be spread uniformly through intergalactic space or else it would produce too strong an x-ray background. Special models are needed to make a massive intergalactic medium consistent with the x-ray observations. From time to time, stories appear in the press claiming that x-ray astronomers have actually found the intergalactic medium and closed the universe. A recent report of measurements by the HEAO satellite (mentioned in Chapter 6) indicates that distant quasars, not the intergalactic medium, probably are the source of the x-ray background. At the present time, this elusive intergalactic gas is not a good candidate for the missing mass, the mass needed to close the universe.

Nonluminous matter

People who like to look really hard for the mass needed to close the universe point out a rather sobering thought for astronomers: There are many things we cannot detect in intergalactic space. We cannot see any large object outside the solar system unless it emits light. Maybe space is full of dark objects that could contribute enough mass to close the universe. What could they be? Planets, dead white dwarfs, black holes—we could never find them if they were there.

Yet when we think about it for a while, this problem looks somewhat less serious. These nonluminous objects could not be small grains, since they would act like smoke in a smoke-filled room and absorb light from distant galaxies. Anything smaller than a good-sized city would have evaporated in the 20 billion years that have passed since the Big Bang. (Smaller objects in clusters of galaxies could survive, but we have seen that the mass in clusters of galaxies is not enough to close the universe.) It is hard to see how dead intergalactic stars (black holes and so on) could be produced in sufficient numbers—remember, we need about fifty times the mass in visible galaxies to close the universe. In summary, we cannot rule out the presence of a cosmologically interesting amount of nonluminous matter, but it is difficult to understand where this stuff might come from.

The mass density of the universe

Figure 16-5 summarizes the immediately preceding thoughts. This long excursion through the matter content of the universe has provided information about what is present in the universe and what is not, although it has not brought us much closer to a direct solution of the cosmological problem. Visible galaxies do not provide enough mass to close the universe. Other luminous material in the form of massive galactic halos or baby galaxies or intergalactic stars could provide the necessary mass only if it were concentrated in the form of low-mass stars. This condition, although possible, is not very plausible. The extra mass present in clusters of galaxies, which may be there to prevent the clusters of galaxies from dissipating. is insufficient to close the universe. The most dramatic limit on a possible constituent of the universe is the limit on the amount of mass present in an un-ionized intergalactic medium, which provides less than 7×10^{-5} (molecules) or 3×10^{-7} (atoms) of the closure density. The best hope for the mass necessary to close the universe lies with an ionized intergalactic plasma. Future work with x-ray telescopes will be very helpful in unraveling the nature of the x-ray background.

This search for the mass necessary to close the universe resembles in some ways the search for the Holy Grail by King Arthur's knights. It was undertaken because of some results from indirect lines of investigation that said that the universe was closed. In the early 1960s, these results caused people to wish to look for the mass needed to close the universe. In the last few years, however, some doubt was cast on these results and some others came to light. So, since we are frustrated in our attempt to close the universe by looking for enough mass, let us consider some indirect methods.

Changes in the expansion rate

If the universe is ever going to stop expanding and start to contract, its expansion must be always slowing down. Is there any way that this slowdown can be detected? When we look out into space, we are also looking backward into time. Thus, if we can measure the redshifts of galaxies sufficiently far into the past, we can perhaps see whether the expansion was faster in the past. Figure 16-6 shows how. Suppose that the universe were practically empty, with a mass density of much less than the closure density. With such little mass, the expansion rate would be constant at all times. When we looked outward into space, the redshifts would fall along the line labeled "no slowdown." If the expansion *is* slowing down significantly, as it heads toward the moment when it turns into a contraction, it would have been more rapid in the past, producing greater redshifts. The relation between distance and redshift would follow the line labeled "slowdown," for when we look at more distant galaxies we are looking at the universe as it was in very early times, long ago. The distant galaxies, or the galaxies that we see as they were in the distant past, would be more highly redshifted, as they are in the graph. This is the reason that the curve bends upward if the universal expansion rate is changing.

If we can turn to the real world and compare real observations with the distance-redshift relation of the left panel of Figure 16-6, we can see whether the expansion of the universe is really slowing down. Yet the distances in the figure are so vast that they cannot be measured directly. The only way to allow observations to enter the picture is to suppose that there are galaxies whose luminosity is the same everywhere in the universe. The apparent brightness, or magnitude, of such galaxies (called *standard candles*) would thus be a good indicator of their distance. An analogous situation might exist on a highway. If all automobile headlights were equally luminous, we could determine the distances of cars quite accurately by looking at the brightness of headlights. The dim ones would be far away and the bright ones close. Thus, if we assume that standard candles exist in

	FORM	DENSITY (g/cm^3)	FRACTION OF CLOSURE DENSITY
	Galaxies	9×10^{-32}	0.02
	Cosmic Light from Galaxy Halos, Intergalactic Stars, etc.	Less than 9×10^{-32} (M/L)	Less than 0.02 (M/L); M/L Must Be 50 If This Is to Close the Universe
	Extra Mass in Clusters	7.2×10^{-31}	0.15 If the Coma Cluster Is Typical
	Neutral Hydrogen	Less than 1.5×10^{-36}	Less than 3×10^{-7}
	Molecular Hydrogen	Less than 3×10^{-34}	Less than 7×10^{-5}
	Intergalactic Ionized Gas	?	Maybe 1? Maybe 0?
	Closure Density	4.7×10^{-30}	

FIGURE 16-5   Matter in the universe

HOW YOU FIND IT	SOURCES OF ERROR	DEPENDENCE ON H
Count Galaxies	Don't Know Galaxy Masses Very Well	—
Look for the Cosmic Light	What Stars, with What Masses, Make Up the Cosmic Light (If It's There)	—
Analyze Motions of Galaxies in Clusters	Limitations of Our Understanding of Cluster Dynamics	$1/H$
Look at Quasar Spectra	Are Quasars Cosmological?	$1/H$
X-Ray Background	What Is the X-Ray Background? Maybe It's Something Else	

the universe, we can compare real observations with the relationship between magnitude and redshift of the right panel of Figure 16-6 to see whether the universal expansion is slowing down, to see whether the universe is closed.

The different curves in Figure 16-6 are labeled with the symbol q, which is called the *deceleration parameter,* and which measures how rapidly the universe is slowing down. Very roughly, it measures the change of the Hubble constant, fractionally, in one Hubble time. The number q is equal to one-half the ratio of the actual density of the universe to the closure density, so that if q is bigger than 0.5, the universe is closed. If q is smaller than 0.5, the universe is open. If the actual density of the universe is the density in visible galaxies, 0.02 of the closure density, then the value of q is 0.01 and the universe is open. If standard candles exist, all we have to do is make observations of their magnitudes and redshifts, enter the figures in the right-hand panel of Figure 16-6, and see what sort of a universe we live in.

In the 1960s, observers discovered that the brightest galaxies in *rich clusters* of galaxies—that is, clusters of thousands of galaxies—tended to have the same luminosity wherever they were found in space. One way to

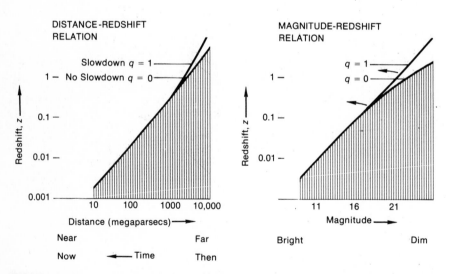

FIGURE 16-6   Is the expansion of the universe slowing down? *Left panel:* If the expansion is slowing down ($q = 1$), the redshifts of galaxies will be larger in the past or at great distances than they would be if the expansion rate did not change ($q = 0$). *Right panel:* For an object whose luminosity is constant, observed magnitude depends on its distance from observer. Near galaxies will be bright and distant galaxies will be dim. The distance-redshift relation of the left panel can thus be transformed into the magnitude-redshift relation of the right panel. The arrows show what happens if galaxies were more luminous in the past.

visualize this is to look at the lower left-hand corner of the graphs in Figure 16-6. No matter what kind of universe we live in, a standard candle with a redshift z of less than 0.1 should fall along the straight line, since the change in the expansion rate of the universe has little effect here. The fact that the observed magnitudes of the brightest elliptical galaxies in rich clusters fall along this line, as is shown in Figure 16-7, supports this view that the brightest galaxies in rich clusters have the same luminosity everywhere, and are thus standard candles.

There are some problems, though, which are indicated by the complicated label on the horizontal axis of Figure 16-7. Astronomers can measure redshifts of galaxies in a straightforward manner. We just obtain a spectrum of light from the galaxy, identify the lines in it, measure their wavelength, see how much they have shifted, and convert to z = wavelength shift per wavelength when the photon left the galaxy (or $\Delta\lambda/\lambda$). The magnitudes are more difficult. The symbols on the label for the horizontal axis all refer to different corrections. I left this caption on the figure not to torture you with all sorts of symbols to remember but to point out how complex the correction procedure is. For those of you who want to know what these symbols mean, here are some brief indications.

The galaxy absorbs some light, so we have to take that out, A_V. When we measure magnitudes of galaxies of different redshifts, we look at different wavelengths as the spectrum of the galaxy is redshifted. This correc-

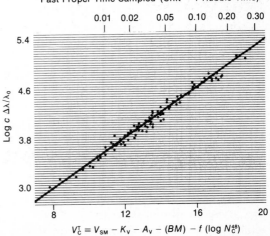

FIGURE 16-7   Magnitude-redshift diagram for different galaxies. The vertical axis plots redshifts, and the horizontal axis plots magnitudes, the same as in the right panel of Figure 16-6. K_V, A_V, (BM), and f (log N_c^{48}) are all correction terms. (From A. Sandage and E. Hardy, *Astrophysical Journal*, vol. 183, p. 755, 1973, published by the University of Chicago Press. Copyright © by the American Astronomical Society. All rights reserved.)

tion and corrections for the geometry of the universe are called K_V. The last two symbols, (BM) and f (log $N_C{}^{48}$), sound horrible, and indeed they are. As we look backward into time, we are looking at fainter and fainter galaxies, and we tend to pick out the brightest galaxies to observe, so we are no longer looking at standard candles. As we shall see later, the effects of such selections may be large.

For the moment, let us assume that we have made all the corrections unerringly and blast ahead to see what happens. A graph like Figure 16-7 is not too illuminating, as the overall slope of the data to the right obscures the effects from evolution of the universe. Figure 16-8 shows the same data replotted in the sense that deviations from the $q = 1$ line are shown on the horizontal axis. At first glance, the data seem to indicate that the universe is closed, because we can use curve-fitting theory to see which theoretical curve best fits the observations. Mathematically, we find $q = 1.13$. If we leave out the three most distant clusters, however—for which the data are not very certain—the value of q becomes 0.65, with a large uncertainty. It might seem that the universe is closed. But wait.

We must be careful not to take these accurate-sounding values for q too literally. Look again at Figure 16-8. The uncertainties in the data, as shown by the scatter of the observations around the curves, are tremendous. One nice feature of curve-fitting theory is that it enables us to estimate our possible error. Including the observations of the three most distant galaxies, it turns out that there is a 50 percent probability that q is somewhere between 0.77 and 1.46. Fine, you say; so the universe is closed. You should remember that the nice-sounding 50 percent probability means also that there is a 50 percent probability that q is *not* within that range, or that if you interpret this data literally, there is a 10 percent probability that the universe is open, even if you assume that all the corrections have been applied correctly. The data seem to favor a closed universe.

Allan Sandage of the Hale Observatories has published data of this sort for some time now. Earlier conclusions from these data, indicating that we live in a closed universe, spurred the search for the missing mass, chronicled in the last section. Yet, when we accept the data literally, there is a crucial assumption involved. Remember, we have been assuming that the brightest galaxies in clusters of galaxies are standard candles, equal in luminosity wherever and whenever they are seen in the universe. We are looking backward in time to see changes in the redshift. Perhaps there are changes in the luminosity of elliptical galaxies too? Early work indicated that these brightest galaxies, generally ellipticals, had not changed in luminosity recently, over the last two billion years or so. It was only in the 1970s that the real magnitude of the evolutionary effects was understood.

In general, we expect that galaxies would have been more luminous in the past. As a galaxy evolves, massive stars in it die, and are no longer seen. If galaxies had been more luminous in the past, they would be moved

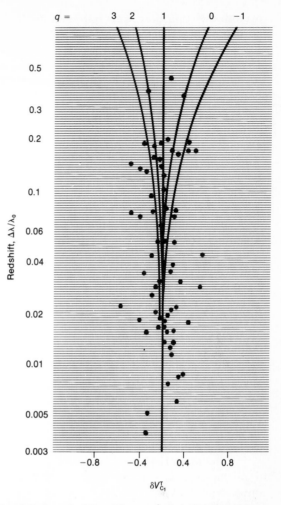

FIGURE 16-8 Figure 16-7 replotted, so that the difference $|\delta V_{c_1}^{\tau}|$ between the observations and the $q = 1$ line is on the horizontal axis. (From A. Sandage and E. Hardy, *Astrophysical Journal,* vol. 183, p. 755, 1973, published by the University of Chicago Press. Copyright © by the American Astronomical Society. All rights reserved.)

along the arrows shown in the right panel of Figure 16-6. If we ignored the effects of this evolution, the universe would seem more closed than it really is.

Beatrice Tinsley of the University of Texas calculated what the effects of galactic evolution would be. Her work indicated that they were unpleasantly large. She took a model of an elliptical galaxy, including different numbers of different types of stars. She let these stars evolve, according to the evolutionary tracks of the stellar evolutionists, and watched, through the computer, what happened to her model galaxy. She found that the evolutionary effects are as large as the effects that come from the changes in the expansion rate. It is not completely certain what these effects are, as the theory that goes into them is not completely worked out. But it seems that the results obtained from diagrams like Figure 16-8 should be viewed with caution.

It is even possible that the effects of the evolution of galaxies go in the opposite direction. Scott Tremaine and Jeremiah Ostriker of Princeton postulate that two galaxies that collided might cannibalize each other, coalesce, and form a larger, more luminous galaxy. Such cannibalism by galaxies would be particularly pervasive in dense clusters of galaxies. It is just those dense clusters that are used for this test of the future of the universe. Do galaxies in fact eat each other? What happens when they do? Once again, our efforts to find out whether the universe is going to keep on expanding are confounded by the complexity of galaxy evolution.

There is yet another problem with this once-classic cosmological test. James Gunn and J. B. Oke of Caltech have shown that the galaxies that one selects to put on graphs like Figures 16-7 and 16-8 critically affect the results. We tend to pick out the brightest and hence the most luminous galaxies to observe because those are the ones we can see. The less luminous ones are too faint to observe. In the case of large redshifts, the most luminous galaxies are the only ones we can see. Gunn and Oke obtained their own sample of galaxies in a less biased way, plotted their galaxies on a diagram like Figure 16-7, and found that their data indicated that the universe was open and would expand forever. They reached their conclusion without considering the possible effects of evolution of galaxies. In late 1978, Sandage and Jerome Kristian made another effort and obtained $q = 1.6 \pm 0.4$, again without taking into account evolutionary effects.[2]

The main reason that this test is so difficult to implement in the case of galaxies is that galaxies with redshifts exceeding 0.3, where the open-universe and closed-universe curves start to diverge considerably (see Figures 16-6 and 16-8), are at the limit of visibility. Astronomers have long hoped to use quasars to probe the future of the universe, since quasars are much more powerful than galaxies and can be seen at far greater distances. To use quasars as a cosmological probe, we must be able to use some intrinsic property of the quasars to estimate their luminosity. Observations of nearby quasars indicate that the intensity of a carbon emission line seems to be a good indicator of luminosity. A hydrogen emission line may also be a possible indicator of luminosity. We then take a sample of quasars with different redshifts but similar line strengths and enter their statistics on a diagram like Figure 16-6, since we hope that quasars with similar line strengths have similar luminosities. Recent applications of this test produced evidence favoring a closed universe. But if this test is to prove useful, we shall have to allow for evolutionary effects. So far the effects of evolution of quasars have been ignored because no one knows how to calculate what they are.

The use of magnitude-versus-redshift diagrams like Figure 16-6 has a long history. In the 1920s Edwin Hubble discovered Hubble's Law by plotting galaxies on such a diagram. His work involved just a tiny little bit of the lower left-hand corner of diagrams like Figure 16-6. A major part of observational cosmology has involved the extension of the measurements of galaxy redshifts into the upper right-hand corner of the diagram, where changes in the expansion rate become apparent. The need to extend measurements on this diagram was one of the principal motivations for the construction of the 200-inch telescope and was a major part of Hubble's life work. This effort is therefore sometimes called the *Hubble program*. It now appears that the Hubble program tells us more about the evolution of galaxies and less about cosmology. It is tempting, perhaps, to dismiss the Hubble program as a cosmological test. If we can learn more about the evolution of galaxies, or learn how to apply this same diagram to quasars, it may turn out to be one of the ways that we can find out whether or not the universe is going to expand forever.

Thus the direct approach to the question of the infinite expansion of the universe has failed to produce an unequivocal answer. A search for the mass needed to close the universe did not turn up the mass needed, but did indicate several places in which it could be hiding. Examination of possible changes in the expansion rate has, so far, produced no results because of the likely effects of the evolution of galaxies on the measurements. There are some indirect approaches to the question of the expansion of the universe. Let us now turn to these.

Deuterium

Deuterium is heavy hydrogen. Ordinary hydrogen has just one proton in its nucleus, whereas deuterium has one proton and one neutron. In 1973 and 1974, with the first discoveries of deuterium in astronomical objects, a new way of determining whether the universe is closed or open appeared. Deuterium can serve as a probe into the mass density of the early universe. We can then use the mass density obtained from such a study to determine whether the universe now contains enough mass to be closed.

When the early universe fused hydrogen to form helium, deuterium was the first step on the road. Most of the deuterium that was made subsequently became helium, for the universe was dense enough at that time for the deuterium to fuse with neutrons and protons, building up to two protons and two neutrons—helium. A small fraction of the deuterium managed to survive the early helium-burning epoch and survive to the present day. How much deuterium managed to survive depends on the density of the universe. Directly related to the density of the universe was the likelihood that a primeval deuterium atom would collide with other particles and become something heavier. A dense, closed universe would not have much cosmological deuterium, since all the deuterium would have become helium. In an open, rarefied universe, the deuterium would still be around. These results are displayed quantitatively in Figure 16-9. By measuring how much deuterium is around, we can then determine the present density of the universe by reading it from a curve like Figure 16-9. Here is another way to answer the cosmological question: Is the universe closed?

Applying this idea was seriously limited by the complete absence, before 1972, of any discovery of deuterium anywhere in the universe except on the earth. In the next four years, however, discoveries of deuterium popped up in the literature like mushrooms, as six discoveries of deuterium in five objects were made. Yet finding deuterium is not the same thing as measuring its cosmic abundance. The only straightforward measurement of abundance was made by Princeton astronomers on the Copernicus satellite, which measured absorption lines from the interstellar medium in the star Beta Centauri. The satellite was needed for the observations because the deuterium lines lie in the ultraviolet part of the spectrum, which is absorbed by our atmosphere. The Princeton group found a deuterium abundance of 1.5×10^{-5}.

On the basis of the idea that this deuterium is really a relic of the Big Bang, all we have to do is go to Figure 16-9 and read the present density of

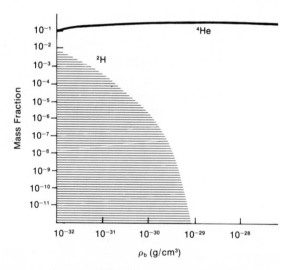

FIGURE 16-9   The abundance of helium and deuterium produced in the Big Bang versus the present density of the Universe ρ_b. The closure density is 4.7×10^{-30}, corresponding to a deuterium abundance of 10^{-6}. (Adapted from R. V. Wagoner, *Astrophysical Journal*, vol. 179, p. 349, 1973, published by the University of Chicago Press. Copyright © by the American Astronomical Society. All rights reserved.)

the universe from the curve. Try it. It turns out to be about 10^{-30} g/cm³, or about 20 percent of the closure density. Nuclear reactions in the Big Bang would go further in a closed universe, producing less deuterium. Therefore, if the deuterium we see was made in the Big Bang, the density of the universe is one-fifth of the closure density. The deuterium evidence indicates, but does not prove, that the universe will expand forever.

But this test, like all the other tests of the cosmological question in this chapter, contains a trap (see Table 16-1). We do not know where this deuterium came from. It may have been made since the beginning of the universe, and as a result the universe could be a good bit denser than a straightforward interpretation would indicate. A closed universe would have a primordial deuterium abundance of 10^{-6}, and the deuterium abundance could have been built up to the observed value of 1.5×10^{-5} by processes occurring subsequently. Theorists are searching for alternative explanations of how this deuterium could have been made. A couple of early ideas turned out not to work on closer examination. In view of the absence of knowledge about the source of the deuterium in the universe, it is too early to tell much about the future of the universe from deuterium. Deuterium has now been discovered in a number of other places in the interstellar medium. Observational and theoretical astronomers are slowly piecing together the picture of just where this deuterium came from. At the moment, the deuterium test seems to indicate that the universe is open and will continue to expand forever.

TABLE 16-1   COSMOLOGICAL TESTS

TEST	VERDICT	QUALIFICATIONS
Measuring the mass	Open	Maybe there is some mass we do not see
Changing rate of expansion (galaxies)	Closed (1970) ??? (1975)	Effects of evolution and selection of galaxies are superimportant
Changing rate of expansion (quasars)	Closed?	Can we use emission lines to indicate luminosities of quasars?
Deuterium	Open	Maybe the deuterium came from some source other than the Big Bang
Ages	Open	Substantial uncertainties

Cosmic ages

Another indirect approach to determining whether the universe will expand forever involves measuring the ages of objects in the universe. The Hubble time is equal to the age of the universe only if the rate of expansion has been constant. Were the expansion faster in the past, it would have taken less time for galaxies to reach their present locations and the universe would be younger than the Hubble time. It turns out that a universe that is older than two-thirds of a Hubble time must be open. Less than two-thirds of a Hubble time must have elapsed since the Big Bang if the rate of expansion has changed enough since then to ultimately bring the expansion to a halt.

We can't measure the age of the universe directly. No one wound up a clock and dropped it into the Big Bang so that we can now go out, read it, and see whether the age of the universe is greater or less than two-thirds of a Hubble time. However, there are some natural clocks that can tell us the age of the oldest stars and when the heavy elements were synthesized. Chapter 14 described these clocks. For definiteness, take a Hubble constant of 50. Two-thirds of a Hubble time is 13 billion years. The ages of the oldest clusters of stars lie between 14 and 16 or 14 and 21 billion years, depending on whose data we pick. Atoms are between 7 and 20 billion years old. These values are just barely consistent with a closed universe and a Hubble constant of 50. They tend to favor an open universe, but uncertainties are so large that this test cannot be called definitive. The Hubble constant—and thus the Hubble time—is itself uncertain. Most people tend to believe that if the Hubble constant isn't 50, it is larger rather than smaller. A Hubble constant larger than 50 shortens the Hubble time and strengthens the verdict in favor of an open universe. With the uncertainties, this cosmological test is a very equivocal one, but it does provide another piece of evidence, even if weak, in favor of the idea that the universe is going to continue to expand forever.

This chapter ends, as this book does, on a note of uncertainty. The results presented here are most uncharacteristic of one view of science as a nice, neat picture, but most characteristic of the confused current state of astronomy. At the moment, our knowledge of the future of the universe is clouded.

Determining the density of the mass of the universe provides a very tentative conclusion that the universe is open. The mass contained in galaxies is only 2 percent of the mass necessary to close the universe, and the extra mass present in the Coma cluster would—if real and if present in other clusters of galaxies—provide only 15 percent of the closure density. The elusive intergalactic medium could close the universe, but has not been found yet. If the intergalactic medium is there, it exists in the form of an ionized gas.

Looking backward to determine the rate of change of the universal expansion was at one time thought to be a way of answering the question as to whether the universe is closed or not. At the present time, the roles of the evolution of galaxies and the effects of selection of galaxies remain to be elucidated. Preliminary indications show that these effects, when included, turn a closed-universe interpretation of the data into an open one. The principal impact of recent work on these effects is to indicate that they are large.

If the deuterium that has recently been found in interstellar space is a relic of the primeval fireball, then the universe is open, since it was sufficiently rarefied in the Big Bang to allow the deuterium to remain and not be processed further. But since it is not yet clear where the deuterium was made, the deuterium test is not yet conclusive.

Some other, recently applied tests also provide equivocal results. The intensities of emission lines may indicate how luminous quasars are, and may make it possible for us to use these powerful beacons for cosmology. If we can use these emission lines for this purpose, and if we can neglect the effects of the evolution of galaxies, then observations indicate that this test tentatively provides evidence that the universe is closed. But on the contrary, the observed ages of the oldest star clusters, combined with current values for the rate of expansion of the universe, indicate that the expansion rate has not changed and will not change, and that the expansion will go on forever.

Although we have the tools to analyze the future of the universe, our application of them is limited by the intervention of many other effects. We have to unravel a complex observational situation before we can decide the question: Is the universe open or closed? Although the trend of the results in the 1970s was to reverse the trend of the 1960s, it is dangerous to rely on trends that may reverse again. The present trend of the observations in the direction of an open universe may not last. But then again it may.

If cosmology is to be a branch of science instead of a philosophical recreation, there must be some data that enable the real world to enter. In the 1920s and 1930s, the discovery of the expansion of the universe gave us our first handle on the evolution of the universe as a whole. The logical interpretation of that expansion was the Big Bang theory. Until 1960, however, it was impossible to decide between the Big Bang and Steady State theories, because the expansion of the universe was the only piece of information that we had on its evolution. We could measure the expansion rate and determine the age of the universe, but we could not do so accurately.

Subsequent developments have greatly refined our knowledge of cosmology. The discovery of the primeval fireball, the background radiation filling the universe, which is a Big Bang relic, confirmed the Big Bang picture. The existence of a cosmic abundance of helium strengthens the Big Bang theory, but the interpretation of the helium data is still not certain. If the radio sources and quasars are very distant objects and we are seeing them as they were when the universe was still young, their abundance in the early universe rules out the Steady State theory in its original, strictly interpreted form. As a result we are fairly confident about our ideas of the past history of the universe.

We are much less certain of the future. Will the universe go on expanding forever? A search for the necessary mass to slow down the expansion did not succeed in uncovering it, but there are places that undetected mass could be hiding. Efforts to look to the past to detect changes in the expansion rate are clouded by problems of interpretation. Recent discoveries of deuterium seem to be pointing in the direction of an open universe, but it is not yet clear where this deuterium came from.

The Big Bang theory leaves one unanswered question. Who *created* the material that exploded as the Big Bang? For this, the astronomer has no answer. We may be able to look back to the early seconds of the evolution of the universe, but our vision stops there. This book ends by leaving the problems of creation to the philosopher and the theologian.

OBSERVATIONAL FACT	A microwave background with a blackbody spectrum exists Everywhere we have looked so far, the universe seems to be about 30 percent helium Deuterium exists in the universe The universe is expanding
CONCRETE THEORY	Big Bang universes make primeval fireball radiation and primordial helium; the Steady State universe contains neither
INFORMED OPINION	Because the microwave background is radiation from the primeval fireball and because helium was made in the first 20 minutes, the Steady State theory is wrong Other alternative cosmologies also conflict with observation The Hubble constant is 50 km/sec per megaparsec, so the universe is less than 20 billion years old Currently the evidence indicates that the universe is open, but all tests relative to this question are slightly ambiguous
UNANSWERED QUESTIONS	Did the deuterium come from the primeval fireball? (If it did, the universe is open.) Can we overcome the uncertainties due to the effects of evolution of galaxies and the effects of selection and measure changes in the expansion rate of the universe? Is there enough mass to close the universe? (We haven't found it so far.)
SPECULATION	What preceded the Big Bang? The cosmological singularity? Is there any connection between this singularity and black holes?

GLOSSARY; SUGGESTIONS FOR FURTHER READING; NOTES

GLOSSARY

Terms marked with an asterisk (*) are discussed in more detail in the Preliminary section.

ABSORPTION LINE
: A feature of a spectrum of electromagnetic radiation in which there is a shortage of photons of a particular wavelength

ACCRETION DISK
: Disk of gas surrounding a black hole. Swirling currents in this disk cause the gas to heat up and emit x rays.

*ANGSTROM UNIT
: 10^{-8} centimeter

ANOMALY
: Discrepancy between theory and observation, between the model world and the real world. One that persists may lead to a crisis.

ANTIMATTER
: Stuff that, on being brought into contact with matter, causes both matter and antimatter to disappear, leaving many gamma rays. Antimatter has so far been found only in laboratories.

*APPARENT BRIGHTNESS (APPARENT MAGNITUDE)
: Measures of how bright something appears in the sky

*ASTRONOMICAL UNIT
: Mean distance between the earth and the sun (1.495985×10^8 kilometers)

B STAR, B SUPERGIANT
: A hot star. A supergiant is an extremely large and luminous star. The letter B refers to the star's spectral class.

BINARY STAR (BINARY SYSTEM)
: A pair of stars orbiting each other

*BLACKBODY RADIATION
: Radiation emitted by an object with no intrinsic color. It has a characteristic spectrum, or distribution of intensity in different wavelengths, that is independent of the nature of the object emitting it.

BLINK MICROSCOPE
: Device that allows alternate viewing of two photographs of the same part of the sky. It is used to discover variable stars and moving objects like asteroids.

CEPHEID VARIABLES	Stars that vary in luminosity in a regular, periodic fashion. If you know the period of a Cepheid variable, you can determine its luminosity.
CONTINUUM RADIATION	Radiation that is not concentrated at particular wavelengths or photon energies
COSMOLOGICAL REDSHIFT	Shift in spectrum lines due to the expansion of the universe
CRISIS	Stage in a scientific revolution at which an anomaly has grown important enough to cause people to question the validity of the prevailing paradigm
DECELERATION PARAMETER	Number that measures rate of change of the expansion of the universe. Sometimes called q_0. If the value is larger than 0.5, the universe is closed and will eventually stop expanding.
DECLINATION	Celestial coordinate used to measure star positions, roughly the equivalent of latitude
*DEGREE	Unit of measure of angle. The sun and moon are each half a degree across in the sky.
DENSITY	Amount of matter per unit volume. Not to be confused with mass, which measures amount of matter.
DEUTERIUM	A type of hydrogen atom, with a nucleus containing a neutron in addition to the bare proton nucleus of ordinary hydrogen. The abundance of deuterium in the universe has been used as a cosmological test.
DISCORDANT REDSHIFT	Hypothetical redshift in the spectrum lines of a quasar that arises from no known physical cause. (Chapter 12 discusses the evidence for and against the existence of discordant redshifts.)
DISTANCE MODULUS	Difference between apparent and absolute magnitudes of an object. (See Chapter 14.)
DOPPLER SHIFT	Change in wavelength that photons make when they travel from an object moving relative to the observer (See Chapter 5.)
*ELECTRON	One of the three types of particles found in an atom. It has a negative electrical charge and is found around the nucleus of the atom.
*ELECTRON VOLT	Unit of energy, equal to 1.60207×10^{-12} erg. Atomic energy levels are typically separated by a few electron volts.
EMISSION LINE	A feature of a spectrum of electromagnetic radiation in which there is an excess or concentration of photons at a particular wavelength
*ERG	Fundamental unit of energy. A two-gram insect crawling along at a speed of 1 centimeter per second has an energy of motion of 1 erg.

GLOBULAR CLUSTER	Group of old stars, generally found in the halo of a galaxy
HOMOGENEOUS	Uniform, not lumpy
HUBBLE'S LAW	Relation between velocity of recession of an object and its distance. Hubble's Law, produced by the expanding universe, can be mathematically expressed as $v = HD$, where H is Hubble's constant, v the velocity of recession, and D the distance.
*KELVIN	Unit for measuring temperature; degrees Celsius above absolute zero
KEPLER'S THIRD LAW	Relation between period, separation, and mass of two objects orbiting each other. It is generally used to measure the mass of the orbiting objects; you can determine the characteristics of the orbit by observation.
*KILOPARSEC	1000 parsecs
LUMINOSITY	Amount of energy that an object radiates into space every second
*MAGNITUDE	Scale astronomers use to measure brightness of objects
MAIN-SEQUENCE STAR	Star that supports itself by fusion of hydrogen in its center. The sun is a main-sequence star.
*MEGAPARSEC	10^6 (one million) parsecs
*MICROMETER	10^{-6} meter
*MICROSECOND	10^{-6} second
*MICROWAVE RADIATION	Radio-frequency radiation of extremely short wavelength, and infrared radiation of long wavelength, generally radiation with wavelengths roughly between 100 micrometers and 1 centimeter
*MINUTE OF ARC	Unit of measure of angle, 1/60th of a degree
NEUTRINO	Particle that travels at the speed of light and rarely reacts with anything
*NEUTRON	One of the particles found in an atomic nucleus. Although roughly equal in mass to the proton, it has no electrical charge.
NORMAL SCIENCE	The usual state of science, with all research based on a prevailing paradigm
PARADIGM	A set of scientific laws that is a basis for scientific work. See Chapter 1.
*PARSEC	3.2615 light-years, 206264.8 astronomical units, or 3.1×10^{13} km. In our part of the galaxy, stars are a parsec apart, on the average.
PHOTOGRAPHIC PLATE	Piece of glass coated with photografhic emulsion. Film is celluloid coated with emulsion. Astronomers generally use plates instead of film because glass does not stretch.
*PHOTON	A light particle
PHOTOSPHERE	Visible surface of a star. This term is generally used to describe the sun.

PLASMA	An ionized gas, composed of electrons and atoms missing one or more electrons
POSITRON	A positively charged antielectron. When a positron and an electron collide, they annihilate each other.
PROTON	One of the particles found in the nucleus of an atom. It has positive electrical charge.
RECOMBINATION	Process in which an electron and an ion collide, stick together, and form an atom (or less highly charged ion). This term is also used to refer to the time that matter in the universe recombined, 1,000,000 years after the beginning.
RIGHT ASCENSION	Celestial coordinate used to measure star positions, roughly the equivalent of longitude
SCHMIDT TELESCOPE	Type of telescope that has a wide field of view, useful in survey work
*SECOND OF ARC	Unit of measure of angle, 1/60 minute of arc, 1/3600 degree
SINGLE-LINED SPECTROSCOPIC BINARY	Star that orbits an invisible object. The star's motion shows up through changing Doppler shifts in the star's spectrum.
*SPECTRAL CLASSIFICATION	Scheme that enables one to determine the general properties of a star by examination of its spectrum. Spectral classes are denoted by letters: O, B, A, F, G, K. M.
*SPECTRUM	Graph or photograph of the intensity of radiation from an object as it varies with wavelength

SUGGESTIONS FOR FURTHER READING

Items marked by asterisks (*) should be intelligible to anyone who has read this book. Other items are more technical.

PRELIMINARY

*Abell, George O. Exploration of the Universe, 3rd ed. Holt, Rinehart, and Winston, New York, 1975.

*Asimov, Isaac. The Universe: From Flat Earth to Quasar. Avon, New York, 1966. (In paperback)

*Hartmann, William K. Astronomy: the Cosmic Journey. Wadsworth, Belmont, CA, 1978.

*Jastrow, Robert, and Malcolm Thompson. Astronomy: Fundamentals and Frontiers, 3rd ed. Wiley, New York, 1977.

*Pasachoff, Jay M. *Contemporary Astronomy.* Saunders, Philadelphia, 1977. (Available in one-year, one-semester, and with-calculus versions.)

*Shipman, Harry L. *The Restless Universe: an Introduction to Astronomy.* Houghton Mifflin, Boston, 1978.

*Zeilik, Michael. *Astronomy: The Evolving Universe,* 2nd ed., Harper and Row, 1979.

These are all basic texts. There are many, many others.

CHAPTER 1   Introduction: the Violent Universe

*American Astronomical Society. *A Career in Astronomy.* (A 24-page brochure describing astronomical careers, available for 25 cents from the Education Office, American Astronomical Society, University of Delaware, Newark, DE, 19711.)

Avrett, E. H., ed. *Frontiers of Astrophysics.* Harvard University Press, Cambridge, 1974. (Advanced undergraduate level)

*Calder, Nigel. *The Violent Universe.* Viking, New York, 1969. (The scientific results are a little old, but this book contains excellent descriptions and photographs of the observatories at which the discoveries described in this book are made.)

*Friedman, H. *The Amazing Universe.* National Geographic Society, Washington, D.C., 1975.

*Gingerich, Owen, ed. *New Frontiers in Astronomy.* Freeman, San Francisco, 1975.

*Gingerich, Owen, ed. *Cosmology + 1.* Freeman, San Francisco, 1977.

*Golden, F. *Quasars, Pulsars, and Black Holes.* Scribners, New York, 1975.

*Kuhn, Thomas S. *The Structure of Scientific Revolutions.* University of Chicago Press, Chicago, 1967. (The classic exposition)

*Synge, J. L. *Talking About Relativity.* Elsevier, New York, 1970.

*Toulmin, Stephen. *Foresight and Understanding.* Harper, New York, 1961. (He argues that prediction is not the essence of science.)

CHAPTER 2   Stellar Evolution: To the White-Dwarf Stage

Chiu, Hong-Yee, and Amador Muriel, eds. *Stellar Evolution.* MIT Press, Cambridge, 1972. (Technical)

Clayton, Donald D. *Principles of Stellar Evolution and Nucleosynthesis.* McGraw-Hill, New York, 1968. (Graduate-level text)

*Greenstein, Jesse L. "Dying Stars." In *Frontiers in Astronomy,* ed. Owen Gingerich. Freeman, San Francisco, 1971.

*Jastrow, Robert. *Until the Sun Dies.* Harper, New York, 1977.

*Meadows, A. J. *Stellar Evolution,* 2nd ed. Pergamon, Oxford, 1978.

CHAPTER 3   Supernovae, Neutron Stars, and Pulsars

*Bell-Burnell, S. Jocelyn. "Petit Four." In M. D. Papagiannis, ed., *Eighth Texas Symposium on Relativistic Astrophysics, Annals of the New York Academy of Sciences, Vol. 302,* 1977. (Jocelyn Bell-Burnell's account of the discovery of pulsars)

*Clark, David H., and F. Richard Stephenson. *The Historical Supernovae.* Pergamon, Oxford, 1977.

*Hewish, Anthony. "Pulsars and High Density Physics." Nobel Prize Lecture, 1974. *Science 188* (June 13, 1975), 1079–1083.

*Kirshner, Robert P. "Supernovas in Other Galaxies." *Scientific American* (December 1976), 89–101.

*Lovell, A. C. B. *Out of the Zenith.* Harper, New York, 1973. (Historical)

Manchester, Richard N., and Joseph H. Taylor. *Pulsars.* Freeman, San Francisco, 1977. [Technical. A shorter treatment by the same authors is "Recent Observations of Pulsars," by Taylor and Manchester, *Annual Review of Astronomy and Astrophysics* 15 (1977), 19–44.]

Smith, F. G. *Pulsars.* Cambridge University Press, Cambridge, England, 1977.

Shklovsky, I. S. *Supernovae.* Wiley, New York, 1968.

*Verschuur, Gerrit L. *The Invisible Universe.* Springer-Verlag, New York, 1974. (Radio astronomy for the nonspecialist)

CHAPTER 4   Journey into a Black Hole

*Asimov, Isaac. *The Collapsing Universe: The Story of Black Holes.* Chicago, Walker, 1977. (Not just about black holes; this book is an excellent introduction to astronomy. Also available in paperback.)

*Kaufmann, William J., III. *Relativity and Cosmology,* 2nd ed. Harper, New York, 1977. (Includes special and general relativity, black holes, and speculative ideas; written for the nonspecialist.)

Misner, C. W., K. S. Thorne, and J. A. Wheeler. *Gravitation.* Freeman, San Francisco, 1973. (A superior text for a senior or graduate-level course on Einstein's theory of gravitation. This book, and Kip Thorne's course based on it, has both inspired and enlightened me.)

*Penrose, Roger. "Black Holes." *Scientific American* (May 1972), 38–46.

*Penrose, Roger. "Black Holes." In *Cosmology Now,* ed. Laurie John. BBC Publications, London, 1974. (The book is a good reference on black holes and especially on cosmology.)

*Thorne, Kip S. "Gravitational Collapse." *Scientific American* (November 1967), 88–98. (Excellent summary)

*Wald, Robert M. *Space, Time, and Gravity.* University of Chicago Press, Chicago, 1977.

CHAPTER 5   The Search for Black Holes

*Gursky, Herbert, and Edward P. J. van den Heuvel. "X-Ray Emitting Double Stars." *Scientific American* (March 1975), 24–35.

*Jones, Christine, William Forman, and William Liller. "X-Ray Sources and Their Optical Counterparts." A three-part series in *Sky and Telescope* 48 (November 1974), 289–291; 48 (December 1974), 372–375; 49 (January 1975), 10–13. Part 1 of this series is most directly relevant to this section.

Thorne, Kip S. "The Search for Black Holes." *Scientific American* (December 1974), 32–43.

These three articles are the most recent reviews at a nontechnical level. The first and third are available in the collection *New Frontiers in Astronomy,* listed above under Chapter 1. Reviews at the more technical level are:

G. R. Blumenthal and W. H. Tucker, "Compact X-ray Sources, *Annual Review of Astronomy and Astrophysics* 12 (1974), 23–46.

M. Oda, "Cygnus X-1: a Candidate of the Black Hole," *Space Science Reviews* 20 (1977), 757–813.

Many of the fascinating details of the stories of Cygnus X-1 and Epsilon Aurigae have been confined to the technical literature. You nonspecialists should not be put off by the fact that the papers below are in technical journals. If you are willing to accept that you will not understand all of each paper, you can learn a good bit from reading them. I list some of the highlights.

Epsilon Alistair G. W. Cameron, "Evidence for a Collapsar in the Binary System
Aurigae Epsilon Aur," *Nature* 229 (1971), 178–179, originally proposed the black-hole idea. R. Stothers, "Collapsars, Infrared Disks, and Invisible Secondaries of Massive Binary Systems," *Nature* 229 (1971), 180–183, agreed. For counterarguments, see P. Demarque and S. C. Morris, "Is There a Black Hole in Epsilon Aurigae?" *Nature* 230 (1971), 516–517. Recent models of Epsilon Aurigae are R. E. Wilson, "A Model of Epsilon Aurigae," *Astrophysical Journal* 170 (1971), 529–539, and Su-Shu Huang, "Interpretation of Epsilon Aurigae, II," *Astrophysical Journal* 187 (1974), 87–92. Huang argues that there is evidence for condensation in the disk that may indicate the formation of a planetary system. Two somewhat old descriptions of this system appeared in Otto Struve's long series of articles in *Sky and Telescope* about strange stars: "The Coming Eclipse of Epsilon Aurigae," *Sky and Telescope* (February 1953), 99–101; "The Story of Epsilon Aurigae," *Sky and Telescope* (March 1962), 127–129.

Cygnus An almost complete set of references to the literature on Cygnus X-1 can be
X-1 found in Oda's review paper, cited above. Some highlights from the technical literature are listed here. Two recent references to the orbital properties and masses of stars in the system are C. T. Bolton, "Orbital Elements and an Analysis of Models for HDE 226868 = Cygnus X-1," *Astrophysical Journal* (hereafter *ApJ*) 200 (1975), 269–279; Y. Avni and J. N. Bahcall, "Masses for Vela X-1 and Other X-Ray Binaries," *ApJ (Letters)* 202 (1975), L131–L134. The x-ray absorption events that link the x-ray source to the companion of the visible star are reported in K. O. Mason et al., "X-Ray Absorption Events in Cygnus X-1 Observed with Copernicus," *ApJ (Letters)* 192 (1974), L65–L70; F. K. Li and G. W. Clark, "Observation of an Absorption Dip in the X-Ray Intensity of Cygnus X-1," *ApJ (Letters)* 191 (1974), L27–L30. Recent models of accretion disks appear in D. M. Eardley, A. P. Lightman, and S. L. Shapiro, "Cygnus X-1: a Two-Temperature Accretion Disk Model which Explains the Observed Hard X-Ray Spectrum," *ApJ (Letters)* 199 (1975), L153–L156, and K. S. Thorne and R. H. Price, "Cygnus X-1: an Interpretation of the Spectrum and Its Variability," *ApJ (Letters)* 195 (1975), L101–L106.

V 861 Only the discovery paper has appeared at this time: R. S. Polidan, G. S. G.
Scorpii Pollard, P. W. Sanford, and M. C. Locke, "X-Ray Emission from the Companion to V 861 Scorpii," *Nature* 275 (28 September 1978), 296–297.

I know it's a long list. But as long as you don't let yourself, as a nonspecialist reader, be intimidated by the equations, a look at some of these papers will further impress upon you the uncertain nature of the scientific adventure.

CHAPTER 6 Frontiers and Fringes
*Calder, Nigel. *The Key to the Universe: a Report on the New Physics.* The Viking Press, New York, 1977.
Davies, Paul C. W. *Space and Time in the Modern Universe.* Cambridge University Press, Cambridge, England, 1977.
Eardley, D. M., and W. H. Press. "Astrophysical Processes Near Black Holes." *Annual Review of Astronomy and Astrophysics* 13 (1975), 381–422.

Freedman, Daniel Z., and Pieter van Nieuwenhuizen, "Supergravity and the Unification of the Laws of Physics," *Scientific American* (February 1978), 126–143.

*Hawking, S. W. "The Quantum Mechanics of Black Holes," *Scientific American* (January 1977), 34–40.

*Hoffman, Banesh. *The Strange Story of the Quantum,* Dover, New York, 1963, 2nd ed., Peter Smith, Magnolia, Mass., n.d.

*Kaufmann, William J., III. *The Cosmic Frontiers of General Relativity.* Little Brown, Boston, 1977.

*Lewin, W. H. G. "X-Ray Outbursts in Our Galaxy." *American Scientist* 65 (September/October 1977), 605–613.

*Overbye, Dennis. "Out from Under the Cosmic Censor: Stephen Hawking's Black Holes." *Sky and Telescope* 54 (August 1977), 84–90. (Probably the best starting point on this difficult topic)

*Will, Clifford. "Gravitation Theory." *Scientific American* (November 1974), 24–34.

CHAPTER 7   Galaxies

*Bok, B. J., and P. F. Bok. *The Milky Way,* 4th ed. Harvard University Press, Cambridge, 1974.

Burbidge, E. M., and G. R. Burbidge. *Quasi Stellar Objects.* Freeman, San Francisco, 1967. (At the intermediate level)

Hodge, P. W. *Galaxies and Cosmology.* McGraw-Hill, New York, 1966.

*National Academy of Sciences Astronomy Survey Committee. *Astronomy and Astrophysics for the 1970s.* National Academy of Sciences, Washington, 1972. (This superb report, prepared by a panel of astronomers under the leadership of Jesse L. Greenstein, looks ahead at the future of astronomy.)

O'Connell, D. J. K. *Nuclei of Galaxies.* Elsevier, New York, 1971. (Proceedings of a study week held at the Vatican on the subjects of Part 2; a good review and general reference.)

Oort, J. H. "The Galactic Center." *Annual Review of Astronomy and Astrophysics* 15 (1977), 295–362.

Sandage, A., M. Sandage, and J. Kristian, eds. *Galaxies and the Universe.* University of Chicago Press, Chicago, 1975. (A compendium of articles by specialists, some of which are more up-to-date than others)

Sanders, R. H., and G. T. Wrixon. "The Center of the Galaxy." *Scientific American* (April 1974), 67–78.

Ulfbeck, O., ed. "Quasars and Active Nuclei of Galaxies." *Physica Scripta* 17 (1978), no. 3. (Proceedings of a symposium held in June 1977 at Copenhagen; a good general reference for Part 2)

CHAPTER 8   Radio Waves from High-Speed Electrons

de Young, D. S. "Extended Extra-galactic Radio Sources." *Annual Review of Astronomy and Astrophysics* 14 (1976), 447–474.

Sciama, D. W. *Modern Cosmology.* Cambridge University Press, Cambridge, England, 1971. (Good review, less mathematical than Peebles)

Many of these references also deal with the topics of Chapters 14, 15, and 16.

*Hey, J. S. *The Evolution of Radio Astronomy.* Neale Watson, New York, 1973. (Historical.)

*Lovell, A. C. B. *The Story of Jodrell Bank.* Harper, New York, 1968. (Historical.)

*Pander, Henk. "An Artist's Astronomical Odyssey." *Sky and Telescope* 57 (January 1979), 7–12. (Listed because it contains some beautiful watercolors of the VLA)
*Sky and Telescope Staff. "The VLA Takes Shape." *Sky and Telescope* 52 (November 1976), 320–326.

CHAPTER 9   Radiation from Quasars: Infrared through Gamma-Ray
Osterbrock, D. "Physical Conditions in the Active Nuclei of Galaxies and Quasi-Stellar Objects," in D. J. K. O'Connell, ed., *Nuclei of Galaxies,* Elsevier, New York, 1971.
Strittmatter, P. A., and Williams, R. E. "The Line Spectra of Quasi-Stellar Objects." *Annual Review of Astronomy and Astrophysics* 14 (1976), 307–338.
Weedman, D. W. "Seyfert Galaxies." *Annual Review of Astronomy and Astrophysics* 15 (1977), 69–95.

CHAPTER 10   Active Galaxies
Burbidge, G. R. "Nuclei of Galaxies." *Annual Review of Astronomy and Astrophysics* 8 (1970), 369–460. (Fairly advanced, but comprehensive)
*Jones, Christine, William Forman, and William Liller. "Optical Counterparts of X-Ray Sources—III." *Sky and Telescope* 49 (January 1975), 10–13.
*Kellermann, K. I. "Extragalactic Radio Sources." *Physics Today* (October 1973), 38–47.
*Metz, William D. "Double Radio Sources: Energetic Evidence that Galaxies Remember." *Science* 188 (June 27, 1975), 1289–1292.
*Metz, William D. "New Light on Quasars: Unraveling the Mystery of BL Lacertae," *Science* 200 (2 June 1978), 1031–1033.
*Strom, Richard G., G. K. Miley, and J. H. Oort. "Giant Radio Galaxies." *Scientific American* (August 1975), 26–36.

CHAPTER 11   The Energy Source
*Metz, William D. "Violently Active Galaxies: the Search for the Energy Machine," *Science* 201 (25 August 1978), 700–702.
*Pacini, Franco, and Martin Rees. "Rotation in High-Energy Astrophysics." *Scientific American* (February 1973), 98–105.
*Rees, Martin. "Quasar Theories." *Eighth Texas Symposium on Relativistic Astrophysics, Annals of the New York Academy of Sciences* 302 (1977), 613–636. [Review articles by Rees on this topic have also appeared in other places, if your library doesn't have this publication. See, for example, *Quarterly Journal of the Royal Astronomical Society* 18 (1977), 429–442 and *Physica Scripta* 17 (1978), 193–200.]
*Schmidt, Maarten, and Francis Bello. "The Evolution of Quasars." *Scientific American* (May 1971), 54–69. (Also in *Cosmology + 1,* referred to under Chapter 1.)

The second part of D. J. K. O'Connell's *Nuclei of Galaxies,* listed under Chapter 7, contains many discussions of quasar theories.

CHAPTER 12   Alternative Interpretations of the Redshift
Arp, H. C. "Anomalous Redshifts in Galaxies and Quasars." *Decalages vers le rouge et Expansion de l'Univers: l'Evolution des Galaxies et*

ses *Implications Cosmologiques.* C. Balkowski and B. E. Westerlund, eds. Centre Nationale de la Recherche Scientifique, 1977. (Do not be put off by the French title, for virtually all the papers in this conference proceedings are in English, with French abstracts. This conference contains many papers supporting the view that the redshifts of some quasars are not produced by the expansion of the universe. This book is a good thing to look at if your local librarian has overcome the barrier of the high price (180 francs) and French title. If you can't obtain this volume, some earlier places where Arp reviewed his position are *Confrontation of Cosmological Theories with Observational Data,* IAU Symposium No. 63, Reidel, Dordrecht, Holland, p. 61, and *Science* 174 (1971), 1189–1199.)

*Field, George B., Halton Arp, and John N. Bahcall. *The Redshift Controversy.* Benjamin, Reading, MA, 1973. (A gold mine. Here you have summary presentations of both points of view along with reprints of the critical papers in the controversy. It's a little dated; Stockton's paper which shows that some quasars have cosmological redshifts was published five years after this volume was. The asterisk is not a mistake. Nonspecialists who do not let themselves be intimidated by equations can understand most of the text in this book.)

Stockton, Alan M. "The Nature of QSO Redshifts." *Astrophysical Journal* 223 (1978), 747–757.

CHAPTER 13   Life Cycle of the Universe: A Model

*Callahan, J. J., "The Curvature of Space in a Finite Universe," *Scientific American* (August 1976), 90–101.

*Ferris, Timothy. *The Red Limit.* Morrow, New York, 1977. (A fascinating, journalistic account focusing on the people doing cosmological research in the twentieth century)

*Gamow, George. *The Creation of the Universe.* Mentor, New York, 1952. (Old, but still good. The numbers are no longer correct, but the general picture is still valid.)

*Groth, E. J., P. J. E. Peebles, M. Seldner, and R. Soneira, "The Clustering of Galaxies," *Scientific American* November 1977), 76–97.

Harrison, E. R. "Standard Model of the Early Universe." *Annual Review of Astronomy and Astrophysics* 11 (1973), 155–186. (A comprehensive review at the advanced undergraduate level. He sets the historical record straight, among other things.)

*John, Laurie, ed. *Cosmology Now.* BBC Publications, London, 1974. (A good review of cosmology, based on a television series)

Peebles, P. James E. *Physical Cosmology.* Princeton University Press, Princeton,. 1971. (A good text, the basis of much of Part 3)

*Schramm, David N. "The Age of the Elements." *Scientific American* (January 1974), 69–77.

Sciama, D. W. *Modern Cosmology.* Cambridge University Press, Cambridge, England, 1971. (Good review, less mathematical than Peebles)

*Singh, Jagjit. *Great Ideas and Theories of Modern Cosmology,* 2nd ed. Dover, New York, 1970. (Sometimes the going is a bit rough, but this is a good overview of cosmology. Singh devotes more attention to unconventional cosmologies than I do.)

*Weinberg, Steven. *The First Three Minutes: a Modern View of the Origin of the Universe.* Basic Books, New York, 1977.

Weinberg, Steven. *Gravitation and Cosmology.* Wiley, New York, 1972. (Advanced undergraduate or graduate-level text)

*Wheeler, John A. "The Universe as Home for Man." *American Scientist* 62 (November/December 1974), 683–691.

Many of these references also deal with the topics of Chapters 14, 15, and 16.

CHAPTER 14   The Cosmic Time Scale

*Iben, I., Jr. "Globular Cluster Stars." *Scientific American* (July 1970) 26–39.

*Kraft, R. P. "Pulsating Stars and Cosmic Distances." In *Frontiers in Astronomy,* ed. Owen Gingerich, Freeman, San Francisco, 1971.

Sandage, Allan. "Distances to Galaxies, the Hubble Constant, and the Edge of the World." *Quarterly Journal of the Royal Astronomical Society* 13 (1972), 282–296.

CHAPTER 15   Relics of the Big Bang

*Burbidge, G. R. "Was There Really a Big Bang?" In W. C. Saslaw and K. C. Jacobs, *The Emerging Universe.* University of Virginia Press, Charlottesville, 1972.

Danziger, I. J. "The Cosmic Abundance of Helium." *Annual Review of Astronomy and Astrophysics* 8 (1970), 161–178.

*Field, G. B. "Big Bang Cosmology: The Evolution of the Universe." In W. C. Saslaw and K. C. Jacobs, *The Emerging Universe.* University of Virginia Press, Charlottesville, 1972.

*Guillen, Michael A. "Galaxy of New Ideas." *Science News* 114 (August 26, 1978), 144–156. (Summary treatment of unconventional cosmologies)

*Motz, Lloyd. *The Universe: Its Beginning and End.* Scribners, New York, 1975.

Muller, R. A. "The Cosmic Background Radiation and the New Ether Drift," *Scientific American* (May 1978), 64–74.

*Pasachoff, Jay M., and William A. Fowler. "Deuterium in the Universe." *Scientific American* (May 1974), 108–118.

*Peebles, P. J. E., and Wilkinson, D. T. "The Primeval Fireball." In *Frontiers in Astronomy,* ed. Owen Gingerich. Freeman, San Francisco, 1971.

*Ryle, M. "Radio Telescopes of Large Resolving Power." Nobel Prize Lecture, 1974. *Science* 188 (June 13, 1975), 1071–1078.

Thaddeus, P. "The Short Wavelength Spectrum of the Microwave Background." *Annual Review of Astronomy and Astrophysics* 10 (1972), 305–334.

*Van Flandern, T. C. "Is Gravity Getting Weaker?" *Scientific American* (February 1976), 44–52. (See footnote 2, Chapter 15, for references to the technical papers that place the uncertainties of van Flandern's work in perspective. His article contains a good discussion of cosmologies based on the Large-Numbers Hypothesis.)

CHAPTER 16   The Future of the Universe

Gott, J. R. III, J. E. Gunn, D. L. Schramm, and B. M. Tinsley, "Will the Universe Expand Forever?" *Scientific American* (March 1976), 62–79.

GENERAL   A complete astronomical bibliography is *A Guide to the Literature of Astronomy,* by Robert A. Seal, Libraries Unlimited, Littleton, Colorado, 1977. The topics discussed in this book are also discussed in "Texas" conferences on relativistic astrophysics. The proceedings of these conferences (which are not always held in Texas) are published in the *Annals of the New York Academy of Sciences.* Recent volumes are 224 (the sixth conference), 262 (the seventh), and 302 (the eighth). The ninth was held in Munich, Germany, in late 1978. Articles on these topics appear in magazines like *Astronomy, Mercury, Scientific American,* and *Sky and Telescope.*

NOTES

CHAPTER 3   1. I. S. Shklovsky, *Supernovae,* London, Wiley, 1968, p. 51 (translation by Duyvendak).

2. Ibid., p. 44.

3. David H. Clark and F. Richard Stephenson, *The Historical Supernovae,* Oxford, Pergamon, 1977.

4. G. A. Tammann, in D. N. Schramm, ed., *Supernovae,* Dordrecht (Holland), Reidel, 1977, pp. 95–116.

5. W. Baade and F. Zwicky, "Supernovae and Cosmic Rays," *Physical Review* 45 (1934), p. 128.

6. N. D. Mermin, *Space and Time in Special Relativity,* New York, McGraw-Hill, 1968. E. F. Taylor and J. A. Wheeler, *Space-Time Physics,* Freeman, San Francisco, 1963, is at a higher, more abstract mathematical level, but still uses practically no calculus. A good popular account is Lincoln Barnett, *The Universe and Dr. Einstein,* New York, Bantam, 1974.

7. The maximum mass of neutron stars has been an intensely studied subject in recent years. At this writing (January 1979), the following conclusions have been reached. (a) Using reasonable equations of state (relations between the pressure and density of matter) in the density region in which these equations of state can be regarded as known (density $\rho < 5 \times 10^{14}$ g/cm^3), and using the restriction that the pressure P must be less than ρc^2 (which probably, but not definitively, means that the speed of sound is less than the speed of light, the maximum mass of stable nonrotating neutron stars is 3.2 solar masses (C. E. Rhoades, Jr., and R. Ruffini, "The Maximum Mass of a Neutron Star," *Physical Review Letters* 32 (1974), 324–327], or 4 solar masses [G. Caporaso and K. Brecher, "The Neutron Star Mass Limit in the Bimetric Theory of Gravitation," *Physical Review D* 15 (1977), 3536–3542], or 5 solar masses [J. B. Hartle and A. B. Sabbadini, "The Equation of State and Bounds on the Mass of Nonrotating Neutron Stars," *Astrophysical Journal* 213 (1977), 831–835]. This conclusion is probably closest to the truth, as long as you accept general relativity (GRT). (b) Using GRT but making fewer restrictions on the equation of state, the maximum mass increases to 8 solar masses [D. Hegyi, "The Upper Mass Limit for

Neutron Stars Including Differential Rotation," *Astrophysical Journal* 217 (1977), 244–247], or 11 solar masses [Hartle and Sabbadini, cited above]. (c) Massive stars made of quarks, the constituents of protons and neutrons, probably do not exist [R. L. Bowers, A. M. Gleeson, and R. D. Pedigo, "On the Possibility of Stable Quark Stars," *Astrophysical Journal* 213 (1977), 840–848]. (d) If you do not assume the validity of general relativity, you can make large neutron stars with equations of state that are both reasonable at low densities and have $P < \rho c^2$. In Rosen's bimetric theory, the limit is 81 solar masses [Caporaso and Brecher, already cited; N. Rosen and J. Rosen, "The Maximum Mass of a Cold Neutron Star," *Astrophysical Journal* 202 (1977), 782–787]. In Ni's theory, there is no limiting mass [D. R. Mikklesen, "Very Massive Neutron Stars in Ni's Theory of Gravity," *Astrophysical Journal* 217 (1977), 248–251]. A good technical review is "'Neutron' Stars within the Laws of Physics," by K. Brecher and G. Caporaso, *Annals of the New York Academy of Sciences* 302 (1977), 471–481. Inclusion of rotation adds about 30% to these maximum masses [Hegyi, cited above; A. P. Lightman and S. L. Shapiro, "Rapidly Rotating Post-Newtonian Neutron Stars," *Astrophysical Journal* 207 (1976), 263–278].

8. K. A. van Riper and W. D. Arnett, "Stellar Collapse and Explosion: Hydrodynamics of the Core," *Astrophysical Journal Letters* 225 (1978), L129–L132.

9. R. F. Green, Ph.D. thesis, California Institute of Technology, 1977.

CHAPTER 4   1. The times in Table 4-1 are given for a particle falling from infinity, taking as time 0 the instant that the particle is 1 AU away from the ten-solar-mass hole.

CHAPTER 5   1. J. de Loore and J. de Greve, *Astrophysics and Space Science* 35 (1975), 241–247.
2. S. L. Shapiro, private communication.

CHAPTER 6   1. J. Grindlay, "New Bursts in Astronomy," *Harvard Magazine* (March/April 1977), 23–27, 81–83. See also H. Gursky, "The Nature of X-Ray Bursters," *Annals of the New York Academy of Sciences* 302 (1977), 197–209.
2. I. B. Strong and W. Klebesadel, "Cosmic Gamma-Ray Bursts," *Scientific American* (October 1976), 66–75.
3. E. B. Newell, G. S. da Costa, and J. Norris, "Evidence for a Central Massive Object in the X-Ray Cluster Messier 15," *Astrophysical Journal Letters* 208 (1976), L55–L60. G. Illingworth and I. King, in "Dynamical Models for Messier 15 Without a Black Hole," *Astrophysical Journal Letters* (1977), 218, L109–L112, show that Newell et al's observations do not require a black hole.
4. See P. J. Young, "The Black Tide Model of QSOs: Destruction in an Isothermal Sphere," *Astrophysical Journal* 215 (1977), 36–52, especially the appendix, and S. L. Shapiro, "The Dissolution of Globular Clusters Containing Massive Black Holes," *Astrophysical Journal* 217 (1977), 281–286.
5. B. Hoffman, *The Strange Story of the Quantum,* 1st ed., Dover, New York, 1963; 2nd ed., Peter Smith, Magnolia, MA, n.d.
6. Bernard J. Carr, "The Primordial Black Hole Mass Spectrum," *Astrophysical Journal* 201 (1975), 1–19. Do not misinterpret the statement in W. J. Kaufmann, III, *The Cosmic Frontiers of General Relativity,* Little Brown,

Boston, 1977, p. 287, as indicating that Carr's calculations show that primeval black holes *must* exist; Carr shows only that under certain circumstances they *may* exist.

7. Don N. Page, "Particle Emission Rates from a Black Hole," *Physical Review D* 13 (1976), 198–206; N. A. Porter and T. C. Weekes, "An Upper Limit to the Rate of Gamma-Ray Bursts from Primordial Black Hole Explosions," *Astrophysical Journal* 212 (1977), 224–226.

8. R. A. Oriti, "The Tunguska Event," *The Griffith Observer,* November 1975; W. H. Beasley and B. A. Tinsley, "The Tungus Event Was Not Caused by a Black Hole," *Nature* 250 (1974), 555–556; J. O. Burns, G. Greenstein, and K. L. Verosub, "The Tungus Event as a Small Black Hole: Geophysical Considerations," *Monthly Notices of the Royal Astronomical Society* 175 (1976), 355–357. A negative review of a recent book treating the more spectacular aspects of the subject is W. McCrea, "Black Hole, Meteorite, or Spacecraft?" (Review of J. Baxter and T. Atkins, *The Fire Came By* (New York, Doubleday, 1976), *Nature* 264 (1976), 683.)

9. C. M. Will, "Gravitational Radiation from Binary Systems in Alternative Metric Theories of Gravity: Dipole Radiation from the Binary Pulsar," *Astrophysical Journal* 214 (1977), 826–839. Footnote 4 of this article contains references to recent solar system tests of general relativity.

10. J. Taylor, et al., 9th Texas Symposium on Relativistic Astrophysics, December 1978, Munich, Germany; *Proceedings of the New York Academy of Sciences* 1979.

11. C. Will and D. Eardley, "Dipole Gravitational Radiation in Rosen's Theory: Observable Effects in the Binary System PSR 1913 + 16," *Astrophysical Journal Letters* 212 (1977), L91–L94.

12. J. Gribbin, *White Holes: Cosmic Gushers in the Universe,* New York, Delacorte (Eleanor Friede), 1976; A. Berry, *The Iron Sun: Crossing the Universe through Black Holes,* Warner paperback, New York, 1977. Roman Znajek exposes the speculative nature of these books in *Nature* 267 (1977), 867 [Berry] and *Nature* 270 (1977), 133 [Gribbin].

13. W. J. Kaufmann, III, *The Cosmic Frontiers of General Relativity,* Little Brown, Boston, 1977. Chapters 10–12 and 14 of Kaufmann's book cover the speculative material in spectacular detail. Be sure to read page 247, the last page of Chapter 14, so that you will appreciate the relationship of these speculations to reality.

14. D. M. Eardley, "Death of White Holes in the Early Universe," *Physical Review Letters* 33 (1974), 442–444.

15. S. W. Hawking, "The Quantum Mechanics of Black Holes," *Scientific American* (January 1977), p. 40.

CHAPTER 7 1. Various names for the Milky Way are discussed in the source of this quotation: R. H. Allen, *Star Names: their Lore and Meaning,* New York, Dover, 1963 (originally published in 1899), p. 476.

CHAPTER 8 1. The information on 4C 39.25 comes from Dave Shaffer's Ph.D. thesis at Caltech, 1973. I thank him for discussions on VLB interferometry.

2. K. Y. Lo, M. H. Cohen, R. T. Schilizzi, and H. N. Ross, "An Angular Size for the Compact Radio Source at the Galactic Center," *Astrophysical Journal* 218 (1977), 668–670. K. Kellermann, D. Shaffer, B. Clark, and B. Geldzahler, "The Small Radio Source at the Galactic Center," *Astrophysical Journal Letters* 214 (1977), L61–L62.

3. Readers interested in the algebraic details might be interested in the two classic papers in this field: M. J. Rees, "Appearance of Relativistically

Expanding Radio Sources," *Nature* 211 (1966), 468–470; "Studies in Radio Source Structure, I," *Monthly Notices of the Royal Astronomical Society* 135 (1967), 345–360. R. D. Blandford, C. F. McKee, and M. J. Rees, "Superluminal Expansion in Extragalactic Radio Sources," *Nature* 267 (1977), 211–216, is a recent review article on this topic.

CHAPTER 9   1. Data from R. H. Hildebrand et al., "Submillimeter Photometry of Extraga-lactic Objects," *Astrophysical Journal* 216 (1977), 698–705; A. F. Davidsen, G. F. Hartig, and W. G. Fastie, "The Ultraviolet Spectrum of the Quasi-Stellar Object 3C 273," *Nature* 269 (1977), 203–206; B. Margon, S. Bowyer, and M. Lampton, "Limits on Intergalactic Helium from the 3C 273 X-Ray Spectrum," *Astrophysical Journal* 174 (1972), 471–475; B. N. Swanenburg et al., "COS B Observation of High Energy Gamma Radiation from 3C 273," *Nature* 275 (1978), 298. I thank Dr.Michal Simon for discussions regarding this figure.
2. See, for example, G. R. Burbidge, T. W. Jones, and S. L. O'Dell, "Physics of Compact Nonthermal Sources. III. Energetic Considerations," *Astrophysical Journal* 193 (1974), 43–54.

CHAPTER 10   1. A. C. S. Redhead, M. H. Cohen and R. D. Blandford, "A Jet in the Nucleus of NGC 6251," *Nature* 272 (1978), 131–134.
2. Spiegel's and Kellermann's remarks were reported by W. D. Metz, "New Light on Quasars: Unraveling the Mystery of BL Lacertae," *Science* 200 (2 June 1978), 1031–1033.
3. Eric R. Craine, *A Handbook of Quasistellar and BL Lacertae Objects,* Tucson, Pachart Publishing House, 1977; Donald Hamilton, William Keel, and J. F. Nixon, "Variable Galactic Nuclei," *Sky and Telescope* 55 (May 1978), 372–374.

CHAPTER 11   1. Quoted in W. D. Metz, "Violently Active Galaxies: the Search for the Energy Machine," *Science* 201 (25 August 1978), 700–702. Ptolemy was the theorist who, flourishing in the second century, believed that the earth was at the center of the universe and used geometrical devices called epicycles to explain the motion of the planets.

CHAPTER 12   1. This phrase is due to Lawrence H. Auer.
2. G. R. Burbidge, E. M. Burbidge, P. M. Solomon, and P. A. Strittmatter, "Apparent Associations between Bright Galaxies and Quasi-Stellar Objects," *Astrophysical Journal* 170 (1971), pp. 233–240.
3. J. N. Bahcall, in G. B. Field, H. C. Arp, and J. N. Bahcall, *The Redshift Controversy,* Benjamin, Reading, MA, 1973, pp. 88–89, 190–194.
4. A. M. Stockton, "The Nature of QSO Redshifts," *Astrophysical Journal* 223 (1978), 747–757.
5. John Faulkner and Martin Gaskell, "Astronomers Licked by QSOs and Active Galactic Nuclei," *Nature* 275 (1978), 91–92. This is a report of a workshop held at the Lick Observatory. The title is more of an indication of the British sense of humor in headlining articles in *Nature* than an indication of the bafflement of the astronomical community.

CHAPTER 13   1. Quoted by G. Gamow, *My World Line,* Viking, New York, 1970, p. 44.
2. Ivan King, "The Evolution of Galaxies: the View on a Cloudy Day," in R. B. Larson and B. M. Tinsley, eds., *The Evolution of Galaxies and Stellar Populations,* New Haven, Yale University Press, 1977.
3. Jamal N. Islam, "The Ultimate Fate of the Universe," *Sky and Telescope* 57 (1979), 13–18. A more technical version of the same article is "Possible

Ultimate Fate of the Universe," *Quarterly Journal of the Royal Astronomical Society* 18 (1977), 3−8.

4. J. J. Callahan, "The Curvature of Space in a Finite Universe," *Scientific American* (August 1976), 90−101.

CHAPTER 14   1. D. N. Schramm, "The Age of the Elements," *Scientific American* (January 1974), 69−77.

2. Most of these numbers come from J. Heidmann, "Expansion of the Universe," in W. Yourgrau and A. Breck, eds., *Cosmology, History, and Theology,* New York, Plenum, 1977.

CHAPTER 15   1. A. Penzias, "Cosmology and Microwave Astronomy," in F. Reines, ed., *Cosmology, Fusion, and Other Matters,* Boulder, Colorado Associated Universities Press, 1972.

2. Most versions of the Large-Number Hypothesis require that the fractional rate of change in the gravitational constant be $3H$ or 15×10^{-11} parts per year. In some versions the change is one-third of this value. The lunar occultation result (T. van Flandern, "Is Gravity Getting Weaker?" *Scientific American* (February 1976) 44−59) was only marginally significant at $7.2\pm 3.7 \times 10^{-11}$ parts per year. In his technical paper, the rate of change is slightly different and the uncertainty is larger. T. van Flandern in "A Determination of the Rate of Change of G," *Monthly Notices of the Royal Astronomical Society* 170 (1975), 333−342, gives the change as $(8 \pm 5) \times 10^{-11}$ parts per year. Neither of these figures is generally acceepted as being significantly different from zero. A detection that is twice the quoted error has a 5% chance of being spurious because of coincidence alone. Generally people seek measurements that are 3 or 4 times the quoted uncertainties if they are to accept the measurements as real. Discussions of these cosmologies and the microwave background can be found in G. Steigman, "A Crucial Test of the Dirac Cosmologies," *Astrophysical Journal* 221 (1978), 407−411, and V. Canuto and S. H. Hsieh, "The 3K Blackbody Radiation, Dirac's Large-Numbers Hypothesis, and Scale-Covariant Cosmology," *Astrophysical Journal* 224 (1978), 302−307.

3. Gary Steigman, "Observational Tests of Antimatter Cosmologies," *Annual Review of Astronomy and Astrophysics* 14 (1976), 339−372.

CHAPTER 16   1. D. W. Sciama, *Modern Cosmology,* Cambridge University Press, Cambridge, England, 1971, Chapter 8.

2. J. E. Gunn and J. B. Oke, "Spectrophotometry of Faint Cluster Galaxies and the Hubble Diagram: An Approach to Cosmology," *Astrophysical Journal* 195 (1975), 255−268; J. Kristian, A. Sandage, and J. A. Westphal, "The Extension of the Hubble Diagram, II. New Redshifts and Photometry of Very Distant Galaxy Clusters: First Indication of a Deviation of the Hubble Diagram from a Straight Line," *Astrophysical Journal* 221 (1978), 383−394.

Planets, origin, 257–259
Pleiades (star cluster), 29, 33
Polarized radiation, 169–171
Polidan, Robert, 109
Pollard, G. S. G., 109
Powers-of-ten notation, 13
Primeval atom, 250
Primeval fireball radiation, 251–252, 280, 282–289, 294–295
Procyon (star), 29, 40
Proton, 2, 327
PSR 0833–45 (pulsar), 55–56
PSR 1913+16, 57, 123–127
PSR 1919+21, 52, 55
Pulsars
    discovery, 49–53
    gravitational radiation, 124–127
    interpretation, 53–57
    in quasars, 218–219
    x-ray, 57–58, 116
    See also neutron stars
Pygmalion syndrome, 17, 66, 77, 81, 119

Quantum mechanics, 116–120
Quark, 3, 224
Quasars, 14–15, 156–239
    absorption lines, 188
    and active galaxies, 190–211
    and BL Lac objects, 182
    comprehensive model, 188–190
    continuum, 177–182
    cosmological test, 316
    defined, 150
    discovery, 150–153
    emission lines, 182–190
    energy source, 213–225
    evolution, 215–216, 293–294
    expansion, 186
    infrared, 178–181
    radio, 153–154, 167–170, 172–180
    redshift, 150, 182–183, 226–239
    variation, 178–180

Racine, René, 274, 278
Radio radiation, 4–7
    Cygnus X-1, 100
    galaxies, 144, 164–166, 194–205
    interferometry, 157–168
    pulsars, 51–55
    quasars, 151–154, 167–180, 211
Reber, Grote, 158, 165
Recombination, 185, 254, 282, 327
Red giant stars, 30–35, 39–41, 62–63, 260
Redshift
    cosmological, 148–150, 183, 226, 248–250, 309–317, 325
    discordant, *see* Redshift, non-cosmological
    Doppler, 89–92
    gravitational, 72–74, 76

non-cosmological, 227–232, 236–239, 325
    relativistic formula, 152
    z describes, 72
Rees, Martin, 204, 217, 220, 223
Reichley, Paul, 55
Reifenstein, Edward, 52
Right ascension, 327
Roll, Peter G., 282
Rose, William, 39
Rosen, Nathan, 67, 123, 127

Sandage, Allan, 152, 267, 273–278 *passim,* 314
Sanders, R. H., 217
Sanford, P. W., 109
Sargent, Wallace, 227, 235
Schmidt, Maarten, 153, 215
Schmidt telescope, 47, 50, 327
Schmitt, J., 208
Schramm, David, 268
Schwarzschild, Karl, 66, 81, 86, 128
Schwarzschild radius, 71–73, 220
Scientific method, 16–19, 236–239
Scientific models, 16–21, 27–35, 56, 95, 104–108, 121–123, 187, 215–216
Scientific revolutions, 18–21, 236–239
Second of arc, 12–13, 327
Seielstad, George, 171
Shaffer, David, 167
Shane, Donald, 256
Shapiro, Stuart, 104
Shields, Gregory, 220
Shklovsky, Iosif, 169
Single-lined spectroscopic binary, 87–93, 327
Singularity, 81, 116–119, 127, 131–132
Sirius A (star), 28–35 *passim,* 39
Sirius B (star), 35–40
Slipher, Vesto, 247
SMC X-1 (x-ray source), 58
Snyder, Hartland, 66, 86
Solinger, Alan, 195
South Polar Cap group (of galaxies), 275
Space telescope, 273
Space warps, 128–130
Spectra (stellar), 9–11, 88–89, 327
Spectroscopic binary star, 87–92, 327
Spiegel, Edward, 210
Spinar, 218–219, 222
Spitzer, Lyman, 216
SS 433, 133
Staelin, David, 52
Standard candles, 309–314
Stars
    collisions, 216–217
    evolution, 24–64
    general properties, 7–11
    origin, 256–258
    spectral classes, 88–89